# Understanding Metals in the Environment and Human Health

Metals play diverse roles in human health and environmental systems, with both positive and negative impacts. The challenges posed by metal pollution and toxicity require a multifaceted approach, combining technological innovation with policy, education, and public health initiatives. This book discusses all metals and their roles in human health and includes the most advanced techniques used in detection, quantification, and remediation of metals, both in human and environmental health related issues. A comprehensive understanding of these advancements enables readers to harness the good in metals, mitigate the bad, and minimize the ugly consequences of metal toxicity for a safer, healthier world.

Features:

- Provides a comprehensive exploration of metals in human health and ecology.
- Covers cutting-edge detection and quantification techniques.
- Presents innovative solutions for environmental metal remediation.
- Addresses personalized health approaches for managing metal levels.
- Includes many case studies showing up-to-date research in different countries and their regulatory aspects and legislation.

This book is an excellent resource for health and environmental professionals, researchers and academics, and master's and graduate students involved in projects related to environmental science, environmental chemistry, ecology, and environmental pollution and remediation.

**Emergent Environmental Pollution**

*Series Editor: Jörg Rinklebe, University of Wuppertal, Germany*

The nature of pollutants in the environment depends largely on their interactions in the soil, water, air, and organisms. Understanding the fate and transport of pollutants in the environment is essential for humans and animals, plants, water bodies, air, soils, and sediments. This series provides timely information on the biological, chemical, and physical processes governing the interactions of pollutants in the environment. Such knowledge is needed for minimizing the negative impact of pollutants on human health and the health of ecosystems. Books from this series are of value to professionals and researchers, students, industrial and mining engineers, government regulators, authorities, and consultants in the environmental arena.

*Competitive Sorption and Transport of Heavy Metals in Soils and Geological Media*
Edited by H. Magdi Selim

*Nickel in Soils and Plants*
Edited by Christos Tsadilas, Jörg Rinklebe, and Magdi Selim

*Phosphate in Soils: Interaction with Micronutrients, Radionuclides and Heavy Metals*
Edited by H. Magdi Selim

*Permeable Reactive Barrier: Sustainable Groundwater Remediation*
Edited by Ravi Naidu and Volker Birke

*Nitrate Handbook: Environmental, Agricultural, and Health Effects*
Edited by Christos Tsadilas

*Vanadium in Soils and Plants*
Edited by Jörg Rinklebe

*Maritime Accidents and Environmental Pollution - The X-Press Pearl Disaster: Causes, Consequences, and Lessons Learned*
Edited by Meththika Vithanage, Ajith Priyal de Alwis, and Deshai Botheju

*Methylmercury Accumulation in Rice: Process and Regulation*
Edited by Xinbin Feng, Jianxu Wang, and Jörg Rinklebe

*Green Technologies for Pollutants Remediation and Recovery from Water and Soil*
Edited by Jakub Zdarta, Teofil Jesionowski, and Jörg Rinklebe

*Emerging Organic Contaminants in Soil: Properties, Translocation, and Remediation*
Edited by Ming Zhang and Mahtab Ahmad

*Understanding Metals in the Environment and Human Health*
Edited by Carlos Lodeiro and Elisabete Oliveira

For more information on this series, please visit: https://www.routledge.com/Emergent-Environmental-Pollution/book-series/CRCEEP

# Understanding Metals in the Environment and Human Health

Edited by
Carlos Lodeiro and Elisabete Oliveira

CRC Press
Taylor & Francis Group
Boca Raton London New York

CRC Press is an imprint of the
Taylor & Francis Group, an **informa** business

Designed cover image: © iStock

First edition published 2027
by CRC Press
2385 NW Executive Center Drive, Suite 320, Boca Raton FL 33431

and by CRC Press
4 Park Square, Milton Park, Abingdon, Oxon, OX14 4RN

*CRC Press is an imprint of Taylor & Francis Group, LLC*



ISBN: 978-1-041-05270-8 (hbk)
ISBN: 978-1-041-05271-5 (pbk)
ISBN: 978-1-003-63209-2 (ebk)

DOI: 10.1201/9781003632092

Typeset in Times
by KnowledgeWorks Global Ltd.

# Contents

# Preface

Metals are among the most pervasive and consequential chemical entities in our natural world. From the iron coursing through our bloodstream to the mercury accumulating in marine ecosystems, their influence—beneficial or harmful—permeates every level of biological organization and environmental complexity. As global industrialization, urbanization, and intensive agricultural practices continue to accelerate, so too the urgency of understanding how metals interact with living systems and the environment at large.

This book, *Understanding Metals in the Environment and Human Health,* was conceived from a recognition that the roles of metals in health and ecological systems cannot be reduced to simple narratives. Essential metals such as iron, zinc, copper, and magnesium are indispensable to life—mediating enzymatic catalysis, immune responses, and cellular metabolism—yet their dysregulation, whether through deficiency or excess, can precipitate serious clinical and environmental consequences. Conversely, non-essential toxic metals such as lead, mercury, cadmium, and arsenic pose well-documented threats to neurological, renal, and cardiovascular health, even at trace exposure levels, with their non-biodegradable nature enabling insidious bioaccumulation across trophic levels in food chains.

Addressing these challenges demands innovation at every frontier. This volume brings together advances in detection and quantification—from classical atomic absorption spectrometry and ICP-MS to cutting-edge electrochemical biosensors, fluorescent chemosensors, quantum dot probes, and paper-based analytical devices—with a focus on sensitivity, portability, and real-world applicability. Equally, it explores the expanding landscape of metal remediation, encompassing bioremediation, phytomining, nanomaterial-based strategies, and electrokinetic approaches, reflecting the field's shift toward sustainable and cost-effective solutions.

Beyond environmental remediation, the book addresses the emerging medical and theragnostic dimensions of metal science. Advances in selective chelation therapy, the comprehensive profiling offered by metallomics, and the transformative potential of gene-editing technologies such as CRISPR-Cas9 for correcting dysregulated metal metabolism signal a new era of precision medicine for metal-related disorders. Cutting across all these themes is the growing role of artificial intelligence (AI) and machine learning in optimizing detection protocols, predicting contamination hotspots, and accelerating bioremediation strategies.

It is the editors' hope that this book will serve as a valuable reference for researchers, clinicians, environmental scientists, students, and policymakers alike, fostering the multidisciplinary dialogue essential for building a future where the benefits of metals are maximized and their adverse consequences minimized. The road ahead is difficult, but the scientific knowledge and resolve at our disposal are plentiful.

**Carlos Lodeiro and Elisabete Oliveira**
*BIOSCOPE Research Labs, LAQV-REQUIMTE*
*NOVA School of Science and Technology, FCT NOVA*
*Caparica, Portugal*

The authors declare no competing financial interest. The authors acknowledge the use of AI-based tools, specifically Grammarly and ChatGPT5.2, solely for language refinement, including grammar correction, spelling, and improvements in clarity. No part of the scientific content, data analysis, interpretation, or conclusions was generated by these tools. The authors take full responsibility for the final text and its accuracy.

# About the Editors

**Carlos Lodeiro** graduated in chemistry in 1995, received his PhD in chemistry in 1999 from the University of Santiago de Compostela, Spain. In 1999, he moved to the NOVA University Lisbon (UNL), Portugal as a European Marie Curie postdoctoral researcher integrated in a project concerning molecular devices and machines, and in 2004 he became a fellow researcher and invited assistant lecturer at the REQUIMTE-CQFB, Chemistry Department (UNL).

In 2008 Dr. Lodeiro got the habilitation in chemistry in Spain, and a year later in 2009 he moved to the University of Vigo, Faculty of Sciences of Ourense (FCOU), Spain as IPP (Isidro Parga Pondal) researcher-lecturer. In 2012, he became Assistant Professor at the Chemistry Department and researcher at the UCIBIO-REQUIMTE Laboratory in the NOVA Science and Technology School, UNL. In 2017, Professor Lodeiro move as research to the LAQV-REQUIMTE Associate Laboratory with his team at the Caparica Campus, NOVA FCT, and in in July 2017 got the habilitation in Inorganic Analytical Chemistry at the NOVAFCT-UNL (Portugal) and became Associate Professor in the Chemistry Department NOVAFCT-UNL., becoming Full Professor in January 2024. At present, he co-leads the BIOSCOPE research group (www.bioscopegroup.org), he is Associated and co-Founder of the PROTEOMASS Scientific Society, and since 2023 Associated and co-Founder of the Nanoarts Scientific Association for the development of chemistry and nanomaterials. Dr. Lodeiro is Fellow of the Royal Society of Chemistry since 2014 and member of the Portuguese Chemistry Society since 2002, and the American Chemical Society since 2016. His research interest include: (i) physical-organic and physical-inorganic chemistry of dyes and chemosensors; (ii) synthesis of functionalized nanoparticles, nanocomposites and nanomaterials; (iii) applications of nanomaterials and sensors in environmental research, detection, and quantification of pollutants; (iv) application of nanomaterials in biomedical research; (v) supramolecular analytical proteomics; and (vi) onco- and nanoproteomics. Dr. Lodeiro has mentored 16 PhDs and has published more than 350 articles, several books, and book chapters. He has three patents to his name.

**Elisabete Oliveira** graduated in 2006, in Applied Chemistry from FCT-Nova University Lisbon, Portugal, in 2007 obtained a master's in biotechnology and completed a PhD degree in Biotechnology in 2010 at the same University. In 2013, she obtained a second PhD degree in "Food Science and Technology" by the Science Faculty of Ourense Campus at the University of Vigo, Spain. Currently, she is Assistant Professor at the Chemistry Department of the NOVAFCT - LAQV-REQUIMTE (Portugal). In 2008, Oliveira received the prize in Creativity and Quality in Research Activity in sensors area, attributed by Foundation Calouste Gulbenkian, Portugal and in 2016 she was awarded with the Prize for Women in Science, "Medalhas de Honra L'Oréal Portugal para as Mulheres na Ciência" in the field of health Sciences.

Her scientific interests focus on chemistry, nanotechnology, and biomedicine, with an emphasis on the synthesis of innovative nanomaterials and their applications. The main research areas include: (i) the development of bioinspired fluorescent chemical sensors, (ii) the creation of nanoparticles and polymers for environmental remediation and antibacterial coatings, (iii) drug delivery studies using mesoporous silica nanoparticles, with a focus on targeted drug delivery to cancer and bacterial cells, and (iv) the design of smart labels and sustainable packaging for the detection and prevention of bacterial contamination. Dr. Oliveira has mentored 4 PhDs and has published close to 100 articles, and several book chapters.

# List of Contributors


**Mario Luigi Antenozio**
Research Institute on Terrestrial Ecosystems
National Research Council (IRET-CNR)
Montelibretti, Rome, Italy

**A. Arias-Borrego**
Research Center of Natural Resources, Health and the Environment (RENSMA)
Department of Chemistry, Faculty of Experimental Sciences
University of Huelva
Huelva, Spain

**Marco Aurélio Zezzi Arruda**
Group of Espectrometry, Sample Preparation and Mechanization-GEPAM
University of Campinas
Institute of Chemistry, University of Campinas
Campinas, São Paulo, Brazil
and
National Institute of Science and Technology in Bioanalytics – Lauro Kubota – INCTBio-LK University of Campinas
Institute of Chemistry
Campinas, São Paulo, Brazil

**Bruno Barbosa**
MEtRICs
Instituto Superior de Educação e Ciências
Lisbon, Portugal

**María Carmen Barciela-Alonso**
Trace Element, Spectroscopy and Speciation Group (GETEE)
Instituto de Materiais (iMATUS)
Faculty of Chemistry
University of Santiago de Compostela
Santiago de Compostela, Spain

**Pilar Bermejo-Barrera**
Group of Trace Elements, Spectroscopy and Speciation (GETEE)
Materials Institute (iMATUS)
Department of Analytical Chemistry, Nutrition and Bromatology
Faculty of Chemistry
University of Santiago de Compostela
Santiago de Compostela, Spain

**Patrizia Brunetti**
Research Institute on Terrestrial Ecosystems
National Research Council (IRET-CNR)
Montelibretti, Rome, Italy

**Joanna Burger**
Rutgers University
The State University of New Jersey
New Brunswick, New Jersey, USA

**R. Callejón Fernández**
Department of Microbiology and Parasitology
University of Seville
Seville, Spain

**B. Callejón-Leblic**
Research Center of Natural Resources, Health and the Environment (RENSMA)
Department of Chemistry, Faculty of Experimental Sciences
University of Huelva
Huelva, Spain

**José Luis Capelo-Martínez**
BIOSCOPE Research Group, LAQV-REQUIMTE
Chemistry Department
NOVA School of Science and Technology, NOVA FCT
Universidade NOVA de Lisboa
Caparica, Portugal
and
PROTEOMASS Scientific Society
Costa de Caparica, Portugal

**Mariko A. Carneiro**
LSRE-LCM, ALiCE
Faculty of Engineering
University of Porto
Porto, Portugal

**Barbara Rachele Ciaramella**
University of Messina
Messina, Sicily, Italy

**Clara C. Costa**
LSRE-LCM, ALiCE
Faculty of Engineering
University of Porto
Porto, Portugal

**Jorge Costa**
MEtRICs
Instituto Superior de Educação e Ciências
Lisbon, Portugal

**Cidália. M. S. Botelho**
LSRE-LCM, ALiCE, Faculty of Engineering, University of Porto
Porto, Portugal

**Frederico Duarte**
BIOSCOPE Research Group, LAQV-REQUIMTE, Chemistry Department, NOVA FCT Universidade NOVA de Lisboa, 2829-516 Caparica, Portugal

**Elena Peña-Vázquez**
Group of Trace Elements, Spectroscopy and Speciation (GETEE)
Materials Institute (iMATUS)
Department of Analytical Chemistry, Nutrition and Bromatology
Faculty of Chemistry
University of Santiago de Compostela
Santiago de Compostela, Spain

**Javier Fernández Lodeiro**
BIOSCOPE Research Group, LAQV-REQUIMTE – NOVA FCT
Chemistry Department
NOVA School of Science and Technology, NOVA FCT
Universidade NOVA de Lisboa
Caparica, Portugal

**Ana Luisa Fernando**
MEtRICs, LAQV REQUIMTE
Associated Laboratory for Green Chemistry
Universidade NOVA de Lisboa
Lisbon, Portugal

**Ricardo M. Ferreira**
LSRE-LCM, ALiCE
Faculty of Engineering
University of Porto
Porto, Portugal

**Jürgen Gailer**
Department of Chemistry
University of Calgary
Calgary, AB, Canada

**Joana Galhano**
BIOSCOPE Research Group, LAQV-REQUIMTE
Chemistry Department
NOVA School of Science and Technology, NOVA FCT
Universidade NOVA de Lisboa
Caparica, Portugal

**T. García-Barrera**
Research Center of Natural Resources, Health and the Environment (RENSMA)
Department of Chemistry, Faculty of Experimental Sciences
University of Huelva
Huelva, Spain

**Michael Gochfeld**
Rutgers University
The State University of New Jersey
New Brunswick, New Jersey, USA

**Leandro A. Gomes**
MEtRICs
University of Catania
Catania, Sicily, Italy

**J.L. Gomez-Ariza**
Department of Chemistry. Faculty of Experimental Sciences. University of Huelva. Huelva. Spain.
Research Centre on Natural Resources, Health and the Environment. University of Huelva. Spain

**C. Gomez-de las Heras**
University Hospital Nuestra Señora de Valme
Sevilla, Spain

**Andreas Gradwohl**
University of Vienna
Department of Inorganic Chemistry
Vienna, Austria

**Franz Jirsa**
University of Vienna
Department of Inorganic Chemistry
Vienna, Austria

**Ketolly Natanne da Silva Leal**
Institute of Chemistry, Chaperone & Protein Biochemistry Laboratory- LaBiC
University of Campinas
Campinas, São Paulo, Brazil
and
Group of Espectrometry, Sample Preparation and Mechanization-GEPAM
Institute of Chemistry
University of Campinas
Campinas, São Paulo, Brazil

**Carlos Lodeiro**
BIOSCOPE Research Group, LAQV-REQUIMTE
Chemistry Department
NOVA School of Science and Technology
NOVA FCT
Universidade NOVA de Lisboa
Caparica, Portugal

**Davide Marzi**
Research Institute on Terrestrial Ecosystems
National Research Council (IRET-CNR),
Montelibretti, Rome, Italy
and
National Biodiversity Future Center (NBFC)
Palermo, Italy

**Juan José López-Mayán**
Group of Trace Elements, Spectroscopy and Speciation (GETEE)
Materials Institute (iMATUS)
Department of Analytical Chemistry, Nutrition and Bromatology
Faculty of Chemistry
University of Santiago de Compostela
Santiago de Compostela, Spain

**Antonio Moreda-Piñeiro**
Trace Element, Spectroscopy and Speciation Group (GETEE)
Instituto de Materiais (iMATUS)
Faculty of Chemistry
University of Santiago de Compostela
Santiago de Compostela, Spain

**C. Morro-Blazquez**
Faculty of Experimental Sciences
University of Huelva
Huelva, Spain

**Elisabete Oliveira**
BIOSCOPE Research Group, LAQV-REQUIMTE
Chemistry Department
NOVA School of Science and Technology, NOVA FCT
Universidade NOVA de Lisboa
Caparica, Portugal
PROTEOMASS Scientific Society
Costa de Caparica, Portugal
Canterbury Christ Church University
School of Human and Life Sciences, Life, Sciences Industry
Liaison Lab
Sandwich, UK

**Inês Pereira-Gomes**
BIOSCOPE Research Group, LAQV-REQUIMTE
Chemistry Department
NOVA School of Science and Technology, NOVA FCT
Universidade NOVA de Lisboa
Caparica, Portugal

**Ariana M. A Pintor**
LSRE-LCM, ALiCE
Faculty of Engineering
University of Porto
Porto, Portugal

**Carlos Henrique Inácio Ramos**
Institute of Chemistry, Chaperone & Protein Biochemistry Laboratory- LaBiC
University of Campinas
Campinas, São Paulo, Brazil

**Isabela Rocha**
Institute of Chemistry, Chaperone & Protein Biochemistry Laboratory- LaBiC
University of Campinas
Campinas, São Paulo, Brazil

**S. Sánchez-Espirilla**
Research Center of Natural Resources, Health and the Environment (RENSMA)
Department of Chemistry, Faculty of Experimental Sciences
University of Huelva
Huelva, Spain,
Research Foundation of Malaga in Biomedicine and Health (FIMABIS)
Málaga, Spain,
Department of Chemistry, Faculty of Sciences
National University of San Antonio Abad of Cusco
Cusco, Peru

**Hugo M. Santos**
BIOSCOPE Research Group, LAQV-REQUIMTE
Chemistry Department
NOVA School of Science and Technology, NOVA FCT
Universidade NOVA de Lisboa
Caparica, Portugal

**Alan J. Stewart**
School of Medicine, Medical and Biological Sciences Building
University of St Andrews
North Haugh, St Andrews, United Kingdom

**R.F Vélez-Pérez**
Research Center of Natural Resources, Health and the Environment (RENSMA)
Department of Chemistry, Faculty of Experimental Sciences
University of Huelva
Huelva, Spain

**Isabella T. Tomasi**
LSRE-LCM, ALiCE
Faculty of Engineering
University of Porto
Porto, Portugal

**Ana Justo-Vega**
Group of Trace Elements, Spectroscopy and Speciation (GETEE)
Materials Institute (iMATUS)
Department of Analytical Chemistry, Nutrition and Bromatology
Faculty of Chemistry
University of Santiago de Compostela
Santiago de Compostela, Spain

**Jakob Windisch**
University of Vienna
Department of Inorganic Chemistry
Vienna, Austria

**Massimo Zacchini**
Research Institute on Terrestrial Ecosystems
National Research Council (IRET-CNR)
Montelibretti, Rome, Italy
and
National Biodiversity Future Center (NBFC)
Palermo, Italy

**Gerald J. Zagury**
Department of Civil, Geological, and Mining Engineering
Polytechnique Montréal
Montreal (QC), Canada

# Introduction: Metals in Health and Environment
## *The Good, the Bad, and the Ugly*

*Carlos Lodeiro and Elisabete Oliveira*

Metals are abundant in the environment and play complex roles in health and ecological systems. These not only have a critical role of essential metals in biological processes to the toxic effects of heavy metal exposure, but their interaction with health and the environment present both benefits and risks. Understanding these dynamics is crucial, especially as industrialization, urbanization, and agricultural practices increase the occurrence of metal pollution. As such, this book aims to offer some tips regarding metal detection, quantification, and remediation in environmental and biological samples.

Advances in detection, quantification, and removal technologies are vital for safeguarding human health and the environment. This book explores the multifaceted impacts of metals, highlighting the importance of innovative techniques to detect, characterize, and mitigate negative effects while maximizing beneficial outcomes, their role in health, and their toxicological profiles in our environment.

## 0.1 THE GOOD: ESSENTIAL METALS AND THEIR ROLE IN HEALTH

Essential metals such as iron, zinc, copper, molybdenum, and magnesium are vital to human physiology. They are involved in a wide range of biological processes, including enzymatic reactions, immune responses, and cellular metabolism. Iron, for example, is essential for hemoglobin formation and oxygen transport, while zinc plays a central role in the immune system and wound healing. Copper contributes to the formation of connective tissue, whereas magnesium is crucial for energy production, neuromuscular activity, and cardiovascular health. Although these metals are required only in trace amounts, an inadequate supply can cause serious health problems such as anemia (iron deficiency), immune dysfunction (zinc deficiency), and cardiovascular issues (magnesium deficiency).

Iron is quite an interesting element from an astrophysical and thermodynamic viewpoint, occupying a remarkable position in the evolution of matter, as it represents the most energetically stable configuration of atomic nuclei. The binding energy per nucleon reaches its maximum around the iron–nickel group, meaning that energy is released when lighter elements fuse toward iron, whereas heavier elements release energy when they undergo fission or decay toward lighter nuclei. As a result of these phenomena, stellar nucleosynthesis naturally drives matter toward

DOI: 10.1201/9781003632092-1

iron-group elements. Massive stars sequentially fuse hydrogen, helium, carbon, oxygen, and silicon until an iron-rich core is formed. At this point, additional fusion mechanisms are no longer energetically favorable and gravitational collapse occurs. Conversely, heavier elements arise only during extreme, short-lived events such as supernova explosions or neutron-star mergers, where enormous energy fluxes push nuclei from their stable status toward more complex and heavy atoms. Over a cosmic timescale, many of these heavy nuclei undergo processes such as radioactive decay, or can also be recycled through successive stellar generations, progressively evolving toward more stable configurations. From a thermodynamic perspective, this trend reflects the universal tendency of physical systems to minimize free energy and maximize entropy. Iron-group nuclei represent a nuclear "energy minimum," meaning that these exist in a state from which no further net energy can be extracted through fusion or fission. As such, while metals do not literally decay into iron in a chemical or terrestrial contexts, the large-scale evolution of matter in the universe tends toward iron as the final, energetically exhausted product of nuclear processing.

## 0.2 CHALLENGES IN MONITORING AND QUANTIFYING ESSENTIAL METALS

The accurate monitoring of essential metals within the body is vital to ensure homeostatic levels are maintained without reaching toxic or non-sufficient concentrations. Traditional methods, such as atomic absorption spectrometry (AAS) and inductively coupled plasma mass spectrometry (ICP-MS), remain the primary methods for the determination of metal concentrations. However, these have associated limitations such as high operational costs and the need for sophisticated laboratory equipment.

Recent advancements in chemosensor and biosensor technologies have updated the processes for detection and quantification of metals in environment and biological samples. For instance, electrochemical and fluorescent bio(sensors) can offer portable, cost-effective, and highly sensitive solutions. Highly sensitive sensors have been described in recent literature reaching detection at nano- and even picomolar levels, providing precise measurements with minimal sample preparation. Moreover, the development of wearable biosensors improves real-time monitoring, particularly for individuals with health conditions who require constant metal level monitoring such as iron deficiency anemia or Wilson's disease (copper overload).

## 0.3 THE BAD: TOXIC METALS AND THEIR HEALTH IMPLICATIONS

While some metals are essential for health, others, such as lead, mercury, cadmium, and arsenic, are highly toxic toward humans and animals even in trace amounts. These metals pose significant health risks, with the potential to induce severe acute neurological, cardiovascular, and renal damage. Moreover, due to their non-biodegradable nature, these metals can accumulate in tissues over time, leading to chronic health issues as well as potential carcinogenic effects.

Environmental exposure is the primary source for toxic metal exposure in humans and animals, through industrial activities, mining, and improper waste disposal, all of which release significant amounts of metals into soil, water, and air. Lead, for example, is a persistent contaminant, especially in older urban environments due to the previous extensive use of lead-based paints and aging infrastructure. Lead exposure is mainly associated with neurotoxicity, particularly in children, leading to consequences such as cognitive impairments and subsequent developmental delays. Another example of environmentally derived heavy metal exposure is mercury. Mercury exposure often derived the consumption of methylmercury contaminated fish. This specific mercury species is highly neurotoxic and can impair cognitive and motor functions. Also, cadmium and arsenic exposure can occur mainly through contaminated water and agricultural products, leading to an increased risk of occurrence of kidney damage and cancer.

## 0.4 ADVANCES IN DETECTION AND QUANTIFICATION OF TOXIC METALS

An accurate determination of the presence of toxic metals accurately and sensitively is crucial for mitigating their impact. Although traditional techniques used to this end be reliable and have proven themselves over the years, their integration with new nano-based technologies and fluorescence-based sensors, can improve the overall sensitivity and specificity of the process. For example, through the combination of traditional fluorescence-base techniques with quantum dots, a type of nanomaterial, the sensitivity of the process has improved significantly. When quantum dots bound to a specific metal, even at low concentrations, a measurable fluorescent signal is produced, making it easier to determine and quantify it even at minimal concentrations.

Additionally, to improve the portability, sustainability and applicability of these sensors, paper or polymer-based analytical devices (PADs) have also had an increased interest in recent research, representing an emerging, low-cost, and portable detection method. These devices become ideal for field testing, providing fast results, requiring none to very low sample preparation. Subsequently, it can provide real-time information, essential for informing and triggeringimmediate field interventions, especially in areas with limited laboratory resources. Additionally, combining PADs with mobile applications enables real-time data collection, analysis and monitoring, providing a critical tool for tracking heavy metal contamination in both clinical and environmental settings.

## 0.5 THE UGLY: ENVIRONMENTAL CONTAMINATION AND ITS LONG-TERM CONSEQUENCES

Heavy metal contamination in the environment presents a severe, long-lasting and often irreversible ecological problem, disrupting ecosystems, affecting soil microorganisms, plants and animals. Among the various sources, industrial waste, mining, and improper agricultural practices are the major contributors to metal pollution in soil and water. Metals such as cadmium, lead, and mercury can persist in ecosystems, and their effects can suffer from biomagnification processes up the food chain, causing harm to wildlife and eventually human populations.

In soil, toxic metals can inhibit microbial activity, reducing soil fertility and disrupting the normal nutrient flow. Some direct consequences of this exposure involve stunted growth and reduced photosynthetic activity by plant species, which leads to severe chain consequences such as decreased agricultural yields. Through the consumption of these contaminated fishes, the nefarious effects of mercury are then propagated to other species, eventually, reaching and posing risks to human consumers. Bioaccumulation and biomagnification intensify the toxic effects of metals, with top predators, including humans, being exposed to highest concentration levels of these metals.

Addressing environmental contamination requires not only the detection of these toxic species but also requires the development of effective metal removal techniques. Traditional methods such as chemical precipitation and electrochemical treatment are widely used to this end, but have limitations, such as high costs, high energy consumption, and secondary waste generation. As a result, research has shifted its focus toward the development of sustainable and efficient alternatives.

### 0.5.1 Bioremediation

Bioremediation takes advantage of the capacity of microorganisms and plants to remove metals from the environment. Certain bacteria, fungi, and algae have evolved toward the development of inherent mechanisms to sequester and accumulate heavy metals from the outer environment. Phytoremediation, a plant-based approach, uses hyperaccumulator plants to absorb metals from soil and water, which can then be harvested and safely disposed of. Although these methods are cost-effective and environmentally friendly, they are often slow and limited to shallow soil contamination.

### 0.5.2 Nanoremediation

Nanotechnology has opened new possibilities for heavy metal removal. Several types of nanomaterials, such as magnetic nanoparticles or carbon nanotubes, can be used for metal remediation processes. Due to their high surface areas and unique properties, these materials are highly effective in adsorbing and removing metals from contaminated sites. Additional functionalization can be performed to these basal materials, for example, through the coating magnetic nanoparticles with ligands that bind selectively to metal ions, allowing for a selective removal process. Once bound, they can be easily separated from water using a magnetic field, making them particularly useful for water purification.

### 0.5.3 Electrokinetic Remediation

Electrokinetic remediation is the process by which, through the application of a low-voltage electric current to contaminated soils or sediments, metal ions are mobilized toward electrodes where they can be collected. This method is effective for the removal of metals from fine-grained soils, making it suitable for use in specific environments, such as land contaminated by mining or industrial waste. Although energy-intensive, recent developments in renewable energy integration are making electrokinetic remediation more sustainable and accessible.

### 0.5.4 Phytomining

Phytomining is an innovative approach that combines phytoremediation with economic value. It involves the cultivation of metal-accumulating plants in contaminated soils, followed by their harvesting and processing, obtaining metal-rich biomass. Through this approach two major benefits can be derived: (1) the cleaning of contaminated land while simultaneously (2) recovering valuable metals for reuse. Although it is still currently in development, phytomining holds promise for sustainable metal recovery, especially in regions with limited access to conventional remediation technologies.

Beyond environmental remediation, the monitoring of metal concentrations in the human body is also another crucial focal point, as it directly impacts human health, and excesses can be traced back to environmental contaminations. Excessive levels of essential metals can be harmful, while simultaneously, their deficiencies also pose health risks. As such, there is an increasing need for methods to regulate metal levels in patients and improve therapeutic interventions for metal-related conditions.

## 0.6 ADVANCES IN METAL CHELATION THERAPY

The standard treatment for heavy metal contamination in the body is chelation therapy. This therapeutic approach involves the administration of chelating agents that bind to the excess metals in the body and subsequently eliminate them through normal excretion pathways. Traditional chelators, like ethylenediaminetetraacetic acid (EDTA), are useful for a variety of heavy metal poisoning cases such as lead poisoning. However, severe side effects have also been reported, including depletion of essential metals. As such, research is focusing on the development of more selective chelators that target toxic metals without affecting beneficial ones. Several examples of novel chelators designed for mercury and arsenic poisoning have been reported in literature, showing improved selectivity while simultaneously reducing the risk of nutrient metal depletion.

## 0.7 METALLOMICS: A COMPREHENSIVE APPROACH TO METAL REGULATION

Metallomics is an emerging omics field that focuses on the role and effect of metals in biological systems on a comprehensive level, considering interactions with relevant biological agents, pathways, and their subsequent effects on health. It employs advanced analytical techniques, such as MS and nuclear magnetic resonance (NMR) spectroscopy, to provide detailed profiles of metal ions in biological samples. By understanding metal distribution patterns and interactions within the body, metallomics has the possibility to offer insights into disease mechanisms and help design targeted therapies for metal-related disorders. This comprehensive approach is particularly valuable for addressing complex conditions like Alzheimer's disease, where metal dysregulation is implicated.

## 0.8 GENE EDITING FOR METAL REGULATION

Cutting-edge genetic technologies, such as CRISPR-Cas9, can also provide innovative ways to control metal metabolism at the genetic level. For instance, through the modulation of expression of proteins involved in metal transport and storage. By enhancing or suppressing the expression of specific genes, it may be possible to correct or help to regulate the dysregulated metal levels, offering a new, personalized therapeutic approach for genetic conditions involving metal metabolism, such as Wilson's disease. Though it is still in the early stages of development, gene editing for metal regulation holds transformative potential for personalized medicine.

## 0.9 THE IMPORTANCE OF FUTURE RESEARCH AND TECHNOLOGICAL DEVELOPMENT

As metal pollution and metal-related health issues continue to pose challenges globally, ongoing research and technological advancements are essential. A comprehensive approach that integrates environmental monitoring, detection and remediation techniques, and theragnostic health-related approaches is necessary to address the complex interplay between metals, health, and the environment.

Other emerging technologies, such as artificial intelligence (AI) and machine learning (ML), are also being increasingly applied to metal detection and remediation protocols. AI-based models can be trained and specialized in the analysis of large datasets, based on years of already available research, to predict possible contamination hotspots, as well as facilitating and optimizing targeted interventions and resource allocation. ML-based algorithms can also help in the optimization of bioremediation processes by predicting the optimal conditions and effectiveness of the applied mechanisms under varying conditions, accelerating the cleanup and recovery of contaminated sites.

Regardless, beyond technological solutions, robust policies and public awareness are also essential for addressing metal pollution issues and promoting safe heavy metal handling, as it is crucial for a variety of industries. Official regulations that limit industrial emissions, enforce safe and intentional waste disposal standards, and restrict the use of hazardous metals can significantly reduce environmental contamination, leading to a decrease in all remaining subsequent negative consequences. In addition, public education campaigns on the safe disposal of electronic waste and minimization of exposure to toxic metals in consumer products are equally important for preventing health risks and raising awareness about the topic in the general population.

## 0.10 CONCLUSION

Metals play both positive and negative roles in health and environmental systems. Although essential metals are vital for human health, these can also cause negative health consequences due to their presence at concentrations higher than their homeostatic levels. As such, these might require careful monitoring to avoid the occurrence of both deficiencies and toxicity.

Conversely, toxic metals pose severe risks and can induce acute toxicity, even at low concentrations, underscoring the need for advanced highly sensitive detection and remediation techniques to mitigate their effects. Cutting-edge methodologies, including nanotechnology, biosensors, and bioremediation, offer promising solutions for controlling metal contamination in the environment. In a parallel approach, innovations in medical treatments, such as selective chelation therapy, metallomics, and gene editing, provide further theragnostic insights for managing and treating dysregulated metal levels in the body.

The challenges presented by metal pollution and toxicity require a multidisciplinary approach, combining technological innovation with policy, education, and public health initiatives. Nevertheless, continued research into new detection, quantification, and remediation methodologies remains essential for building a future where the beneficial roles of metals are maximized, while their adverse impacts on health and the environment are minimized. Embracing these advancements will enable society to harness the "good" in metals, mitigate the "bad," and minimize the "ugly" consequences of metal toxicity for a safer, healthier world.

# 1 Melding Metals Data for a Food Chain Analysis to Provide a Basis for Assessing Ecosystem and Human Health and Designing Biomonitoring

*Joanna Burger and Michael Gochfeld*

## 1.1 INTRODUCTION

Governmental agencies, Tribal Nations, public policy managers, land and resource managers, and the public are concerned about the health of humans and the environment, and require as much information as possible about several ecological indicators. These include data about geochemical and climatic conditions, population and ecosystems, diseases, and contaminants, to name a few. Contaminant levels are elevated in marine and coastal environments and still continue to increase because of direct pollution and runoff from rivers and the land, as well as due to atmospheric deposition [1, 2]. While lead and cadmium levels have generally declined, others have not [3, 4]. Some contaminants pose a threat to the marine organisms themselves, and to consumers that eat them, including humans [5, 6]. The contaminants bioaccumulate with increasing nodes or links on the food web, resulting in top trophic levels being most contaminated [7]. Assessing heavy metals in a complex food chain, however, is difficult because different agencies and people have different goals and objectives, and compromising among these is often difficult and time-consuming. Yet having such information for local and regional areas is essential for management and formation of policies and regulations, as well as for making remediation and restoration decisions [8–11].

Gathering enough information on different species is often difficult because of time, lack of personnel and funding, and projects are often limited to gathering information on one or two species and in one or two locations. These assessment programs are often essential for a specific management goal. This may be especially true in offshore marine ecosystems where large ships are required to complete the project. For some management needs, in-depth studies are required, in species examined, in spatial patterns, or in temporal patterns, or some combination. Some

 DOI: 10.1201/9781003632092-2

examples are the Great Lakes project that has been examining contaminant levels in birds for decades [11, 12] and the biota and contaminant studies in the Everglades [10]. These examples were designed as biomonitoring programs. Sometimes, however, a broad species assessment of contaminants is required to address an intractable public health concern.

The Department of Energy (DOE) had over a hundred sites to clean up [13, 14] and only about 15 remain today; some will require 50 years or more to complete cleanup at a cost of almost $400 billion [15–17]. In the early 2000s, DOE wanted to close Amchitka Island (in the Bering Sea of Alaska) which was a site of their nuclear tests in the late 1960s. Even though they had completed complicated ground water models and human health risk assessments that showed little risk, governmental and state agencies, Native Alaskans (Aleuts), commercial and subsistence fishers, and the general public were not convinced [18–20]. They were concerned that there might be radionuclide leakage, that the fish and other marine species were not safe to eat, that the ecosystem might be contaminated, and that DOE was not responsive to their concerns. The study was initiated due to this stakeholder pressure on the DOE to assure them that the organisms in the food web around the Amchitka Island were healthy and free of radionuclides and the foods were safe for human consumption. As part of this expedition, biota were also collected for mercury analysis because Consortium for Risk Evaluation with Stakeholder Participation (CRESP) felt mercury might pose a greater risk to people and biota.

Metal levels in more than 30 species representing different nodes on the food web of marine organisms from Amchitka Island and some other nearby islands were synthesized. Objectives of this chapter are to: (1) describe the multiple goals of stakeholders concerning metal threats to the biota living around Amchitka Island; (2) synthesize available data on cadmium, lead, mercury, and selenium that were collected from Nikolski, Adak, Kiska, and Amchitka Islands in the Aleutian Chain in the Bering Sea in 2004; (3) provide a food web approach; and (4) describe and discuss the advantages, disadvantages, and lessons learned from a broad-scale food chain approach to understanding metal levels. The chapter describes the entire process from stakeholders identifying a problem with an agency response, through a stakeholder-driven process that resulted in a scientific expedition involving key stakeholders to gather information on organisms that are part of a complex food web in the Aleutians, through data analysis to a report that allowed DOE managers to close a nuclear facility, with the land reverting to the United States Fish and Wildlife Service (USFWS).

## 1.2 BACKGROUND ON AMCHITKA ISLAND

The Atomic Energy Commission and the Energy Research and Development Administration, predecessors of the DOE detonated three underground nuclear tests (1965–1971) on Amchitka Island. This island (51°N lat, 179°E long) is located in the Bering Sea between Alaska and Russia (Figure 1.1). At the time, there was considerable controversy about nuclear testing at Amchitka because of potential health risks to humans (especially Native Aleuts), injury to marine ecosystems, and possible

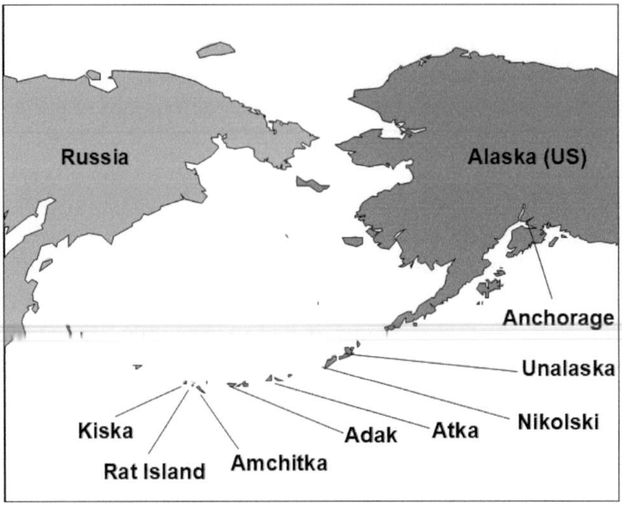


**FIGURE 1.1** Map showing the location of Amchitka and other islands in the Aleutian Chain of Alaska.

generation of tsunamis [21, 22]. People were not concerned about surface release of radiation because the radioactive material was probably vitrified by the intense heat of the blast [19, 23]. Conducting underground tests required construction of facilities as well as developing an airstrip. DOE was responsible for its cleanup, while the USFWS has landowner responsibility for Amchitka Island, which remains part of the Alaska Maritime National Wildlife Refuge.

The metal levels in all the invertebrates we sampled, including king crab, from the Aleutians were well below any established risk levels. Over the years from 1971, there has been concern about the possibility of subsurface transport of radionuclides from the three cavities into the marine waters around Amchitka because the region is one of the most seismically active and dynamic subduction zones on earth [24]. DOE finally removed all structures, remediated the surface contamination, and the surface was closed as part of the Alaska Department of Environmental Conservation's (ADEC) contaminated sites program [25]. When DOE announced that Amchitka remediation was complete and they intended to relinquish the land, years of frustration and anger resurfaced over DOE's lack of concern for people and their lack of understanding or disregard about the science base that stakeholders felt was necessary.

The primary concerns were: (1) whether the subsistence foods of the Aleuts were safe to eat, (2) whether the marine environment was contaminated, and (3) whether the commercial fish and shellfish from the island and region were safe to eat. At the time, Dutch Harbor in the Aleutians, the port for commercial fish in the Bering Sea, had the highest tonnage of fish landings in the world in the past several years, and provided 17% of Alaska's $811 million fish landings (2.3 million metric tons of fish [26]). Understanding baseline concentrations of contaminants is particularly important for the Bering Sea region as it is the site of intense commercial fishing. More than 90% of the world's fish catch comes from 10% of the world's oceans (including the Bering Sea [27]).

## 1.3 FROM RADIONUCLIDES TO METALS

The enormous controversy blossomed when, after finally removing all the buildings, structures, and debris from Amchitka Island (in 2001), DOE wanted to close the site and essentially turn over the site to Alaska and/or the USFWS. DOE always retains responsibility for any radionuclide remaining. The critical mechanism for fostering discussions was a joint meeting of all stakeholders that was held at the University of Alaska where all the major stakeholders and the public met. Stakeholders (ADEC, Aleutian Pribilof Island Association [APIA], USFWS, and DOE) signed an agreement that would: (1) include the four major stakeholder groups, (2) reassure the Aleuts and others that the subsistence and commercial foods were safe to eat and free from radionuclides, (3) reassure all stakeholders that there was not harm to the marine system and its species, (4) DOE would leave the radionuclides in the underground test sites in place, (5) CRESP would develop (and implement) a Science Plan to achieve the objectives (3) and (4), and (6) that would allow DOE to close Amchitka.

Once the first draft of the biological part of the Science Plan was written, the authors visited several villages in the Aleutians', and met first with the elders, then with the rest of the adult community, and finally with the children after the latter came to us and asked if they could also have a meeting. All groups were very informative and made many comments on the Plan, including that Aleuts come on the expedition to collect in their usual manner. This requires considerable effort, but four Aleuts were on the expedition. After visiting the Aleut villages, there were several meetings with key stakeholders, which included DOE-EM, National Nuclear Security Administration, National Oceanographic and Atmospheric Administration (NOAA), USFWS, ADEC, and the APIA, to hammer out the finer details of the expedition. There were also public meetings with Greenpeace and the public [18]. This public meeting was a critical component of the entire Science Plan. It was also essential to collect samples and analyze them for radionuclides. It was also essential to include stakeholders during all phases, including everyone who was affected or expressed an interest in the environmental safety of the organism in the Aleutian/Bering Sea ecosystem. There were also meetings after the expedition, which included a range of stakeholders such as subsistence and commercial fishers (both Aleut and NOAA that regulated fisheries), recreational fishers, resource trustees (USFWS, fauna of the Aleutians), and others in the public. The authors went back to the Aleut villages to report and discuss the findings.

The Science Plan dealt with radionuclide exposure [28], but CRESP insisted that samples should also be collected for analysis of mercury (and other metals) because we felt metals could possibly pose a risk to subsistence and commercial foods and to the local food chains that ultimately led to the Native Alaskans (Aleuts) and commercial fisheries [18]. DOE bore no responsibility for metals and thus analysis costs were not borne by DOE.

The 55-m retrofitted fishing ship would be away from Adak for about 3 weeks, with only limited (once/day or less) contact with shore and without any chance of rescue if something happened. With that in mind, the ship physician (MG) prepared a detailed Safety Plan. There was an established and clear line of authority to make decisions. If either the physician or the expedition leader thought some aspect of sampling was unsafe – it was not done. The divers were always present on the ship, as that was the most dangerous aspect of the trip, and the team on shore always had a tent and supplies to last a day should they get stranded on an island. The advice and knowledge of the Aleuts about winds, waves, storm conditions, and other aspects of these waters were critical to the success of the expedition. The importance of all three points (having a safety plan, a chain of command, and knowledgeable Aleuts) were critical to such an expedition. For instance, in a couple of cases an overenthusiastic person collecting samples wanted to stay on an island when command called them back (and soon it would not have been possible to land someone to retrieve them); in other circumstances people wanted to climb vertical rocks to reach nesting birds to collect eggs (a delicacy for Aleuts).

## 1.4 METHODS: METAL ANALYSIS

Under appropriate Alaska State and federal permits, body tissues were collected from algae and invertebrates, muscle tissues were collected from fish and birds, and one egg each was collected from eiders and gulls. Sufficient samples were collected for both radionuclides and metals. All tissues were frozen for later dissection and analysis. Dissections were conducted in the laboratory retrofitted in the hold of the ship. Collecting pieces of long algae was done on shore where the sometimes 20-m-long blades could be laid out. Dissections were necessary to reduce the biomass of birds and large fish. Samples were collected from Nikolski and Adak islands (currently inhabited by Aleuts), and Amchitka (site of nuclear tests) and Kiska islands (a battleground during the World War II); the latter two are not currently inhabited by humans.

All metal analyses were conducted in the Elemental Analysis Laboratory of the Environmental and Occupational Health Sciences Institute (Rutgers University) in Piscataway. A 2-g (wet weight) sample of tissue was digested in Ultrex ultrapure nitric acid in a microwave (CEM, MDX 2000), using a digestion protocol of three stages of 10 minutes each under pressure of 50, 100, and 150 pounds per square inch (3.5, 7, and 10.6 $kg/cm^2$) at 70% power. Digested samples were subsequently diluted to 25 mL with deionized water. All laboratory equipment and containers were washed in 10% $HNO^3$ solution and rinsed with deionized water prior to each use. Mercury was analyzed by cold vapor technique, and the other elements were analyzed by graphite furnace atomic absorption. All concentrations are expressed in ppm, wet weight. The mercury in avian tissues is usually 90–100% methylmercury [29, 30]. Detection limits for the equipment were: 0.02 ppb for cadmium, 0.15 ppb for lead, 0.2 ppb for mercury, and 0.7 ppb for selenium. All specimens were analyzed in batches with known standards, calibration standards, and spiked specimens. Recoveries ranged from 88 to 102%. Batches with recoveries of less than 85% were reanalyzed. The coefficient of variation on replicate, spiked samples ranged up to 10%. In the sections that follow, we describe particular phases of the research, specifically noting the advantages or problems with each phase. Statistical analyses were generated on Statistical Analysis Systems [31].

## 1.5 COLLECTORS AND COLLECTING

The overall approach was to collect species representing plants, invertebrates, fish, and birds. One of the unique features of this study is that the species collected to examine different nodes on the food chain were collected by three teams, representing the major interests in metal exposure: Aleut fishers/hunters, Rutgers University/University of Alaska scientists (the research vessel team), and a scientist on a NOAA ship collecting to representing the commercial fisheries. The university team comprised a land-based team and a team of divers. The Aleuts were collecting foods that they would gather if they were stranded on each beach or island, while the scientists were collecting samples along transects.

While the initial Science Plan specified species to be collected, there were two very important modifications: (1) After discussions with Aleuts in their villages prior to the expedition, four Aleuts were added to the collecting team, and species were added that were important subsistence foods (e.g., octopus, bird eggs, additional fish were added). (2) Once on the vessel, additional subsistence foods were added that were not on our list. The addition of species after visiting the villages necessitated seeking permits for some species. The possible addition of species once on the ship required the State of Alaska to add this provision to permits. Once it became clear that the target species might vary by both their availability and the need to add more subsistence foods, the biological team planned a week in Adak prior before the expedition to try out the sampling and preparation methods. This turned out to be a very important component that ensured success of the project because it allowed the team to substitute one species for another because of the availability of a given species or the difficulty of collecting them. Further, it allowed time to add additional species to collecting permits.

## 1.6 SPECIES COLLECTED FOR A FOOD CHAIN ANALYSIS

Table 1.1 shows all the species (and their scientific names) collected along with: (1) whether they were initially on the sampling list or were added by the Aleuts as subsistence foods; (2) their trophic level; (3) what team was to collect them; and (4) the maximum and mean mercury levels. Duncan letter groupings for each trophic level are given in the table. Samples were collected from more than 30 species. However, 11 species and eggs from 2 species were not on the original Science Plan (Table 1.1 [32]). Once on the island, the expedition leader (JB) directed the Aleuts to collect what they would collect if they were stranded there, a challenge they had faced in their lives. The presence of the Aleuts on the expedition was critical to collecting their main subsistence foods and proved to be a real advantage in assuring that other Aleuts in the villages approved of the species collected and methods used. Lastly, the initial Science Plan had a standardized sampling regime, the same number of samples of each species from different islands were to be collected. This proved unrealistic. Sometimes the fish or birds were simply not present where we had expected to find them, and sometimes the stormy weather precluded sampling at a particular site.

**TABLE 1.1**
**Samples Collected on the Amchitka Island Expedition, and Mercury Levels (ppm)**

| Common Name | Scientific Name | Trophic Level | Original Collect (A, C, U) | n | Max | Mean ± SE | Duncan Letter Grouping |
|---|---|---|---|---|---|---|---|
| **ALGAE** | | | | | | | |
| Kelps (Brown algae) | *Hedophyllum sessile* | PP | U | 15 | 0.001 | 0.001 ± 0.000 | (C)' |
| Kelps (Brown algae) | *Fucus distichus* | PP | | 22 | 0.002 | 0.001 ± 0.000 | (B,C) |
| Sea lettuce | *Ulva lactuca* | PP | U | 12 | 0.003 | 0.002 ± 0.000 | (A) |
| Kelps (Brown algae) | *Alaria nana* | PP | | 52 | 0.039 | 0.005 ± 0.001 | (A,B) |
| **INVERTEBRATES** | | | | | | | |
| Green sea urchin | *Strongylocentrotus polyacanthus* | GR | U | 30 | 0.016 | 0.005 ± 0.001 | (F) |
| Limpet | *Tectura scutum* | GR | | 10 | 0.022 | 0.009 ± 0.002 | (E) |
| Chiton | *Cryptochiton stelleri* | GR | U | 20 | 0.021 | 0.010 ± 0.001 | (E) |
| Blue mussel | *Mytilus trossulus* | FF | U | 148 | 0.040 | 0.016 ± 0.001 | (D) |
| Rock jingle | *Pododesmus macroschisma* | FF | U | 20 | 0.026 | 0.018 ± 0.001 | (C,D) |
| Oregon triton | *Fusitriton oregonensis* | GR | | 36 | 0.096 | 0.024 ± 0.004 | (C,D) |
| Octopus | *Octopus dofleini* | PR | | 5 | 0.064 | 0.038 ± 0.011 | (B,C) |
| Golden king crab | *Lithodes aequispina* | GR/SC | C | 14 | 0.070 | 0.043 ± 0.005 | (B) |
| Red king crab | *Paralithodes camtschaticus* | GR/SC | C | 5 | 0.205 | 0.101 ± 0.027 | (A) |
| **FISH (muscle)** | | | | | | | |
| Sockeye salmon | *Oncorhynchus nerka* | K/PR | U | 15 | 0.066 | 0.042 ± 0.005 | (H) |
| Atka mackerel | *Pleurogrammus monopterygius* | K/PR | C, U | 19 | 0.077 | 0.046 ± 0.004 | (F,G,H) |
| Pacific ocean perch | *Sebastes alutus* | K/PR | C, U | 17 | 0.139 | 0.048 ± 0.009 | (G,H) |
| Northern rock sole | *Lepidopsetta polyxystra* | P/PR | | 15 | 0.133 | 0.068 ± 0.007 | (D,E,F,G) |
| Walleye pollock | *Theragra chalcogramma* | PR | A, C | 12 | 0.215 | 0.074 ± 0.018 | (E,F,G,H) |
| Rock sole | *Lepidopsetta bilineata* | PR | | 27 | 0.247 | 0.092 ± 0.011 | (C,D,E,F) |

*(Continued)*

**TABLE 1.1** *(Continued)*
**Samples Collected on the Amchitka Island Expedition, and Mercury Levels (ppm)**

| Common Name | Scientific Name | Trophic Level | Original Collect (A, C, U) | n | Max | Mean ± SE | Duncan Letter Grouping |
|---|---|---|---|---|---|---|---|
| Rock greenling | *Hexagrammos lagocephalus* | PR | A | 84 | 0.493 | 0.099 ± 0.010 | (D,E,F,G,H) |
| Dolly Varden | *Salvelinus malma* | PR | C, U | 75 | 0.552 | 0.114 ± 0.013 | (E,F,G,H) |
| Red Irish lord | *Hemilepidotus hemilepidotus* | PR | A | 57 | 0.335 | 0.130 ± 0.010 | (C,D) |
| Pacific halibut | *Hippoglossus stenolepis* | PR | A, C | 24 | 0.928 | 0.158 ± 0.044 | (C,D,E) |
| Black rockfish | *Sebastes melanops* | P/PR | | 68 | 0.506 | 0.167 ± 0.015 | (B,C) |
| Pacific cod | *Gadus macrocephalus* | PR | A, C | 142 | 0.859 | 0.173 ± 0.011 | (B,C) |
| Flathead sole | *Hippoglossoides elassodon* | PR | | 39 | 0.517 | 0.277 ± 0.013 | (A) |
| Yellow Irish lord | *Hemilepidotus jordani* | PR | | 68 | 0.944 | 0.281 ± 0.024 | (A,B) |
| Great sculpin | *Myoxocephalus polyacanthocephalus* | PR | | 27 | 1.096 | 0.366 ± 0.058 | (A) |
| **BIRDS** | | | | | | | |
| Tufted puffin | *Fratercula cirrhata* | PR | A | 21 | 0.257 | 0.121 ± 0.016 | (D) |
| Common eider | *Somateria mollissima* | PR | A, U | 20 | 0.182 | 0.123 ± 0.008 | (C,D) |
| Bald eagle (chicks) | *Haliaeetus leucocephalus* | PR | | 2 | 0.224 | 0.128 ± 0.097 | (D) |
| Common eider egg | *Somateria mollissima* | PR | | 51 | 0.409 | 0.191 ± 0.010 | (C) |
| Glaucous-winged gull | *Larus glaucescens* | PR | U | 32 | 0.851 | 0.329 ± 0.030 | (B) |
| Glaucous-winged gull egg | *Larus glaucescens* | PR | | 21 | 0.808 | 0.364 ± 0.037 | (A,B) |
| Pigeon guillemot | *Cepphus columba* | PR | | 21 | 0.855 | 0.494 ± 0.037 | (A) |
| **MAMMAL** | | | | | | | |
| Sea lion muscle | *Eumetopias jubatus* | PR | | 1 | | 1.03 | |
| Sea lion liver | *Eumetopias jubatus* | PR | | 1 | | 69.5 | |

*Note:* For tropic level, PP = primary producer, GR = grazer, FF = filter feeder, P = plankton, K = krill, SC = scavenger, PR = predator. For collector: A = Aleut, C = commercial fishing trawler, and U = university scientists.

## 1.7 MERCURY, OTHER METALS, AND TROPHIC LEVELS

The sampling plan was to collect primary producers, invertebrates, fish, and birds, making sure that these included typical and subsistence foods, invertebrates, fish that were commercially important, and birds that were scavengers and predators and were consumed by Aleuts. Thus, the species were selected to provide information for subsistence Aleuts, commercial fishers, resource trustees, and the public. A study by Hamrick and Smith [33] about foods eaten in Unalaska (where Dutch Harbor is located) reported that the most frequently eaten foods were halibut, salmon (several species), seal oil, cod, and king crab. Although the initial discussions included some marine mammals, the permitting process did not allow for this option. In the following sections, key points with each node, including primary producers, invertebrates, fish and birds, are discussed with a focus on mercury levels.

## 1.8 PRIMARY PRODUCERS

Kelp are important bioindicators because they are primary producers that are eaten by higher organisms trophic levels, including livestock and people [34]. Algae in the form of dried tablet are sold in health food stores in North America [35]. Algae are also used in phytoremediation because they bioaccumulate metals in vacuoles rich in polyphenols [36, 37]. They are part of the sea urchin and sea otter (*Enhydra lutris*) food chain. Kelp and other algae are eaten not only in the Aleutians, but also in other places in the Arctic [38], Canada [39], and Asia [35]. Dried and tablet algae are appearing in health food stores [35].

The kelp and sea lettuce tested from Amchitka had the lowest levels of mercury (Table 1.1 [40]). Overall level for lead in *Alaria nana* was 1.0 ± 0.1 ppm and for selenium it was 4.4 ± 1.0 ppm. There were significant locational differences in mercury and selenium levels in *Alaria* as well [40]; selenium was significantly higher in one cove than another. We were particularly interested in whether there were differences as a function of different parts of the algae. There were no significant differences for lead, mercury, and selenium, but there were differences for cadmium. Cadmium was significantly higher in the holdfast (1.1 ± 0.2 ppm) compared to other parts (tip and blade were 0.6 ± 0.1 ppm, stipe was 0.4 ± 0.1 ppm, $X^2 = 9.2$ ($p < 0.06$)).

## 1.9 INVERTEBRATES

The Science Plan called for collecting invertebrates in several categories: urchins, mussels, and crabs that occurred around the Aleutian Islands, but it was unclear which species could be collected. The species collected along with their trophic level and mercury levels are shown in Table 1.1. These species were filter feeders, scavengers, grazers, and predators (octopus). Octopus proved to be the hardest to find. Red king crab, a deeper water species, was difficult to find as well. It is noticeable that the mean mercury levels of all the invertebrates were higher than the mean levels for algae.

The largest sample size was for blue mussel, from Adak, Amchitka Island, and Kiska Island, and there were no inter-island differences [32]. However, there were

**TABLE 1.2**
**Mean Metal Levels (± SE, ppm, Wet Weight) as a Function of Sampling Location for Blue Mussels from Adak Island in the Aleutian Islands. All Metals Differed Significantly with p > 0.001 ($X^2$ Test). The Letters after the Mean Indicate Significant Differences (Sample Size was 25 for Each Island)**

| Island | Cadmium | Lead | Mercury | Selenium |
|---|---|---|---|---|
| Adak Harbor | 0.99 ± 0.88 | 0.23 ± 0.021 | 0.024 ± 0.001 | 0.69 ± 0.019 |
| Breakwater/Jetty | 0.81 ± 0.06 | 1.02 ± 0.15 | 2.96 ± 0.19 | 0.70 ± 0.03 |
| Airport Creek | 1.51 ± 0.15 | 0.29 ± 0.11 | 0.02 ± 0.001 | 0.90 ± 0.12 |
| Navfac Creek | 0.76 ± 0.05 | 0.60 ± 0.33 | 0.01 ± 0.0004 | 1.76 ± 0.64 |
| Clam Lagoon | 0.88 ± 0.08 | 0.33 ± 0.21 | 0.01 ± 0.0007 | 0.44 ± 0.06 |

*Source:* After Burger and Gochfeld [12].

significant differences among the different sampling locations on Adak for all metals (Table 1.2 [41]). No one sampling site had the highest or lowest level of all metals. Although the differences were statistically significant, this was due to the small variances at sites, so that relatively small differences were statistically significant. The team had predicted that the highest metal levels would occur in the biota of the Adak, because of prior shipping activity and direct deposition from land-based activities. Clam Lagoon on Adak, distant from human activities, was expected to have the lowest levels; however, this was not the case (Table 1.2). Currents carrying sediment accounted for the high values of lead and mercury at the base of the breakwater jetty.

## 1.10 FISH

The metal levels in fish were a major concern for all stakeholders because fish forms an important part of subsistence diet for Aleuts, and the commercial fishery of the Bering Sea out of Dutch Harbor is one of the largest in the world. The fish sampled represent the major fisheries (e.g., walleye pollock, Pacific cod, Pacific halibut, and salmon). All of the fish sampled in this study are eaten by Aleuts. Several aspects, including sampling, tissue selected, among and within islands, and fish age, that affect metal levels in fish are discussed in subsequent text.

One important question in sampling is always – Were the right species and sizes of species collected from the right locations? Since there were three teams collecting fish, the sizes of the main fish sampled could be compared (Table 1.3). For all fish, there were no significant differences in the mean length of fish collected, except for red Irish lord and rock greenling [42]. The mean weight of fish collected by the university divers, however, was lower for rockfish, rock greenling, and red Irish lord, although the divers collected heavier yellow Irish lords than the Aleuts ([42] Table 1.3).

**TABLE 1.3**
**Mean (± SE) Length (cm) of Fish Collected from Islands in the Aleutian Chain of Alaska. Not All Fish were Collected by the NOAA Team (after [42]). There were No Significant Differences in Length Except for Red Irish Lord and Great Sculpin**

| Species | Diving Team | Aleut Team | NOAA Trawl |
|---|---|---|---|
| Atka mackerel | 44 ± 2 | 42 ± 2 | 10 ± 0.5 |
| Dolly Varden | 32 ± 1 | 31 ± 1 | |
| Flathead sole | 40 ± 1 | 38 ± 1 | |
| Great sculpin | 49 ± 2 | 5 ± 2 | |
| Pacific halibut | 84 ± 40 | 81 ± 7 | 62 ± 15 |
| Pacific cod | 60 ± 3 | 61 ± 2 | 64 ± 5 |
| Rock sole | 33 ± 1 | 35 ± 2 | 37 ± 1 |
| Rockfish | 37 ± 4 | 37 ± 1 | 40 ± 1 |
| Rock greenling | 33 ± 0 | 35 ± 1 | |
| Red Irish lord | 28 ± 1 | 34 ± 1 | |
| Yellow Irish lord | 41 ± 1 | 40 ± 1 | |

### 1.10.1 Interspecific Differences for Fish

There were interspecific differences for all of the metals examined (Table 1.4 [43]). Mercury levels were highest in sculpin and lowest in Dolly Varden. Highest values are shaded, indicating no correlation between cadmium with lead or mercury and selenium. The mean values for mercury and selenium are highly correlated (r = +0.70). This may reflect the high affinity of mercury and selenium for forming insoluble selenides [44].

**TABLE 1.4**
**Mean (± SE) Metal Levels in Fish from the Amchitka Expedition (ppm, Wet Weight [43]). All p Values were < 0.0001 with Kruskal Wallis 1-Way ANOVA**

| Fish Species | Cadmium | Lead | Mercury | Selenium |
|---|---|---|---|---|
| Black rockfish | 0.018 ± 0.011 | 0.002 ± 0.0008 | 0.145 ± 0.018 | 0.527 ± 0.022 |
| Dolly Varden | 0.002 ± 0.0011 | 0.022 ± 0.005 | 0.114 ± 0.013 | 0.349 ± 0.036 |
| Pacific halibut | 0.010 ± 0.004 | 0.028 ± 0.01 | 0.158 ± 0.044 | 0.373 ± 0.029 |
| Great sculpin | 0.004 ± 0.001 | 0.012 ± 0.004 | 0.294 ± 0.054 | 0.613 ± 0.035 |
| Pacific cod | 0.009 ± 0.001 | 0.061 ± 0.006 | 0.173 ± 0.012 | 0.183 ± 0.007 |
| Rock greenling | 0.002 ± 0.0003 | 0.016 ± 0.004 | 0.099 ± 0.014 | 0.188 ± 0.016 |
| Yellow Irish lord | 0.007 ± 0.001 | 0.051 ± 0.012 | 0.294 ± 0.058 | 0.321 ± 0.019 |
| Pink salmon | 0.003 ± 0.001 | 0.027 ± 0.012 | 0.042 ± 0.005 | 0.246 ± 0.026 |
| Flathead sole | 0.004 ± 0.001 | 0.050 ± 0.008 | 0.276 ± 0.013 | 0.398 ± 0.037 |
| Rock sole | 0.009 ± 0.004 | 0.043 ± 0.016 | 0.095 ± 0.023 | 0.374 ± 0.054 |

**TABLE 1.5**
**Metal Levels (ppm, Wet Weight) as a Function of Tissues for Great Sculpin (after [45]). The Letter after Each Value Indicates the Significant Difference. $X^2$ with Kruskal Wallis 1-Way ANOVA**

| Metals | Kidney | Liver | Muscle | $X^2$ |
|---|---|---|---|---|
| Cadmium | 0.063 ± 0.02 (B) | 1.26 ± 0.3 (A) | 0.004 ± 0.001 (B) | 38 (0.01) |
| Lead | 0.048 ± 0.003 (A) | 0.035 ± 0.08 (A) | 0.014 ± 0.004 (B) | 18 (0.001) |
| Mercury | 0.11 ± 0.02 (B) | 0.29 ± 0.05 (A) | 0.36 ± 0.06 (A) | 15 (0.006) |
| Selenium | 1.86 ± 0.2 (A) | 2.08 ± 0.2 (A) | 0.61 ± 0.03 (B) | 32 (0.001) |

### 1.10.2 Tissue Differences for Fish

Muscle tissue was collected for all fish and birds because it is the main tissue consumed by people and other animals. However, for some fish we examined liver tissue as well [45]. As an example, we provide the values for great sculpin (Table 1.5). As is typical of most fish, the metal concentrations are higher in liver than in muscle. These data indicate the importance of examining tissue levels, particularly given that birds and other biota often eat the whole fish (except the skeleton). The great sculpin forages on the bottom over sand and mud [46].

Tissue differences were also examined in Pacific cod. Selenium levels were significantly higher in liver than muscle tissue (1.6 ppm in liver vs. 0.18 ppm in muscle). However, mercury levels were similar in muscle than the liver (0.17 ppm in muscle vs. 0.14 ppm in the liver) [47, 48]. Examining levels in liver and kidney is important for the fish as it is an indication of potential harm to the fish itself, while muscle levels indicate potential harm to consumers of fish (people as well as other biota on the food chain). Once fish were collected, tissues other than muscle can be taken and analyzed. However, it is often unclear which species may have levels that need further examination, such as the high level of mercury in muscle in Pacific cod. In this study, it was logistically impractical to go back after analysis to take more samples to determine whether our result occurred in other years.

### 1.10.3 Locational Differences in Fish

One of our objectives was to determine if there were differences within and among islands in metal levels in the fish. There were some significant differences both within and among islands, but the differences were usually not great. However, the highest levels for cadmium, mercury, and selenium (but not lead) were on Adak, the island with a long history of human occupation and military activity, providing many opportunities for metal contamination [43]. Mean mercury levels in Pacific cod from Adak was 0.34 ppm, which is a cause for concern, partially offset by a Se:Hg molar ratio of 2 (Table 1.6 [32, 49]).

There were some locational differences of note for other species. For example, mercury levels in ocean perch were three times higher at Amchitka than at Kiska; mercury levels in chiton were three times higher at Adak than Amchitka; and levels

**TABLE 1.6**
**Mean Concentrations (ppm, Wet Weight) of Metals in Pacific Cod Muscle from the Aleutians Comparing Islands [48]. $X^2$ with Kruskal Wallis 1-Way ANOVA**

| Metals | Nikolski | Adak | Amchitka | Kiska | $X^2$ (p) |
|---|---|---|---|---|---|
| Cadmium | 0.0001 ± 0.0004 | 0.013 ± 0.004 | 0.012 ± 0.001 | 0.004 ± 0.001 | 41 (0.0001) |
| Lead | 0.060 ± 0.014 | 0.022 ± 0.011 | 0.076 ± 0.009 | 0.038 ± 0.006 | 11.6 (0.009) |
| Mercury | 0.15 ± 0.02 | 0.32 ± 0.07 | 0.20 ± 0.017 | 0.11 ± 0.012 | 18.1 (0.0003) |
| Selenium | 0.12 ± .016 | 0.28 ± 0.067 | 0.18 ± 0.007 | 0.19 ± 0.011 | 16.1 (0.001) |

in Dolly Varden were seven times higher at Amchitka than Adak, likely due to their being in a land-locked lake created during the 1971 Cannikin nuclear explosion [22]. Dolly Varden from Adak were migratory and entered streams to spawn. The main difficulty with examining differences within and among islands derives from the overall difficulty of obtaining samples of each species from every island. This could be remedied by a longer sampling period, more personnel, or reduced sample sizes or number of species. The advantage is that information can be used to inform cleanup decisions (if they are necessary), and to provide information to consumers about what places to avoid for fishing. All of the islands, however, were occupied by Aleut communities for centuries, and Adak, Amchitka, and Kiska saw military occupations in World War II.

### 1.10.4 Fish Size and Age

Although it is generally assumed that for many species of fish, levels of metals such as mercury increase with the length of the fish. However, this is not always the case [50, 51]. There was no significant positive correlation for size and mercury levels for northern rock sole, flathead sole, and Atka mackerel, and the correlations were low for yellow Irish lord (correlation coefficient of [$r = 0.2$, $p > 0.01$]). The highest correlations were for Pacific ocean perch ($r = 0.6$, $p > 0.0006$) and walleye pollock ($r = 0.5$, $p < 0.02$) due to small sample sizes. This suggests that further studies should examine this relationship for pollock as they are one of the major commercial catches from Dutch Harbor. Length is used as a surrogate for age in fish [52]. It is difficult to age fish and usually size is a surrogate for age but examination of otoliths (small ear bones) can be used to do so. Otoliths were removed from a sample of 49 Pacific cod and were aged by Delsa Anderl of NOAA [47]. Multiple regression models (SAS, PROC GLM, 2005) indicated that age affected the levels of mercury and selenium in the liver and muscle tissue (but not cadmium). The otolith age accounted for more metal variation than length. Selenium decreased significantly, while mercury increased [32, 48]. Examining differences in metal levels as a function of age (using otoliths) would be useful, because it provides better information on age than using the surrogate of length. For example, mercury levels positively correlated with age in Pacific cod (0.45 ppm), compared to 0.38 ppm for both weight and length [48]. Selenium levels did not correlate with any of these three measurements. While it is easy to learn how to extract otoliths from the fish (but the process is time-consuming), it is costly to age fish from otoliths and it is not possible for all fish.

## 1.11 BIRDS

Birds and eggs were collected because they are popular subsistence foods on the Aleutians and may represent several levels on the food chain. Gulls are top-level predators, eating a variety of foods including gull chicks in season. Common eiders, feeding mainly on mussels, are mid-level predators [53–55]. Gulls are relatively easy to find on these islands because they nest conspicuously in colonies and because they mob people approaching their nests. Eiders nests are solitary, usually hidden in vegetation on all the islands. Aleuts collected gulls by shooting, and female eiders were grabbed as they struggled through tall grass when flushed from the nest. Mercury levels were high in the eggs of the two bird species examined (Table 1.7). Guillemots were mainly fish eaters and had the highest mercury levels in their flesh. The bald eagle chick was half-grown and had been fed crabs and rats, which explains their low mercury level.

The most surprising observation was that the mercury level in bird eggs was similar to the level in adults. The high levels in eggs are of concern because the Aleuts prize them and likely eat more at a time than is advisable. Cadmium and lead were low in eggs but differed by an order of magnitude from each other. Eiders are lower on the food chain than gulls. Bald eagle chicks had quite low levels compared to the gulls (Table 1.7), but the sample size was low (n = 2) because the authors couldn't stand to kill any. Perhaps the chicks did not have enough time to accumulate metals. The bald eagles were collected at the request of USFWS.

Sea lion was not part of the expedition protocol, but Native Alaskans have the right to hunt marine mammals, and the Aleuts shot one. After the Aleuts donated some of the tissues to the Alaska Museum, the museum requested that CRESP analyze the tissues for mercury. The levels of mercury were the highest of all the other

**TABLE 1.7**
**Mean (± SE) Metal Levels (ppm, Wet Weight) for Birds from the Aleutian Islands [56–59]. Given Are Levels in Their Muscle**

| Species | Cadmium | Lead | Mercury |
|---|---|---|---|
| Common eider | | | 0.12 ± 0.008 |
| Eggs | 0.076 ± 0.013 | 0.21 ± 0.89 | 0.19 ± 0.022 |
| Glaucous-winged gull | | | 0.33 ± 0.03 |
| Eggs | 0.14 ± 0.019 | 0.016 ± 0.012 | 0.36 ± 0.97 |
| Pigeon guillemot | | | 0.49 ± 0.04 |
| Bald eagle (chick) | | | 0.13 ± 0.97 |

species and specially of the liver were exceedingly high (69.5 ppm, wet weight). The high levels found in the liver in the present study were similar to those reported for sea lions in California [60] and Oregon [61]. Interestingly, Buhler et al. [60] reported that only 1.5–3.7% of the mercury in the liver was methylmercury, while almost a 100% of mercury in the muscle was methylmercury. It is methylmercury that has the most toxic effect in animals [62].

## 1.12 LESSONS LEARNED FROM SAMPLING A FOOD WEB

As with any study, there were methodological issues and lessons learned from sampling, and they varied by the node on the food chain. Overall, except for algae and some invertebrates (e.g., mussels) it was difficult to balance sampling from different sites, both within an island and between islands. Every species differed in abundance and accessibility at different sites. The more mobile the organism, the harder it was to find and collect them. This was partly ameliorated by selecting species that were ecological counterparts and played the same role within the food chain. Other difficulties specific to species groups are discussed in subsequent text.

The difficulty of collecting some kelp was their extreme lengths and weights. Pulling large masses of kelp to shore that could be 30-m long and were intertwined was very difficult. Furthermore, kelp beds were often dense and difficult to walk through, and walking in the deeper intertidal spaces requires hip boots and balance. Kelp, however, were abundant and easy to find, and were often close enough to the beach to be collected from shore, which would make them useful for monitoring. However, their extreme length made it cumbersome and time-consuming to lay and sample efficiently. Furthermore, the differences in metals in different sites on the kelp makes it important to first ascertain if there are such differences among species of kelp. Some algae, such as sea lettuce were small enough to easily collect and process.

Some species of invertebrates were very difficult to collect in numbers (e.g., king crab, octopus), while others were much easier. King crabs were caught only by the NOAA trawler. The Aleut team spend many hours trying to capture octopus, which is a delicacy. There was little difficulty in collecting other species (e.g., limpet, chitons, mussels). Blue mussels were collected at low tide when the mussels were exposed and could be collected by wading from shore. The sampling challenge is that the mussels are clumped, and we collected from different parts of a mussel bed to represent possible differences in exposure of these sessile organisms. Thus, mussels were collected from different, isolated rock formations. There are no standard protocols for collection of mussels and the pooling of samples is seldom mentioned. This is unfortunate as there are numerous papers on metals in mussels [e.g., 63–65]. It is difficult to compare levels with the US Mussel Watch because mean levels for individual sites are not readily available [66].

Fish were a problem because it was nearly impossible to collect the same number of each and of the same size or a range of sizes. Without the Aleut team the expedition would not have been able to collect some species that had very specific habitat types, times of day they were active, and other very specific requirements.

## 1.13 HEALTH IMPLICATIONS FOR ALGAE, INVERTEBRATES, AND BIRDS

The importance of examining different nodes on the food chain is to understand which species may be at risk from their internal levels of heavy metals, which scavengers and predator might be a risk from eating plants and other animals, and whether there are groups of species at risk, as well as to develop future management and biomonitoring plans. Mercury, lead, and cadmium are all common contaminants, commonly analyzed in marine waters due to natural sources as well as anthropogenic ones [62, 67].

In many places in the world, algae are an important human food source [34, 38], including the Aleuts [68]. They are also becoming more common in traditional medicine in non-traditional places (such as the US). Thus understanding potential risk from consumption is useful. Algae take up metals in proportion to the concentration in the water [69]. The level of cadmium (mean of 0.7 ppm), lead (mean of 1.0 ppm), and selenium (mean of 4.4 ppm) were sufficiently high to suggest that predators, including humans, may be at risk from consuming *Alaria*, at least from one of the coves [40]. That is, adverse effects have been demonstrated at dietary levels from 0.1 to 1.0 ppm for cadmium [70], from 0.1–0.5 ppm for lead [71], 1.0 ppm for selenium [72], and 0.2 ppm for mercury [73]. The variation within each *Alaria* and among coves suggests that more attention should be given to examining metal levels and human consumption patterns of algae. The differences in metal levels in different parts of each *Alaria* has two implications: (1) metal levels were highest in the holdfast for some, suggesting it not be used for human consumption, and (2) levels of lead, mercury, and selenium did not differ by location on the kelp leaf blade itself.

The king crab industry is an important component of the seafood industry that is processed and shipped from Dutch Harbor. As with all metals, it is important to determine whether levels could pose a risk to the organism themselves, or to their predators (including humans). There are few consumption limits for shellfish, although WHO [74] set a limit of 0.5 ppm (wet weight for mercury in shellfish). The Codex Alimentarius, a joint FAO/WHO agency proposed a standard of 1.0 ppm for cadmium and 0.5 ppm for lead in mussels [73]. Mean lead and cadmium levels in mussels were above this standard for some of the sampling sites, which presents some cause for concern for human consumption. Mercury levels were well below the WHO standards and any established risk levels in all the invertebrates we sampled from the Aleutians, including king crab. Selenium levels of 1.0–2.6 ppm are known to be toxic for some animals [72] and the mussels from Adak were in this range, which could pose concern for both human and other biota.

While the flesh of gulls is not generally eaten, the eggs of gulls are considered a delicacy by the Aleuts and they are easy to collect because they nest on the ground, in colonies, and are very vocal [33, 68, 75, 76]. You can't miss a gull colony. However, their eggs are available for a relatively brief period in late spring. In contrast, eiders nest on all the islands, but nest solitarily, and are silent as they sneak into their nests. Even when incubating on a nest 3 meters away, the camouflaged female eiders are difficult to see. It takes diligence to find many nests. The gulls from the Aleutians could pose some risk to consumers. Lead levels as low as 0.1 to 0.5 have caused some behavioral and reproductive effects in wildlife consumers, suggesting that levels in the eggs and muscle of some birds may place some consumers at risk [3, 77]. There is no safe lead level (or a standard for lead in people), but presumably the lead levels in eggs should be a concern for populations that consume many a week. Cadmium at 1.0 ppm causes effects to predators that eat them, but the tissues from gulls had mean values below this threshold [70].

The glaucous-winged gull eggs from the Aleutians averaged 0.36 ppm of mercury while eider eggs averaged 0.18 ppm (Table 1.7 [32]). Not only are the levels for glaucous-winged gulls over the 0.3 ppm mercury standard used by the European Union and USEPA, but since eggs are collected only during the beginning of the egg laying period, they are available for a limited time (a few weeks), and as a delicacy are eaten frequently. Aleuts in the villages we visited reported eating them daily when available. The effect of eating several eggs a day should be considered. Ginsberg and Toal [78] suggested that eating a single meal of high mercury foods may pose a risk to pregnant women.

## 1.14 HEALTH IMPLICATIONS FOR FISH

Several shellfish and fish species are of considerable commercial interest. Dutch Harbor (Unalaska) is the leading fishing port in North America and one of the largest commercial landings for shellfish and fish in the World (Unalaska 2016)). [33, 68]. Fish are very important in Aleut diets Salmon of several species come up to spawn in the small streams on many of the populated islands, making it easy for them to collect for several months. Many Aleuts live in small communities (30s to 130s) that are isolated with supply ships coming only a couple times a year. Sometimes these ships can't land because of rough seas and small harbors. Thus, Aleuts in these communities are reliant on fish and shellfish, among other subsistence foods. People on these isolated islands often eat fish morning, noon and night, whole, parts (whole fillets, livers), and in soups and casseroles.

The present study identified several factors that affected metal levels, resulting in differences among species, within and among islands (different coves or offshore fishing), and as a function of trophic level, internal tissues, size, weight, and age (for Pacific cod). Usually, the differences between islands were not great, except for cadmium in Pacific cod and mercury levels in chiton and Dolly Varden, but other fish should be examined in more detail. The interspecific differences were significant (Table 1.1). Lead and cadmium were generally below any health standards for consumption of fish, although liver levels of cadmium were within a range to cause some concern.

Mercury, which in fish is largely methylmercury [29, 30] is the main metal of concern in marine environments, especially for subsistence fishing. Mercury is naturally present in seawater, but its levels are augmented nearshore from anthropogenic sources and runoff and offshore by atmospheric deposition. Mercury has significant toxic effects on humans and other biota [62, 79]. People who eat fish frequently, like in the Aleutian communities, are at risk of mercury poisoning. Mean mercury levels in fish from the Aleutians ranged from 0.114 ppm in Dolly Varden to 0.366 ppm in great sculpin (Table 1.1 [32]). Sculpin and sole (species with the highest levels) are bottom dwelling fish. Sculpin are predators in invertebrates and fish. They are not fished commercially and are discarded as by-catch. However, the Aleut team reported eating great sculpin and related species regularly. The fish sampled represent different trophic levels, and forage in different locations. While there is a tendency for mean mercury levels to increase with the mean size of each fish species, there is a great deal of variation in mean mercury level for fish of the same size (Figure 1.2a [49]).

The action level for methylmercury in fish is 1.0 ug/g (ppm, wet weight, [76]), which is really a regulatory action level rather than a risk level. The European

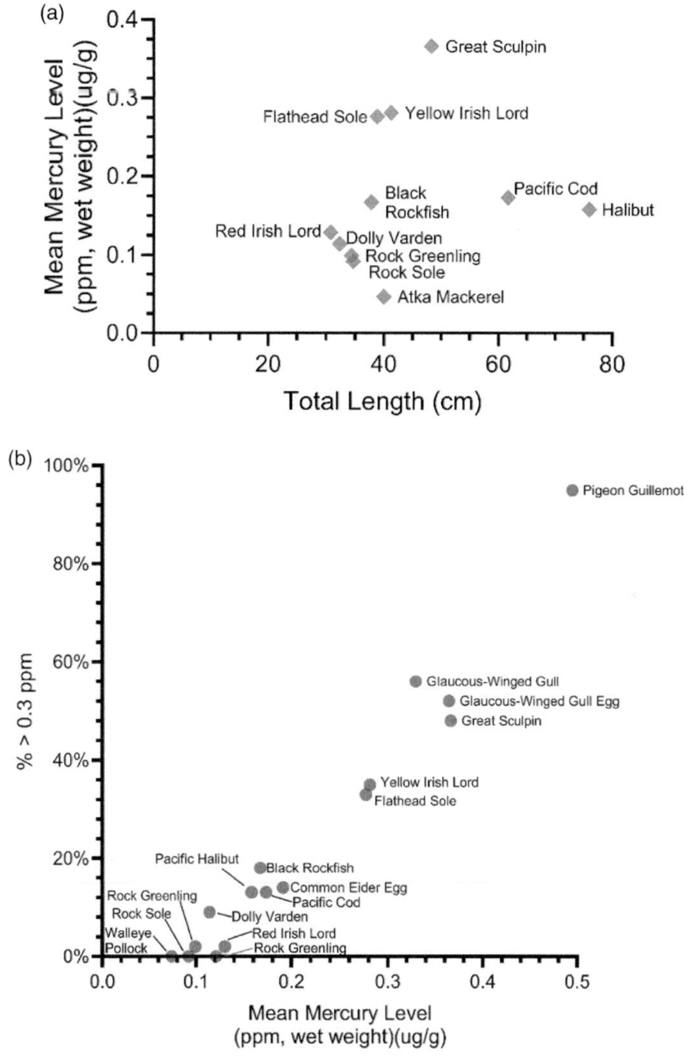


**FIGURE 1.2** Relationships of mercury levels and risk. **(a)** Relationship between total length and mean mercury levels for fish collected on islands in the Bering Sea, Alaska [42, 43]. **(b)** Relationship of mean mercury levels and the percentage of samples that were over 0.3 ppm for each fish and bird. (Developed from [32].)

Commission's environmental quality standard for mercury is that a representative sample of fish shall not exceed 0.3 ppm (wet weight). In 2001, the US Environmental Protection Agency set an ambient freshwater quality standard of 0.3 ppm in fish [80]. These guidance levels are for the levels in fish that are safe to eat.

The effect of eating several fish at the high end of exposure should be considered. Figure 1.2b shows the percentage of fish above 0.3 ppm as a function of mean mercury levels. It is instructive to remember the maximum levels for each fish from Table 1.1; the highest maximum level for any fish was 1.096 pm, well over any action level or guidance (great sculpin). The maximum level for yellow Irish lord and Pacific halibut were both 0.9 ppm. Further, 30% of great sculpin were above 0.5 ppm, and 4% were above 1.0 ppm [32]. Ginsberg and Toal [78], for example, suggested that even eating a single meal of high mercury foods may pose a risk to pregnant women. While it seems logical that any pregnant women would not always have fish that are at the high end of exposure, it may be likely for some fish. For example, when we caught Pacific cod, they were usually about the same size. Fish of the same age school together, thus increasing the likelihood that when Aleuts catch a big fish they may catch several at the same time. Determining risk from consumption of any of the foods eaten by Aleuts or others requires information on not only the species eaten, but the frequency and amount. Our expedition included Aleuts who reported the species eaten, what parts were eaten, and when. Prey types for subsistence people are seasonal: many birds and some fish are migratory, some fish simply move deeper or offshore, making them less available. For example, the Aleuts on the expedition said they ate rock greenling all year but ate sculpin only when nothing else was available. They preferred octopus, halibut, and sea lion over other fish.

Regardless of the availability of the fish and the catchability of the fish, our job was to analyze the metal levels, and to report them to the Aleuts communities. After the samples were all analyzed, we went back to the villages to discuss the results. They, like other fish consumers, make risk decisions of one type or another on a daily basis. And they have risky decisions to make daily as they live on an isolated island, in the middle of rough seas, that they have to navigate to catch fish and collect other subsistence foods and sometimes face winds that are strong they have to string ropes between houses to allow passage between them (by holding on to the ropes). Fishing and consumption of fish create three difficulties: (1) for many parts of the world, including the Aleutians, shellfish and fish are an important and essential source of protein, (2) overfishing results in fish population declines and fishing of smaller and smaller fish [81, 82], and (3) some fish contain methylmercury at high enough levels to cause effects in the fish themselves and in higher trophic level species that consume them, including humans [83, 84]. Fish consumption is the most significant source of mercury (and methylmercury) exposure in human diet [85–87]. It is unlikely that overfishing is a problem in the waters around the isolated Aleutian islands where Aleut populations number less than a couple of hundred people. It is the very isolation, however, that results in fish and shellfish being the major part of their diet.

Determining the risk from fish consumption, however, is complicated by the fact that some other metals can mitigate the toxic effect of mercury. For example, experiments as early as the 1960s with rads showed a protective effect of selenium on mercury [88]. However, laboratory and field studies have demonstrated the toxic effects of selenium [77, 89, 90]. Selenium is an essential trace element and is generally regulated in vertebrates that is toxic both at low and high levels [77]. These observations led Ralston and others [44, 91–93] to propose that a Se/Hg molar ratio of 1 or above would be protective for mercury poisoning. Assuming this were true, then the fish collected from the Aleutians in this study would all be protective because they were above that level but suppose this were true only for a ratio of 5 (Figure 1.3a),

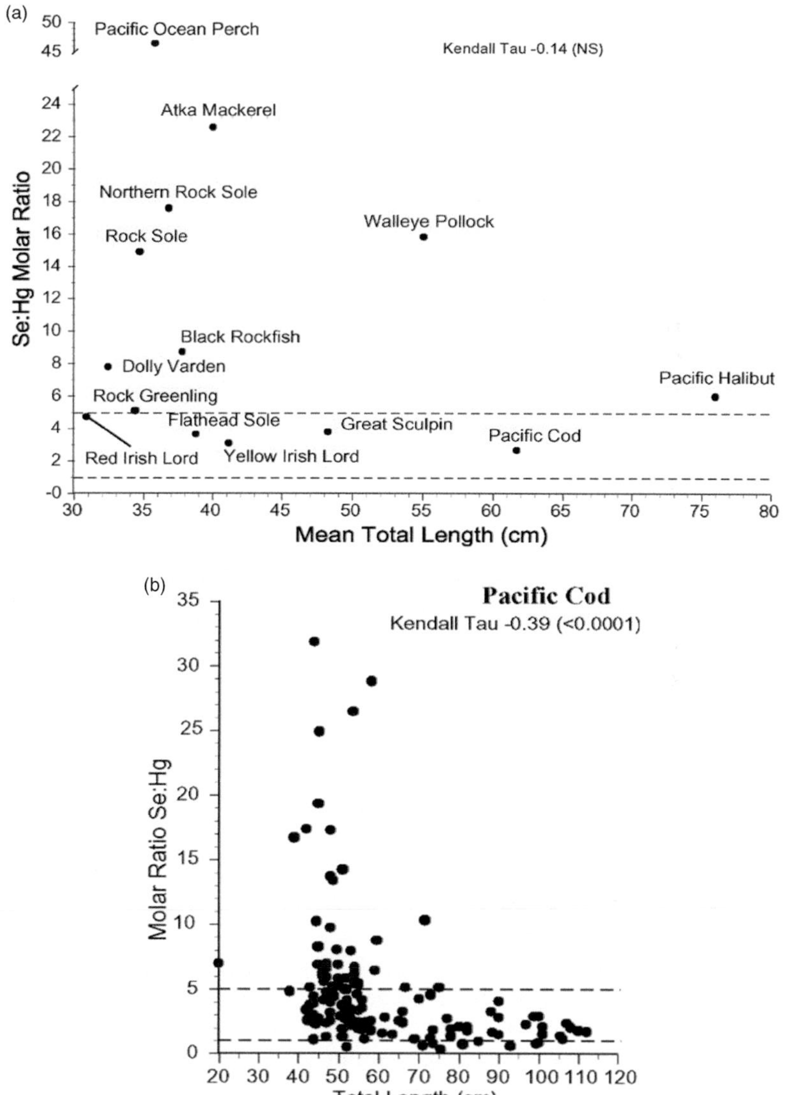


**FIGURE 1.3** Se/Hg molar ratio in human health assessment. **(a)** Relationship of mean total length to the Se/Hg molar ratios, with two "safe levels" of 1 and 5 in many species. **(b)** Variation in levels in Pacific cod by total length. (After [49].)

then several fish would be below that level. We see at least three problems with this: (1) even if it were true that a ratio of 1 was protective, would that be protective for every tissue? (e.g., brain, liver, kidney). The answer to this is unknown. (2) The mean values do not show the whole picture and a fish that has a Se/Hg ratio of 1 may have many individuals below that ratio, and (3) at some point, mercury levels are sufficiently high so that mercury poses a problem because of its toxicity (and that level is unknown). Selenium can be toxic as well.

Figure 1.3b shows the relationship of the ratio to mean total length for Pacific cod. Each dot represents one fish. There are fish below the Se/Hg ratio of 1 (meaning they still pose a risk), especially large fish. Further the bulk of the fish fall between a ratio of 1 and 5. The uncertainties involved in deciding that fish with over a Se/Hg ratio of 1 (or even 5) are sufficient that depending upon that ratio as "safe to consume", especially for pregnant women, may not be worth the risk. Further, in many cases, a fish consumer usually does not know the length of the fish they buy, or where it fits on the curve of mean Se/Hg ratio against size of the fish. And a pregnant women may not know as well. And neither know whether they have a fish with a high or low ratio. In the end, there is significantly more information about the levels of mercury in fish generally (e.g., means) than there is for selenium, making it difficult even for scientists to know the Se/Hg ratios for a given species, much less have this information for fish from many areas of the World. Pacific cod are the species found in supermarkets throughout the US as fillets, as well as in fish sticks.

## 1.15 CONCLUSIONS

We synthesized data on a project to examine levels of cadmium, lead, mercury, and selenium using a food chain approach on biota collected from marine waters and islands in the Aleutian Chain of Alaska where the DOE had detonated underground nuclear tests in the late 1960s. The project was a stakeholder-driven process involving several state and federal regulatory and natural resource agencies, the APIA, Aleuts, university scientists, and the public. All of the species collected (except Bald Eagle) were subsistence foods of the Aleuts, and some were of considerable economic importance for the fishery out of Dutch Harbor; all are part of a complex food web. Mean mercury (the metal of primary concern in marine ecosystems) level increased with every node on the food chain. Levels were lowest in algae (0.001–0.005 ppm), ranged from 0.005 ppm in green sea urchin to 0.043 ppm in golden king crab (invertebrates), ranged from 0.42 ppm in salmon to 0.37 ppm in great sculpin (fish), and ranged from 0.121 to 0.49 ppm in birds. These data show a clear increase in mercury levels at higher trophic levels. Some of the kelp were 30-m long, and some levels varied by location on the algae; algae are becoming a common food in many parts of the world, as well as have medicinal uses. The wide range of invertebrate species examined showed a greater range in mercury levels than any other group. Some of the delicacies of the Aleuts (e.g., octopus, king crab) had the highest mercury levels. Mussels, a species that is sessile and remains in one place, showed locational differences for several metals, with levels varying in different creeks around islands. The fish differed in size and trophic levels, had significant differences in mercury levels, with species showing variations in metals as a function of species, geographical location, trophic level, size, age, and tissue level. Mercury levels increased with increasing weight of the organism for limpets and for 11 of 15 species examined. Of the 15 fish species examined in detail, 9 had some samples over the 0.3 ppm human health standard. For the two species of birds examined, eggs had slightly higher levels of mercury than adults, and gulls and their eggs were above the 0.3 ppm level. The muscle tissue of pigeon guillemot averaged 0.49 ppm; the one sea lion collected by Aleut subsistence hunters averaged 1.03 ppm in muscle and 69 ppm in liver. These species pose a potential risk for humans.

## ACKNOWLEDGEMENTS

Fish were collected under appropriate state permits. Our studies were approved by the Rutgers University Animal Review Board. We thank many people who contributed to the CRESP Amchitka Geophysical and Biological Project, especially C. W. Powers, D. Kosson, B, Friedlander, C. Jeitner, S. Burke, D. Volz, and S. Jewett. We also thank the following for help throughout the project: D. Barnes, L. Duffy, A. Morkill, R. Patrick, D. Rogers, D. Dasher, and the people of the villages of Unalaska, Nikolski, Atka, and Adak in the Aleutians. We thank the entire crew of the *Ocean Explorer* for making our field work possible and pleasant, and for bringing us safely back to port. This research was funded by the Consortium for Risk Evaluation with Stakeholder Participation (CRESP) through the Department of Energy (DE-FG 26-00NT 40938, DE-FC01-86EW07053), the Division of Life Sciences of Rutgers University, and by NIEHS ESO 5022. The results, conclusions, and interpretations reported herein are the sole responsibility of the authors and do not represent the views of the funding agencies.

## REFERENCES

1. Furness, R. W., and P. S. Rainbow. 1990. *Heavy Metals in the Marine Environment.* Boca Raton, FL: CRC Press.
2. Houghton, J. T., B. A. Callander, and S. K. Varney. 1992. *Climate Change 1992.* Cambridge, UK: Cambridge University Press.
3. Agency for Toxic Substances and Disease Registry (ATSDR). 2007. *Toxicological Profile for Lead.* Atlanta, GA: Agency for Toxic Substances and Disease Registry. Accessed January 20, 2025. http://www.atsdr.cdc.gov/toxprofiles/tp.asp?id=96&tid=22.
4. Agency for Toxic Substances and Disease Registry (ATSDR). 2012. *Toxicological Profile for Cadmium.* Atlanta, GA: Agency for Toxic Substances and Disease Registry.
5. Center for Disease Control (CDC). 2022. "One Health." Accessed [insert date]. https://www.onehealth.cdc.gov/onehealth/index.html.
6. Wiener, J. C., D. P. Krabbenhoft, G. H. Heinz, and M. Scheuhammer. 2003. "Ecotoxicology of Mercury." In *Handbook of Ecotoxicology*, edited by D. J. Hoffman, B. A. Rattner, G. A. Burton Jr., and J. Cairns Jr. Boca Raton, FL: Lewis Publishers.
7. Baudrot, V., C. Fritsch, A. Perasso, M. Banerjee, and F. Raoul. 2018. "Effects of Contaminants and Trophic Cascade Regulation on Food Chain Stability: Application to Cadmium Soil Pollution on Small Mammals – Raptor Systems." *Ecological Modelling* 382: 33–42. https://doi.org/10.1016/j.ecolmodel.2018.05.002.
8. Burger, J. 2019. "A Framework for Increasing Sustainability and Reducing Risk to Ecological Resources through Integration of Remediation Planning and Implementation." *Environmental Research* 172: 586–95.

9. Cappuyns, Valérie. 2016. "Inclusion of Social Indicators in Decision Support Tools for the Selection of Sustainable Site Remediation Options." *Journal of Environmental Research* 184: 45–56.
10. Doren, R. F., J. H. Richards, and J. C. Volin. 2009. "Ecological Indicators for System-Wide Assessment of the Greater Everglades Ecosystem Restoration Project." *Ecological Indicators* 9: S1–S16.
11. Evers, D. C., A. K. Jackson, T. H. Tear, and C. E. Osborne. 2012. Hidden Risk: Mercury in Terrestrial Ecosystems of the Northeast. Biodiversity Research Institute, *BRI Report* 2012–07. Gorham, Maine.
12. Fox, G. A. 2001. "Wildlife as Sentinels of Human Health Effects in the Great Lakes–St. Lawrence Basin." *Environmental Health Perspectives* 109 (Suppl. 6): 853–61.
13. Crowley, K. D., and J. F. Ahearne. 2002. "Managing the Environmental Legacy of U.S. Nuclear-Weapons Production." *American Scientist* 90: 514–23.
14. Department of Energy (DOE). 2002a. *A Review of the Environmental Management Program (by Top to Bottom Review Team).* Washington, DC: Department of Energy.
15. Department of Energy (DOE). 2019. *Hanford Lifecycle Scope, Schedule, and Cost Report.* DOE/RL-2018-45 (Rev. 0). Richland, WA: Richland Operations Office.
16. Department of Energy (DOE). 2022. *EM Strategic Vision: 2022–2032. U.S.* Department of Energy. https://www.energy.gov/em/articl/em-strategic-vision.
17. National Research Council (NRC). 2019. *Independent Assessment of Science and Technology for the Department of Energy's Defense Environmental Cleanup Program.* Nuclear and Radiation Studies Board. Washington, DC: National Academy Press.
18. Burger, Joanna, Michael Gochfeld, Charles W. Powers, and David Kosson. 2011. "Amchitka Island: Melding Science and Stakeholders to Achieve Solutions at a Former Department of Energy Nuclear Site." In *Science and Stakeholders: Achieving Implementable Solutions to Energy and Environmental Issues*, edited by Joanna Burger, 1–25. New York: Springer.
19. Department of Energy (DOE). 2002b. *Modeling Groundwater Flow and Transport of Radionuclides at Amchitka Island's Underground Nuclear Tests: Milrow, Long Shot, and Cannikan.* Draft. DOE/NV-11508-51. Nevada Operations Office, Las Vegas, Nevada.
20. Department of Energy (DOE). 2002c. *Screening Risk Assessment for Possible Radionuclides in the Amchitka Marine Environment. Draft.* DOE/NV-857. Nevada Operations Office, Las Vegas, Nevada.
21. Greenpeace. 1996. *Nuclear Flashback: The Return to Amchitka.* Washington, DC: Greenpeace.
22. Kohlhoff, D. W. 2002. *Amchitka and the Bomb: Nuclear Testing in Alaska.* Seattle, WA: University of Washington Press.
23. Faller, S. H., and D. E. Farmer. 1998. *Long-Term Hydrological Monitoring Program: Amchitka, Alaska. EPA-402-R-98-002.* U.S. Environmental Protection Agency, Washington, DC.
24. Eichelberger, J. C., J. Freymueller, G. Hill, and M. Patrick. 2002. "Nuclear Stewardship: Lessons from a Not-So-Remote Island." *Geotimes* 47: 20–23.
25. Department of Energy (DOE). 2008. *Long-Term Surveillance and Maintenance Plan for U.S. DOE Amchitka, Alaska Site.* DOE-LM/1121-2008.
26. National Oceanic and Atmospheric Administration (NOAA). 2005. Dutch Harbor–Unalaska, in Alaska, Top U.S. Port for Landings in 2004. *NOAA Report* 04–096. Available at: www.nmfs.noaa.gov/docs/04-096_top_ports.pdf. Accessed May 26, 2024.
27. Waldichuk, M. 1974. "Coastal Marine Pollution and Fish." *Ocean Management* 2: 1–60.

28. Consortium for Risk Evaluation with Stakeholder Participation (CRESP). 2005. Amchitka Independent Science Assessment: Biological and Geophysical Aspects of Potential Radionuclide Exposure in the Amchitka Marine Environment. Vanderbilt University.
29. Lescord, G. L., T. A. Johnston, B. A. Branfireun, and J. M. Gunn. 2018. "Percentage of Methylmercury in the Muscle Tissue of Freshwater Fish Varies with Body Size and Age and among Species." *Environmental Toxicology and Chemistry* 37 (10): 2682–91. https://doi.org/10.1002/etc.4233.
30. Thompson, D. R., K. C. Hamer, and R. W. Furness. 1991. "Comparison of the Levels of Total and Organic Mercury in Seabird Feathers." *Marine Pollution Bulletin* 20: 577–79.
31. Statistical Analysis Systems (SAS). 2020. Cary, NC.
32. Burger, Joanna, Michael Gochfeld, Christian Jeitner, Sean Burke, Thomas Stamm, Ronald Snigaroff, Daniel Snigaroff, Robert Patrick, and Joshua Weston. 2007a. "Mercury Levels and Potential Risk from Subsistence Foods from the Aleutians." *Science of the Total Environment* 384: 93–105.
33. Hamrick, K., and J. Smith. 2003. *Subsistence Food Use in Unalaska and Nikolski*. Anchorage, Alaska: Aleutian Pribilof Island Association.
34. Garza, D. 2005. *Common Edible Seaweeds in the Gulf of Alaska*. Fairbanks, Alaska: Alaska Sea Grant Program.
35. VanNetten, C., S. A. H. Cann, D. R. Morley, and J. P. VanNetten. 2000. "Elemental and Radioactive Analysis of Commercially Available Seaweed." *Science of the Total Environment* 255: 169–75.
36. Mehta, S. D., and J. P. Gaur. 2005. "Use of Algae for Removing Heavy Metal Ions from Wastewater: Progress and Prospects." *Critical Reviews in Biotechnology* 25: 113–52.
37. Ragan, M. A., O. Smidsrod, and B. Larsen. 1979. "Chelation of Divalent Metal Ions by Brown Algal Polyphenols." *Marine Chemistry* 7: 265–71.
38. Chan, Hing Man, Chul Kim, Kareen Khoday, Olivier Receveur, and Harriet V. Kuhnlein. 1995. "Assessment of Dietary Exposure to Trace Metals in Baffin Inuit Food." *Environmental Health Perspectives* 103: 740–46.
39. Sharp, G. J., H. S. Samant, and O. C. Vaidya. 1988. "Selected Metal Levels of Commercially Valuable Seaweeds Adjacent to and Distant from Point Sources of Contamination in Nova Scotia and New Brunswick." *Bulletin of Environmental Contamination and Toxicology* 40: 724–30.
40. Burger, Joanna, Michael Gochfeld, Christian Jeitner, Maureen Gray, Tara Shukla, Sarita Shukla, and Sean Burke. 2007b. "Kelp as a Bioindicator: Does It Matter Which Part of 5 M Long Plant Is Used for Metal Analysis?" *Environmental Monitoring and Assessment* 128: 311–21.
41. Burger, J., and M. Gochfeld. 2006. "Locational Differences in Heavy Metals and Metalloids in Pacific Blue Mussels (*Mytilus [edulis] trossulus*) from Adak Island in the Aleutian Chain, Alaska." *Science of the Total Environment* 368: 937–50.
42. Burger, Joanna, Michael Gochfeld, Sean Burke, Christian W. Jeitner, Sarah Jewett, Daniel Snigaroff, Ronald Snigaroff, Thomas Stamm, Sharon Harper, Michael Hoberg, Heloise Chenelot, Robert Patrick, Craig D. Voltz, and Joshua Weston. 2006. "Do Scientists and Fishermen Collect the Same Size Fish? Possible Implications for Exposure Assessment." *Environmental Research* 101: 34–41.
43. Burger, Joanna, Michael Gochfeld, Christian Jeitner, and Taryn Pittfield. 2014. "Heavy Metals in Fish from the Aleutians: Interspecific and Locational Differences." *Environmental Research* 131: 119–30.
44. Ralston, N. V. C., C. R. Ralston, J. L. Blackwell, and L. J. Raymond. 2008. "Dietary and Tissue Selenium in Relation to Methylmercury Toxicity." *NeuroToxicology* 29 (5): 802–11. https://doi.org/10.1016/j.neuro.2008.07.007.

45. Burger, Joanna, Michael Gochfeld, Christian Jeitner, Sean Burke, and Thomas Stamm. 2007c. "Metal Levels in Flathead Sole (*Hippoglossoides elassodon*) and Great Sculpin (*Myoxocephalus polyacanthocephalus*) from Adak Island, Alaska: Potential Risk to Predators and Fishermen." *Environmental Research* 103: 62–69.
46. Eschmeyer, W. N., E. S. Herald, and H. Hammann. 1983. *A Field Guide to Pacific Coast Fisheries of North America*. Houghton Mifflin, New York, NY.
47. Burger, J., and M. Gochfeld. 2007a. "Risk to Consumers from Mercury in Pacific Cod (*Gadus macrocephalus*) from the Aleutians: Age and Size Effects." *Environmental Research* 105: 276–84.
48. Burger, Joanna, Michael Gochfeld, Christian Jeitner, Thomas Stamm, Sean Burke, Daniel Snigaroff, Ronald Snigaroff, Michael Donio, and Craig Volz. 2007d. "Heavy Metal Levels in Pacific Cod (Gadus macrocephalus) from the Aleutians: Location, Age, Size, and Risk Effects." *Journal of Toxicology and Environmental Health*, Part A 70: 1–15.
49. Burger, Joanna, Michael Gochfeld, Christian Jeitner, Michael Donio, and Taryn Pittfield. 2012. "Interspecific and Intraspecific Variation in Selenium:Mercury Molar Ratios in Saltwater Fish from the Aleutians: Potential Protection of Mercury Toxicity by Selenium." *Science of the Total Environment* 431: 46–56.
50. Green, N. W., and J. Knutzen. 2003. "Organohalogens and Metals in Marine Fish and Mussels and Some Relationships to Biological Variables at Reference Localities in Norway." *Marine Pollution Bulletin* 46: 362–77.
51. Pinho, A. P., J. R. D. Guimarães, A. S. Marins, P. A. S. Costa, G. Olavo, and J. Valentin. 2002. "Total Mercury in Muscle Tissue of Five Shark Species from Brazilian Offshore Waters: Effects of Feeding Habit, Sex, and Length." *Environmental Research* 89: 250–58.
52. Boening, D. W. 2000. "Ecological Effects, Transport, and Fate of Mercury: A General Review." *Chemosphere* 40: 1335–51.
53. Burger, Joanna, and Michael Gochfeld. 2016. Habitat, Population Dynamics, and Metal Levels in Colonial Waterbirds: A Food Chain Approach. https://doi.org/10.1201/9780429159435.
54. Mendoza-Carranza, M., A. Sepúlveda-Lozada, C. Dias-Ferreira, and V. Geissen. 2016. "Distribution and Bioconcentration of Heavy Metals in a Tropical Aquatic Food Web: A Case Study of a Tropical Estuarine Lagoon in SE Mexico." *Environmental Pollution* 210: 155–65. https://doi.org/10.1016/j.envpol.2015.12.014.
55. Rolfhus, K. R., B. D. Hall, B. A. Monson, M. J. Paterson, and J. D. Jeremiason. 2011. "Assessment of Mercury Bioaccumulation within the Pelagic Food Web of Lakes in the Western Great Lakes Region." *Ecotoxicology* 20 (7): 1520–29. https://doi.org/10.1007/s10646-011-0733-y.
56. Burger, J., and M. Gochfeld. 2007b. "Metals and Radionuclides in Birds and Eggs from Amchitka and Kiska Islands in the Bering Sea/Pacific Ocean Ecosystem." *Environmental Monitoring and Assessment* 127: 105–17.
57. Burger, J., and M. Gochfeld. 2009. "Comparison of Arsenic, Cadmium, Chromium, theLead, Manganese, Mercury, and Selenium in Feathers in Bald Eagle (*Haliaeetus leucocephalus*), and Comparison with Common Eider (*Somateria mollissima*), Glaucous-Winged Gull (*Larus glaucescens*), Pigeon Guillemot (*Cepphus columba*), and Tufted Puffin (*Fratercula cirrhata*) from the Aleutian Chain of Alaska." *Environmental Monitoring and Assessment* 152: 357–67.
58. Burger, Joanna, Michael Gochfeld, Christian Jeitner, Daniel Snigaroff, Ronald Snigaroff, Thomas Stamm, and Craig Volz. 2008. "Assessment of Metals in Down Feathers of Female Common Eiders and Their Eggs from the Aleutians: Arsenic, Cadmium, Chromium, Lead, Manganese, Mercury, and Selenium." *Environmental Monitoring and Assessment* 143: 247–56.

59. Burger, Joanna, Michael Gochfeld, Christian Jeitner, Sean Burke, Craig Volz, Ronald Snigaroff, Daniel Snigaroff, Tara Shukla, and Sarita Shukla. 2009. "Mercury and Other Metals in Eggs and Feathers of Glaucous-Winged Gulls (*Larus glaucescens*) in the Aleutians." *Environmental Monitoring and Assessment* 152: 179–94.
60. Buhler, D. R., R. R. Claeys, and B. R. Mate. 1975. "Heavy Metal and Chlorinated Hydrocarbon Residues in California Sea Lions (*Zalophus californianus californianus*)." *Journal of the Fisheries Research Board of Canada* 32: 2391–97.
61. Buhler, D. R., and B. R. Mate. 1971. "Mercury Levels in California Sea Lions." In *Mercury in the Western Environment: Proceedings of a Workshop*, Portland, Oregon, February 25–26. Corvallis, OR: Continuing Education Publications.
62. Agency for Toxic Substances and Disease Registry (ATSDR). 2013. *Addendum to the Toxicological Profile for Mercury (Alkyl and Dialkyl Compounds)*. Atlanta, GA: Agency for Toxic Substances and Disease Registry.
63. Andersen, V., P. Maage, and J. Johannessen. 1996. "Heavy Metals in Blue Mussels (*Mytilus edulis*) in the Bergen Harbor Area, West Norway." *Bulletin of Environmental Contamination and Toxicology* 57: 589–96.
64. Avelar, W. E. P., F. L. M. Mantelatto, A. C. Tomazelli, D. M. L. Silva, T. Shuhama, and J. L. C. Lopes. 2000. "The Marine Mussel *Perna perna* (Mollusca, Bivalvia, Mytilidae) as an Indicator of Contamination by Heavy Metals in the Ubatuba Bay, São Paulo, Brazil." *Water, Air, and Soil Pollution* 118 (1): 65–72. https://doi.org/10.1023/A:1005109801683.
65. Baraj, B., L. F. Niencheski, and C. Corradi. 2003. "Trace Metal Content Trend of Mussel *Perna perna* (Linnaeus, 1758) from the Atlantic Coast of Southern Brazil." *Water, Air, and Soil Pollution* 145: 205–14.
66. Sarver, D. J., A. Ellis, N. A. Sims, and D. Wise. 2004. "Using Pearl Oysters as Heavy Metal Monitors in Tropical Waters." *Journal of Shellfish Research* 23: 310–15.
67. Fowler, S. W. 1990. "Critical Review of Selected Heavy Metal and Chlorinated Hydrocarbon Concentrations in the Marine Environment." *Marine Environmental Research* 29 (1): 1–64. https://doi.org/10.1016/0141-1136(90)90027-L.
68. Patrick, R. 2002. "How Local Alaska Native Communities View the Amchitka Issue." In *Proceedings of the Amchitka Long-Term Stewardship Workshop*. Fairbanks, AK: CRESP/University of Alaska.
69. Phillips, D. J. H. 1990. "Use of Macroalgae and Invertebrates as Monitors of Metal Levels in Estuarine and Coastal Waters." In *Heavy Metals in the Marine Environment*, edited by R. W. Furness and P. S. Rainbow, 81–100. Boca Raton, FL: CRC Press.
70. Eisler, R. 1985. *Cadmium Hazards to Fish, Wildlife, and Invertebrates: A Synoptic Review*. U.S. Fish and Wildlife Service Report 85(1.4). Washington, DC.
71. Eisler, R. 1988. *Lead Hazards to Fish, Wildlife, and Invertebrates: A Synoptic Review*. U.S. Fish and Wildlife Service. Washington, DC.
72. Lemly, D. A. 1993. "Guidelines for Evaluating Selenium Data from Aquatic Monitoring and Assessment Studies." *Environmental Monitoring and Assessment* 28: 83–100.
73. Codex Alimentarius. 2005. "Draft Maximum Level for Cadmium in Marine Bivalve Mollusks." Joint FAO/WHO Committee on Food Additives and Contaminants of the Codex Alimentarius Commission. *Codex Reference CX* 4/30.2 CL 2005/36 FAC.
74. World Health Organization (WHO). 1976. *Environmental Health Criteria. I. Mercury*. Geneva: World Health Organization.
75. U.S. Fish & Wildlife Service (USFWS). 2004. *Alaska Subsistence Spring/Summer Migratory Bird Harvest*. Anchorage, AK: U.S. Fish & Wildlife Service.
76. U.S. Food and Drug Administration (U.S. FDA). 1994, 2001. *FDA Consumer Advisory*. Available at: http://www.fda.gov/bbs/topics/ANSWERS/2000/advisory.html. Accessed December 1, 2001.
77. Eisler, R. 2000. *Handbook of Chemical Risk Assessment: Health Hazards to Humans, Plants, and Animals*, Three Volume Set. CRC Press. https://doi.org/10.1201/9780367801397.

78. Ginsberg, G. L., and B. F. Toal. 2000. "Development of a Single-Meal Fish Consumption Advisory for Methylmercury." *Risk Analysis* 20: 41–47.
79. Whitney, M. C., and D. A. Cristol. 2018. "Impacts of Sublethal Mercury Exposure on Birds: A Detailed Review." In *Reviews of Environmental Contamination and Toxicology Volume 244*, edited by P. de Voogt, 113–63. Springer International Publishing. https://doi.org/10.1007/398_2017_4.
80. Environmental Protection Agency (EPA). 2001. *Methylmercury Fish Tissue Criterion: Fish Tissue Criterion for Methylmercury to Protect Human Health.* http://www.epa.gov/waterscience/criteria/methylmercury/factsheet.html. Accessed June 10, 2008.
81. Pauly, D., V. Christensen, J. Dalsgaard, R. Froese, and F. Torres. 1998. "Fishing Down Marine Food Webs." *Science* 279 (5352): 860–63. https://doi.org/10.1126/science.279.5352.860.
82. Safina, C. 1998. *Song for the Blue Ocean: Encounters Along the World's Coasts and Beneath the Seas.* 1st ed. Henry Holt & Co.
83. Hightower, J. M., and D. Moore. 2003. "Mercury Levels in High-End Consumers of Fish." *Environmental Health Perspectives* 111 (4): 604–8. https://doi.org/10.1289/ehp.5837.
84. World Health Organization (WHO). 1989. *Mercury: Environmental Aspects.* Geneva, Switzerland: WHO. https://iris.who.int/handle/10665/40021.
85. Eisler, R. 1987. *Mercury Hazards to Fish, Wildlife, and Invertebrates: A Synoptic Review.* Contaminant Hazard Reviews (Report 10; Biological Report 85(1.10)). U.S. Department of the Interior, Fish and Wildlife Service. https://pubs.usgs.gov/publication/5200073.
86. National Research Council (NRC). 2000. *Toxicological Effects of Methylmercury.* National Academies Press. https://doi.org/10.17226/9899.
87. Rice, G., J. Swartout, K. Mahaffey, and R. Schoeny. 2000. "Derivation of U.S. EPA's Oral Reference Dose (RfD) for Methylmercury." *Drug and Chemical Toxicology* 23 (1): 41–54. https://doi.org/10.1081/DCT-100100101.
88. Lindh, U., and E. Johansson. 1987. "Protective Effects of Selenium against Mercury Toxicity as Studied in the Rat Liver and Kidney by Nuclear Analytical Techniques." *Biological Trace Element Research* 12 (1): 109–20. https://doi.org/10.1007/BF02796669.
89. Ohlendorf, H. M. 2000. "Ecotoxicology of Selenium." In *Handbook of Ecotoxicology*, edited by D. J. Hoffman, B. A. Rattner, G. S. Burton, and J. Cairns, 465–500. Boca Raton: CRC Press.
90. Ohlendorf, H. M., and R. L. Hothem. 1995. "Agricultural Drain Water Effects on Wildlife in Central California," 577–95. https://pubs.usgs.gov/publication/5210575.
91. Peterson, S. A., N. V. C. Ralston, P. D. Whanger, J. E. Oldfield, and W. D. Mosher. 2009a. "Selenium and Mercury Interactions with Emphasis on Fish Tissue." *Environmental Bioindicators* 4 (4): 318–34. https://doi.org/10.1080/15555270903358428.
92. Peterson, S. A., N. V. C. Ralston, D. V. Peck, J. V. Sickle, J. D. Robertson, V. L. Spate, and J. S. Morris. 2009b. "How Might Selenium Moderate the Toxic Effects of Mercury in Stream Fish of the Western U.S.?" *Environmental Science & Technology* 43 (10): 3919–25. https://doi.org/10.1021/es803203g.
93. Raymond, L. J., and N. V. C. Ralston. 2009. "Selenium's Importance in Regulatory Issues Regarding Mercury." *Fuel Processing Technology* 90 (11): 1333–38. https://doi.org/10.1016/j.fuproc.2009.07.012.

# 2 Indispensable Role of LC-Based Metallomics Tools to Unravel the Mechanisms Which Causally Link Human Exposure to Toxic Metal(loid) Species to Organ Damage

*Jürgen Gailer*

## 2.1 INTRODUCTION

Industrial activities, including metal and coal mining [1], metal refining and the consumption of unprecedented amounts of fossil fuels [2] emit quantities of toxic metal(loid) species into the environment which rival or exceed those released by natural processes (e.g. chemical/microbial weathering of minerals, volcanic activity) at local, regional and global scales [3, 4]. The concomitant deposition of these inorganic pollutants into agricultural soils [5] represents the predominant pathway for the chronic exposure of humans, including babies, children and pregnant women to these pollutants. The increasing levels of these pollutants in the diet [6, 7] is being recognized as a global public health threat [1, 5, 8]. This undesirable situation is further exacerbated by the climate change-related increased incidence of forest fires which deposit additional amounts of toxic metal(loid) species (e.g. Hg) into ecosystems that are used for food production [9, 10], thus further compromising food safety and security [11].

The indigenous people that inhabit northern Canada – the Inuit – are particularly affected by the aforementioned increased anthropogenic transport of toxic metal(loid)s, such as mercury, via the air and ocean currents to the Arctic [12]. Their inherent reliance on the ingestion of freshwater fish laden with methylmercury ($MeHg^+$) [12] has resulted in a discernible seasonal pattern of mercury concentrations being detected in their hair [13], which is higher in the summer months

DOI: 10.1201/9781003632092-3 

presumably owing to increased fish consumption. Unsurprisingly, the adverse health effects of the increased dietary exposure to $MeHg^{+}$ and persistent organic pollutants is most apparent in Inuit children who exhibit an increased attention deficit/hyperactivity disorder symptomatology [14].

From a global point of view, human populations in developed countries are similarly exposed to gradually increasing levels of toxic metal(loid) species and organic pollutants in their diet [15]. To this end and most disconcertingly, the analysis of baby foods has recently revealed elevated concentrations of Pb, Cd [16], As [17] and Hg [7]. Compared to the United States Food and Drug Administration (US FDA) mandated maximum limits for As, Cd, Pb and Hg for bottled water, the highest concentrations of these metal(loid)s supplied by baby food producers in the United States were between 5 and 177 times above these limits [7]. These facts have prompted a discourse between the food industry and the US FDA [7] which has resulted in the FDA crafting the 'Closer to Zero' action plan. This initiative is intended to reduce exposure to these toxic element species in foods commonly eaten by babies and young children to bring heavy metals levels to acceptable range within two years.

While only time will tell if this goal will be reached, the unsafe levels of these metal(loid)s in baby foods are of immediate concern to parents who may want to know which of these particular metal(loid)s are most concerning. To answer this question, one needs to consider the bioavailability of the metal(loid) species that are present in the baby foods; bioavailability refers to the fraction of the total ingested metal(loid) species that is absorbed into the bloodstream from the human gastrointestinal tract. While studies that address this important question are being experimentally conducted using innovative analytical approaches [18], another way to evaluate the bioavailability of toxic metal(loid) species in food is by looking at their actual blood concentrations in the general population. These latter data are obtained by so-called biomoniotoring studies which are conducted by the Centers of Disease Control and Prevention (CDC) in the United States. Their latest report provides accurate concentrations of environmental chemicals, including As, Cd, Hg and Pb in the human bloodstream for different age groups, sex and ethnicity [19]. To this end, the blood concentration of Cd and Hg in children in the United States between the age of 1 and 6 years is 0.3 ppb and that of Pb is 2.2 ppb [20]. Similar Cd concentrations have been reported for children in Canada [21]. It is when we ask if these blood concentrations are tolerable or causally implicated in the onset of environmental diseases, that we realize we do not have an answer to this public health related question.

Given that it has been recently estimated that 9 million deaths per year are attributed to pollution related causes [22, 23] and that these numbers likely represent a serious underestimation of the actual numbers, the aforementioned blood concentrations leave us with considerable uncertainty about what blood concentrations of toxic metal(loid)s can be considered safe, with regard to not only babies, children and pregnant women but also industrial workers. Accordingly, one of the arguably biggest public health problems in the postgenomic world is to address the exposure-response problem. Owing to the enormous difficulty in answering this seemingly simple question [24], there is an urgent need to establish novel approaches to linking human exposure to multiple toxic metal(loid) species with the etiology of distinct adverse health effects and/or environmental diseases. While many different 'omics' approaches can be employed for tackling this fiendishly complex problem [25], a molecular toxicology approach can be utilized to establish a mechanism-driven integrated biochemical framework to link human exposure to inorganic pollutants with the onset of adverse health effects and/or environmental diseases [6]. Since probing the bioinorganic chemistry of toxic metal(loid) species in the complexity of biological fluids is not an easy task, new instrumental analytical approaches need to be considered, as classical toxicology-based approaches – which work to address the acute toxicity of various chemicals – can no longer provide adequate answers [26].

## 2.2 HOW CAN THE APPLICATION OF HYPHENATED CHROMATOGRAPHIC TECHNIQUES HELP ANSWER QUESTIONS RELATED TO THE BIOINORGANIC/ TOXICOLOGICAL CHEMISTRY OF METAL(LOID)S?

It may at first seem far-fetched to envisage how liquid chromatography (LC)-based separation techniques can be employed to gain new insight into questions related to the toxicological chemistry of metal(loid) species until we realize that essentially all bioinorganic chemistry-related problems that pertain to toxic metal(loid) species involve their interaction with multiple biomolecules (or ligands) in a concerted fashion and in different biological compartments predominantly at near-physiological conditions (pH 7.4, high salt concentration) until an observable toxicological outcome is evident (e.g. organ damage [27]). From a conceptual point of view there are two reasons why LC methods are particularly well suited to answer questions pertaining to the bioinorganic chemistry of toxic metal(loid) species.

1. All toxicology relevant processes within an organism (i.e. in vivo) unfold in an aqueous environment, which makes their observation by established LC techniques inherently amenable as these most frequently employ water-based mobile phases [28]. Biological fluids, however, sometimes contain comparatively high salt concentrations (e.g. blood plasma contains ~105 mM of $Cl^-$), which – in the context of observing metal(loid)s – necessitates the need to carefully select an appropriate atomic spectroscopic detector that can handle this salt concentration to avoid the generation of artifacts [29]. To this end, inductively coupled plasma mass spectrometers (ICP-MS), for example, cannot handle mobile phase salt concentrations >0.5%, which in turn limits the LC separation mechanisms that can be employed to probe any given bioinorganic chemistry-related questions.
2. A large variety of LC-based separation mechanisms are available nowadays from which one can chose the one that appears best suited to observe a relevant bioinorganic process at near-physiological conditions. In this context, choosing a flame atomic absorption spectrometer (FAAS) or an inductively coupled plasma atomic emission spectrometer (ICP-AES) are most desirable as they can accept mobile phases containing >0.5% of salt (ICP-AES, for example, can accept up to 30% of salt). Thus, using these detectors gives the analyst a lot of flexibility to select an appropriate separation mechanism from the arsenal of available LC-based separation mechanisms to address any particular problem.

In terms of bioinorganic chemistry-related questions that remain unanswered pertaining to the metabolism of environmentally ubiquitous As, Cd, Hg and Pb species in mammals, two specific questions stand out. The first one refers to the biomolecular mechanisms by which toxic metal(loid) species that have infiltrated the systemic blood circulation are delivered to their toxicological target organs (e.g. across the blood–brain barrier [BBB]). The second one refers to the type of biomolecular mechanisms which unfold within tissue cells. To this end, the modulation of the metabolism of toxic metal(loid) species by endogenous ligands is pertinent as these processes may be exploited to mitigate adverse health effects by hastening their excretion from organ cells. What each of these questions have in common is that addressing them will require the application of LC based analytical tools which can overcome the main obstacle to gain insight: biological complexity.

As will be illustrated in more detail in the remainder of this text, hyphenated LC-based approaches are ideal to answer the aforementioned toxicological-chemistry-related questions. To design actual experiments to address each of the selected four toxicological questions, an adequate separation mechanism, the chemical composition of the mobile phase and the element-specific detector of choice have to be carefully chosen. It therefore appears useful to illustrate how this was accomplished for each of these investigations to advance the organ uptake of $Cd^{2+}$ (example 1), $MeHg^{+}$ (example 2), $Hg^{2+}$ and $MeHg^{+}$ (example 3) and the mobilization of $MeHg^{+}$ from organs (example 4). It goes without saying that the general approach of tailoring these main components of a hyphenated analytical system to a specific toxicological-chemistry-related problem represents a powerful new means to address pertinent research questions of toxic metal(loid)s that are relevant to public health. In this context, we have to consider a considerable number of inherently toxic metal(loid)s which are essential for the manufacturing of high-tech products and therefore often inadvertently end up in the waste stream [2] and eventually in the ground water [30], the wastewater [31] and/or the diet [7].

The overall thought processes that underlies the concept of 'using LC-based approaches to solve toxicological chemistry related problems' in terms of the execution of actual experiments is conceptually illustrated in Figure 2.1.

### 2.2.1 Structural Characterization of the $Cd^{2+}$ Species Which Cause Nephrotoxicity

While the kidneys have long been known to be one of the major toxicological target organ following the environmental exposure of humans to $Cd^{2+}$ [32, 33], more recently other organs such as the spleen have also been shown to be of toxicological relevance [34, 35]. From a mechanistic point of view, the uptake of any toxic metal(loid) species into a target organ will ultimately be determined by two critical factors. The first one is the structure of the metal(loid) species that is/are formed with biomolecular ligands within blood plasma, which may involve the formation of ternary adducts with plasma proteins and/or small-molecular-weight species [e.g. L-cysteine (Cys), L-homocysteine (hCys), $Cl^{-}$]. The second one is the uptake of the formed metal(loid) complexes into target organ cells by individual ATP-binding biomolecular ABC-cassette transporters [36, 37]. While answering each of these

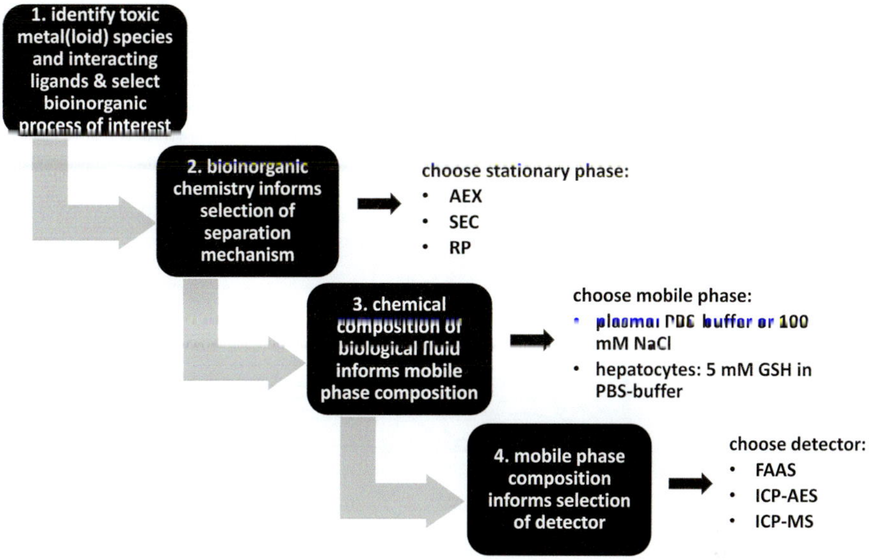


**FIGURE 2.1** Conceptual depiction of the thought processes that play a pivotal role in terms of employing LC-based approaches to solve important public health related toxicological chemistry related problems, including the organ uptake of toxic metal(loid)s and their mobilization from target organs by endogenous or exogenous ligands. (*Abbreviations*: AEX, anion-exchange chromatography; SEC, size-exclusion chromatography; RP, reversed phase chromatography; FAAS, flame atomic absorption spectrometer; ICP-AES, inductively coupled plasma atomic emission spectrometer; ICP-MS, inductively coupled plasma mass spectrometer. FAAS, ICP-AES and ICP-MS are element/metal-specific detectors.

questions is challenging in their own right, the determination of the structure of metal(loid) complexes that are formed at near-physiological conditions in blood plasma represents a feasible strategy to causally link their bioinorganic chemistry in the bloodstream with their uptake into target organs to establish their respective mechanisms of chronic toxicity [24].

While the binding of $Cd^{2+}$ to plasma protein, such as human serum albumin, transferrin and $\alpha_2$ macroglobulin, has historically received a lot of attention [38], several other plasma-binding proteins need to be considered [34]. More recently, the interaction of $Cd^{2+}$ with small-molecular-weight species in plasma, such as Cys [39] (human plasma concentration up to 300 μM [40]) and $Cl^-$ [34] (human plasma concentration ~105 mM), has been shown to be implicated in its translocation to organs [41]. From a purely inorganic chemistry perspective, it is well established that $[CdCl_3]^-$ and $[CdCl_4]^{2-}$ complexes are formed in aqueous solutions that contain $Cl^-$ [42], but the possible involvement of $Cl^-$ as well as Cys in the uptake of $Cd^{2+}$ into target organs has not been systematically investigated to date.

To better understand the roles that $Cl^-$ and hCys in plasma play in the transport of $Cd^{2+}$ to target organs, a 100 mM NaCl mobile phase (this is close to the $Cl^-$ plasma concentration) was employed in conjunction with an anion-exchange chromatography (AEX) high-performance liquid chromatography (HPLC) column. The injection

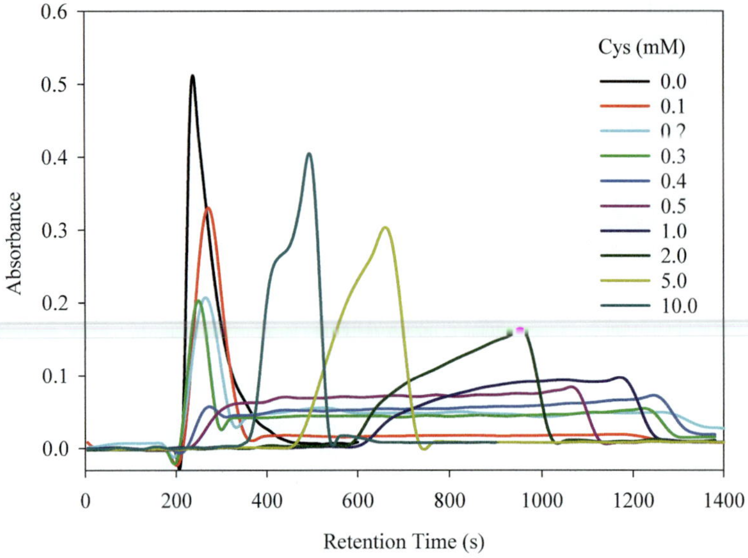


**FIGURE 2.2** Representative Cd-specific chromatograms obtained by HPLC-FAAS for the analysis of $Cd^{2+}$ (50 μg) on a PRP-X100 anion-exchange column (250 mm × 4.1 mm ID; 10 μm particle size) with a 100 mM NaCl mobile phase that contained 5 mM Tris-buffer (pH 7.4). Flow rate: 1.0 mL/min; injection volume: 50 μL (50 μg $Cd^{2+}$); detector: flame atomic absorption spectrometer at 228.8 nm. (Reproduced from [41].)

of 50 μg of $Cd^{2+}$ resulted in the detection of a single, minimally retained Cd peak by means of a FAAS. We then observed the retention behavior of $Cd^{2+}$ as a function of the gradually increasing hCys concentrations (100 μM to 10 mM) in the 100 mM NaCl mobile phase (Figure 2.2).

While the retention behavior of $Cd^{2+}$ could be rationalized based on its competitive interaction with $Cl^-$ and the increasing concentration of Cys in the mobile phase [41], the results obtained with 100 and 200 μM hCys were most relevant from a physiological/toxicological point of view. The small changes of the retention time of $Cd^{2+}$ that were observed with these hCys concentrations implied changes in the coordination environment around the metal ion, which resulted in changes of the overall net negative charge of the metal species that formed on the column. To unravel their structure, Cd fractions of the control experiment (100 mM NaCl) as well as the 100 and the 200 μM Cys in the 100 mM NaCl-containing mobile phase were collected and subsequently analyzed by X-ray absorption spectroscopy (XAS; Canadian Light Source, Saskatoon/SK, Canada). The results revealed that $[Cd(Cys)(H_2O)]^+$ and $Cd(Cys)_2$ species had eluted from the column with 100 and 200 μM Cys in the 100 mM NaCl mobile phase. While it is unclear if these Cd-species which were formed at near-physiological conditions in vitro are also formed in vivo, they certainly need to be considered to possibly represent the actual nephrotoxic Cd species that impinge on biomolecular uptake mechanism at the surface of kidney cells for their subsequent organ uptake.

This example of a successful application of an LC-based hyphenated technique to a bioinorganic chemistry problem illustrates how fundamentally new insight can be obtained regarding the uptake of $Cd^{2+}$ by toxicological target organs. While it was somewhat counterintuitive to choose an AEX-based separation mechanism given that a positively charged metal cation was injected (this cation, however, forms $CdCl_3^-$ and $CdCl_4^{2-}$ species on the column), the choice of the mobile phase (100 mM NaCl and gradual increase of the Cys concentration therein) and the detector (FAAS) were critical to be able to observe the Cd species that eluted. Since the latter could be structurally characterized by XAS, the amount of $Cd^{2+}$ that was injected had also to be carefully selected as well (XAS requires a minimum metal concentration to achieve signal-to-noise results to obtainhigh quality structural data). While one limitation of the LC-based hyphenated approach is that the employed mobile phase does not contain plasma proteins (e.g. human serum albumin), the results that were obtained suggest that the competitive interaction of plasma protein-bound $Cd^{2+}$ with $Cl^-$ and Cys may play an important role in its translocation from the bloodstream to known ABC transporters which are located of the surface of target organ cells [37]. In principle, this proposed $Cd^{2+}$ uptake mechanism is somewhat reminiscent of a possible translocation mechanism of $MeHg^+$ from its Cys-34 binding site on rabbit serum albumin (RSA) toward the BBB which is mediated by hCys [43].

### 2.2.2 Structural Characterization of the $MeHg^+$ Species Which Causes Neurotoxicity

Minamata disease refers to a poisoning epidemic which unfolded in a small fishing village in Japan in the mid 1950 where several hundred people consumed $MeHg^+$-contaminated fish and eventually displayed overt neurotoxic symptoms [44]. While recent studies have challenged the notion that $Hg^{2+}$ in the waste effluent was methylated by anaerobic microbes to $MeHg^+$ which then caused this poisoning epidemic [45], it is fairly well established that the latter mercury species ultimately caused the observed severe neurotoxic effects [46]. The biomolecular mechanism by which this neurotoxic organomercury species is translocated from the bloodstream to the BBB, however, has remained elusive ever since [47]. To unravel the biomolecular 'players' that are present in plasma and orchestrate this translocation process, rabbit blood plasma that had been spiked with a detectable amount of $MeHg^+$ was analyzed with a hyphenated chromatographic technique [43]. The technique used was size-exclusion chromatography (SEC) that was operated in conjunction with a PBS buffer mobile phase (to simulate the electrolyte of blood plasma) and an ICP-AES, which served as the Hg-specific detector. The results that were obtained from the analysis of MeHg-spiked rabbit plasma revealed a single Hg peak (Figure 2.3A) which eluted shortly before the 44 kDa molecular weight size calibration marker and had an identical retention time as a pure RSA–MeHg complex.

When PBS buffer mobile phases which contained 50, 100, 150 or 200 μM of hCys were used to analyze aliquots of the same MeHg-spiked rabbit blood plasma, the obtained Hg-specific chromatograms revealed a Cys concentration-dependent increase of the retention time of the major Hg peak (Figure 2.3A). This observation was rationalized by the mobilization of $MeHg^+$ from RSA at these near-physiological

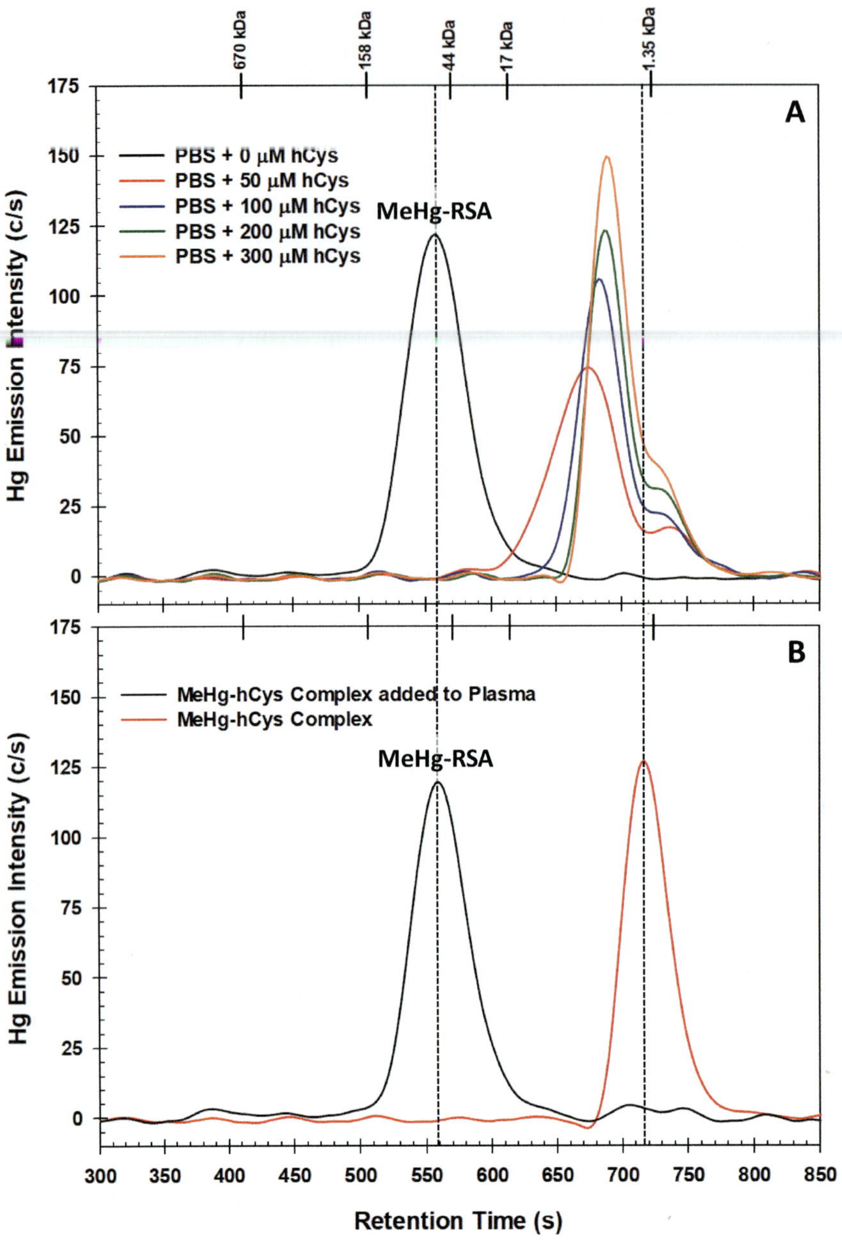


**FIGURE 2.3** Representative Hg-specific chromatograms obtained by SEC-ICP-AES for the analysis of rabbit blood plasma which had been spiked with MeHgOH (A) or an equimolar hCys-MeHg solution in PBS buffer and after its addition to rabbit blood plasma (B)'. Sepax SRT-10C SEC column (7.8 mm ID, 300 mm length, 10 μm particle size, 300 Å pore size, fractionation range: 5–1,250 kDa); mobile phase: PBS buffer (pH 7.4); temperature: 20°C; flow rate: 1.0 mL/min; injection volume: 500 μL (250 μL blood plasma and 250 μL of PBS buffer containing 5 μg Hg, 50 μM); Hg detection at 253.652 nm. The retention times of molecular weight standards are indicated at the top of the figure. (Reproduced from [43]).

hCys conditions (similar results were obtained with a PBS buffer mobile phase which contained 50 μM Cys). To gain insight into the nature of the SMW Hg species that had eluted from the SEC column, the experiment with the 50 μM hCys mobile phase was repeated using a 50 mM Tris buffer (pH 7.4) mobile phase. The analysis of the first major Hg peak that eluted by electrospray ionization mass spectrometry (ESI-MS) conclusively revealed a hCys–MeHg complex which eluted with a slightly different retention time compared to a pure hCys–MeHg complex (Figure 2.3B), most likely owing to the matrix of plasma. These results are important because previous studies have shown that the structurally closely related Cys–MeHg complex is a substrate for L-amino acid transporters 1 and 2 [48] which are located at the BBB. The obtained in vitro results, therefore, provide the first direct experimental evidence that a hCys–MeHg complex which is formed in the bloodstream will be delivered to the BBB for the subsequent translocation of this neurotoxic cargo into the brain [49].

### 2.2.3 Formation of Complexes between $Hg^{2+}$ and $MeHg^{+}$ with Cys: Implication for Organ Uptake

The toxicological chemistry-related translocation of mercury species from the bloodstream to their respective target organs is likely to involve dynamic changes of their underlying coordination chemistry. While these mercurials have been shown to be bound to plasma proteins (e.g. human serum albumin), the role of small-molecular-weight species, such as Cys, that are also present in plasma in this translocation process is much less understood [50]. To evaluate this proposition, the retention behavior of $Hg^{2+}$ and $MeHg^{+}$ was investigated in the pH range 5.0–8.0 using mobile phases comprising 10 mM Cys in 100 mM phosphate buffer using AEX-HPLC and using a FAAS as the Hg-specific detector [51]. The results revealed that at pH 7.5 $Hg^{2+}$ produced a single Hg peak, while $MeHg^{+}$ was undetectable, most likely because it was strongly retained. The fact that the largest $k'$ value for both mercury species was observed at pH 7.0 was rationalized by the putative on-column formation of $[MeHg(Cys)_2]^-$ and $[Hg(Cys)_3]^-$ species, which were strongly retained owing to their increased negative charge and/or their larger hydrophobicity compared to the MeHgCys and $Hg(Cys)_2$ species, which eluted at lower pH values (pH 5.0–6.5). The formation of the putative trinuclear Cys species of $Hg^{2+}$ is congruent with previous results obtained with $^{13}C$-nuclear magnetic resonance (NMR) spectroscopy at similar conditions and at physiological pH [52]. Further evidence for the formation of a tridentate $Hg(Cys)_3$ species at physiological pH comes from studies which identified a trinuclear $Hg^{2+}$ species in the metalloregulatory bacterial protein (MerR) [53] and related studies which reported on the formation of $Hg(NAC)_4$ at physiological pH using XAS [54]. Taken together, the findings obtained by using a hyphenated LC approach based on AEX-HPLC, a mobile phase comprising 10 mM Cys in 100 mM phosphate buffer (pH 7.5) and FAAS as the Hg-specific detector, provided relevant insight into the coordination chemistry of both mercury species with Cys at near-physiological conditions (pH 7.4). Correspondingly, the formed metabolites are implicated in their uptake to target organs. The results that were obtained by AEX-HPLC represent a starting point to systematically investigate the effect of other relevant species that either are present in the bloodstream [e.g. ~25 mM hydrogen carbonate ($HCO_3^-$)] or may enter the bloodstream (e.g. small-molecular-weight dithiols that can be administered orally when intoxicated people are treated with 2,3-dimercaptopropionic-1-sulfonate or DMPS) on the formation of toxicologically relevant novel Hg–X complexes at near-physiological conditions of plasma which will need to be structurally characterized.

### 2.2.4 Formation of MeHg$^+$-N-Acetyl-L-Cysteine Complexes at Simulated Cell Cytosolic Conditions: Implication for Its Mobilization from Target Tissues/Organs

The starting point for these investigations were results from experiments with mice that used radiolabeled MeHg$^+$ [55]. It was observed that after the exposure of mice to this toxic metal species, the latter could be effectively mobilized from their brain, liver and bloodstream when the animals were administered with drinking water that contained N-acetyl-L-cysteine (NAC; Figure 2.4b) [55]. Interestingly, simultaneously executed experiments with $Hg^{2+}$ did not remove the latter compound from mouse tissues, suggesting the bioinorganic chemistry to be unique to MeHg$^+$. Follow-up studies eventually revealed that the organic anion transporter-1 (OAT-1) is possibly involved in translocating a putative NAC–MeHg complex from the blood to proximal tubular cells (i.e. kidney cells) [56] and that the apical multidrug resistance-associated protein-2 (MRP-2) expels this putative complex into urine [57]. The formation of this complex, however, has never been investigated experimentally at the near-physiological conditions of mammalian cell cytosol and it has also not been qualitatively identified in biological excretory fluids such as urine.

To employ an LC-based hyphenated technique to probe the formation of an NAC–MeHg complex at conditions that simulate cell cytosol, one needs to first consider that this mercury compound has a high affinity for glutathione (GSH, Figure 2.4a), which is a thiol group containing tripeptide that is present in essentially all mammalian cell cytosols (e.g. in hepatocytes) at mM concentrations. It is therefore necessary to inject MeHg$^+$ on an appropriate LC column using a mobile phase that contains GSH, and to demonstrate that a MeHg–SG complex elutes from the column. The addition of increasing concentrations of NAC to the GSH-containing mobile phase should then gradually transform the injected MeHg$^+$ from an MeHg–SG species to a MeHg–NAC species. The observation of this biotransformation thus critically

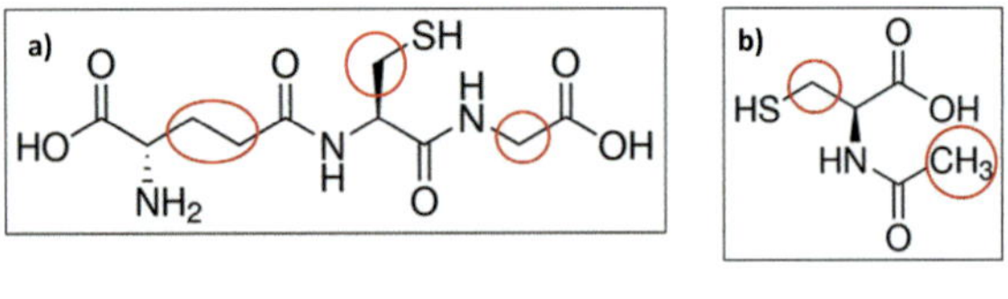


**FIGURE 2.4** (a) The molecular structure of L-glutathione (GSH), and (b) N-acetyl-L-cysteine (NAC) with hydrophobic moieties highlighted in red. Note that the hydrophobic acetyl moiety in NAC makes this thiol comparatively more hydrophobic as its hydrophobic moieties are more solvent accessible, while those of GSH are more buried within its structure. (Reproduced from [58]).

requires an appropriate separation mechanism that can distinguish between these on-column formed Hg species. Since from a structural point of view NAC has considerably more hydrophobic character compared to GSH (Figure 2.4), a reversed phase (RP) HPLC approach was deemed to be a suitable separation mechanism to observe the aforementioned ligand-exchange reaction at near-physiological conditions as the more hydrophobic MeHg–NAC complex should be more strongly retained compared to the less hydrophobic MeHg–SG complex on this separation system.

Evidence that an RP-HPLC-based hyphenated approach should be capable of observing the proposed competitive interaction of $MeHg^+$ with GSH and NAC under near-physiological conditions comes from related previous studies [58].

Based on the aforementioned conceptual considerations, a $C_{18}$ HPLC column was equilibrated with PBS buffer that contained 5.0 mM GSH to simulate protein-free hepatocyte cytosol [59]. After the injection of both mercurials, the obtained Hg-specific chromatograms revealed single peaks for $Hg^{2+}$ and $MeHg^+$ (Figure 2.5a, b) [60].

To determine the nature of the mercurial–GSH complexes that formed at these conditions on the column, the experiments were repeated with a 50 mM Tris buffer mobile phase (instead of PBS buffer) and the collected Hg-containing fractions were analyzed by ESI-MS. The ESI-MS results provided evidence for the formation of $(GS)_2Hg$ and MeHgSG complexes, which corroborates what one would expect based on the biochemistry of these mercurials [60]. Thereafter, the NAC concentration was gradually increased from 0.5 mM to 10 mM in the 5.0 mM GSH-containing mobile phase. The Hg-specific chromatograms that were obtained after the injection of both mercurials revealed gradually increasing retention times for both mercurials (Figure 2.5a, b), which was already apparent with 0.5 mM of NAC in the mobile phase and implied the on-column formation of more hydrophobic Hg species. From a bioinorganic chemistry point-of-view these results are interesting as they can be

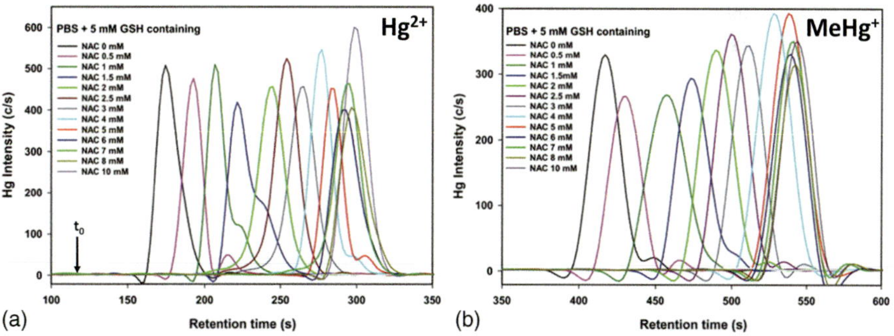


**FIGURE 2.5** (a) Representative Hg-specific chromatograms obtained after the injection of $Hg^{2+}$ on a Gemini $C_{18}$ RP-HPLC column (150 × 4.6 mm I.D.) at 37°C using 5.0 mM GSH with 0–10 mM NAC in PBS buffer (pH 7.4). Flow rate: 1.0 mL/min; detector: ICP-AES (Hg at 253.65 nm); injection volume: 20 µl (5 µg Hg). (b) Representative Hg-specific chromatograms obtained after the injection of $MeHg^+$ on a Gemini $C_{18}$ RP-HPLC column (150 × 4.6 mm I.D.) at 37°C using 5.0 mM GSH with 0–10 mM NAC in PBS buffer (pH 7.4). Flow rate: 1.0 mL/min; detector: ICP-AES (Hg at 253.65 nm); injection volume: 20 µl (5 µg Hg). (Reproduced from [58]).

rationalized by the putative on-column formation of $(GS)_2Hg$–NAC and MeHg–NAC species, which are slightly more hydrophobic compared to the species that are formed in solution using the NAC-free mobile phase. Once the NAC concentration of ~5.0 mM had been reached in the mobile phase, the retention factor of each mercurial had increased by ~1 *k*' unit. Taken together, these observations were attributed to the progressive coordination of NAC to both mercurials, importantly in the presence of a constant physiological GSH concentration. To determine the structure of the mercurial complexes that had formed at these conditions, the experiment with the 5 mM of GSH and 5.0 mM NAC mobile phase was repeated using a 50 mM Tris buffer (pH 7.4) and the collected Hg-containing fractions were analyzed by ESI-MS. The obtained ESI-MS results revealed that with regard to $Hg^{2+}$, the species $Hg(NAC)_2$, Hg(GS)(NAC) and $Hg(GS)_2$ had formed on the column; while in the experiment with $MeHg^+$, the species MeHgNAC and MeHgGS eluted from the column [60]. Altogether, these findings conclusively demonstrate a progressive coordination of NAC molecules via their thiol groups to each of these mercury species in the presence of physiologically relevant GSH concentrations. This observed competitive coordination of NAC to both mercurials establishes a feasible mechanism by which $MeHg^+$ is directed to proximal tubular cells involving OAT-1 and then expelled from mouse tissues via MRP-2 transporters to urine. Although these hyphenated LC-based experiments were conducted with aqueous mobile phases that did not contain any protein, they nevertheless provide a feasible biomolecular explanation as to how the influx of small NAC concentrations into the cell cytosol may facilitate the excretion of $MeHg^+$ in the presence of severalfold higher concentration of GSH [57]. Relatedly, the fact that $Hg^{2+}$ was not mobilized by NAC from mouse tissues is attributed to the lack of existing efflux transporters, which are able to accept Hg(GS)(NAC) and/or $Hg(NAC)_2$ as a substrate once they are formed at near-physiological conditions.

## 2.3 APPLICATION OF LC-BASED APPROACHES TO ADVANCE THE BIOINORGANIC CHEMISTRY OF TOXIC METAL(LOID)S

Based on the illustrated examples it is evident that the chosen LC separation modes may be similarly employed to gain much needed biomolecular insight into other toxicological chemistry-related processes involving other metal species that are of immediate public health concern, such as the Hg-containing compound thimerosal [51], nickel [61] and manganese [62] that unfold at the blood plasma–red blood cell interface [63]. The most recent application of an LC-based approach to probe the competitive interaction of two mercurials between GSH and NAC at conditions that simulate cytosolic conditions glutathione [60] is particularly noteworthy as it implies that highly relevant bioinorganic chemistry interactions between other toxic metal(loid) species can also be probed within target organ cells. As LC-based approaches have inherent limitations, it is important to make readers aware that other powerful instrumental analytical methods, such as synchrotron-radiation-based XAS [64] and/or advanced metalloproteomic approaches [65] can also provide important new insight into biochemical mechanisms that underly the chronic toxicity of metal(loid) species in mammals [66]. Most recently it has been shown that probing the biochemical fate of $MeHg^+$ within microbes that inhabit the gut [67] is highly relevant to delineate its mechanism of chronic toxicity as its demethylation therein can significantly decrease its absorption into the bloodstream.

## 2.4 CONCLUSION

Arguably one of the biggest problems in the post-genomic world is to gain deeper insight into the exposure–response relationship that pertains to the increasing exposure of certain human populations to environmentally ubiquitous As, Cd, Hg and Pb species. While dietary factors have recently been shown to greatly affect the *bioavailability* of mercury species in humans [68], an equally important question that directly pertains to the exposure–response relationship is to delineate the bioinorganic processes that determine the *organ availability* of toxic metal(loid) species, which is perhaps the most critical prerequisite to causally connect human exposure to inorganic pollutants with the etiology of organ-based diseases. The difficulty in gaining insight into these processes which unfold in complex biological fluids can be overcome by the application of LC-based hyphenated techniques that require atomic spectroscopic detectors. Recent results have conclusively demonstrated that specific bioinorganic chemistry-related questions that are of considerable public health relevance (e.g. the organ uptake of a specific toxic metal species) can be addressed by tailoring hyphenated chromatographic techniques to the problem at hand. The latter approach requires one to first choose an appropriate separation mechanism (i.e. the stationary phase), then to identify the biological fluid wherein the bioinorganic chemistry process of interest unfolds (i.e. the mobile phase) and, last but not least, to observe the metal(loid) of interest with an apprrpriate detector. The results that were obtained by using AEX, SEC and RP-HPLC approaches underscore the inherent versatility that unconventional hyphenated LC separation methods offer in terms of obtaining important new insight into highly relevant bioinorganic processes involving other relevant toxic metal(loid) species, such as those used in the high-tech industry. All results that are obtained employing these hyphenated LC approaches, however, need to be interpreted with caution as in vitro results can never be extrapolated to what is unfolding in vivo. Nevertheless, in vitro results should be used to iteratively establish the biochemical mechanisms of chronic toxicity that unfold in the bloodstream–organ system with regard to individual toxic metal(loid) species. In contrast to epidemiological research which inherently cannot establish causal relationships between the exposure of humans to toxic metal(loid)s and the etiology of specific diseases, the proposed molecular toxicology approach offers the inherent potential to establish such causal links which may be used to tighten regulatory standards for toxic metal(loid)s in food and drinking water to foster public health. In terms of relevant research problems that may be addressed using this apprach, one may think of the structural characterization of the $Mn^{2+}$ metabolite(s) that is/are translocated from the bloodstream to the human brain to better understand the etiology of manganism [69], as well as the uptake of other toxic metal species, such as $Pb^{2+}$ and organotin species, into the brain [70] as the obtained insights contribute to gain urgent new insight into the etiology of neurodegenerative diseases [71]. Therefore, the tailoring of hyphenated LC-based techniques to address research problems that pertain to toxic metal(loid)s represents a powerful new approach to uncover the mechanism-integrated framework to link human exposure to multiple toxic metal(loid) species with the etiology of environmental diseases [6].

## ACKNOWLEDGMENTS

Some of the research that is discussed in this book chapter was funded by the Natural Sciences and Engineering Research Council (NSERC) of Canada.

## REFERENCES

1. Macklin MG, Thomas CJ, Mudbhatkal A, Brewer PA, Hudson-Edwards KA, Lewin J et al. Impacts of metal mining on river systems: A global assessment. Science. 2023;381:1345–50.
2. Briffa J, Sinagra E, Blundell R. Heavy metal pollution in the environment and their toxicological effects on humans. Heliyon. 2020;6:e044691.
3. Chowdhury S, Jafar Mazumder MA, Al-Attas O, Husain T. Heavy metals in drinking water: Occurrences, implications, and future needs in developing countries. Sci Total Environ. 2016;569–570:476–88.
4. Gerson JR, Szponar N, Zambrano AA, Bergquist B, Broadbent E, Driscoll CT et al. Amazon forests capture high levels of atmospheric mercury pollution from artisanal gold mining. Nat Commun. 2022;13:559.
5. Hou D, O'Connor D, Igalavithana AD, Alessi DS, Luo J, Tsang DCW et al. Metal contamination and bioremediation of agricultural soils for food safety and sustainability. Nat Rev Earth Envir. 2020;1:366–81.
6. Gailer J. Toward a mechanism-driven integrated framework to link human exposure to multiple toxic metal(loid) species with environmental diseases. Int J Mol Sci. 2024;25:3393.
7. Thomas R. Regulating heavy metals in baby foods: The challenges of food manufacturers and the FDA being on the same page. Spectroscopy. 2021;36:10–6.
8. Steckling N, Tobollik M, Plass D, Hornberg C, Ericson B, Fuller R et al. Global burden of disease of mercury used in artisanal small-scale gold mining. Ann Glob Health. 2017;83:234–47.
9. Kelly EN, Schindler DW St., Louis VL, Donald DB, Vladicka KE. Forest fire increases mercury accumulation by fishes via food web restructuring and increased mercury inputs. Proc Natl Acad Sci USA. 2006;103:19380–5.
10. Lopez AM, Pacheco JJ, Fendorf S. Metal toxin threat in wildland fires determined by geology and fire severity. Nat Commun. 2023;14:8007.
11. Barone G, Storelli A, Meleleo D, Dambrosio A, Garofalo R, Busco A et al. Levels of mercury, methylmercury and selenium in fish: Insights into children food safety. Toxics. 2021;9:39.
12. Donaldson SG, Van Oostdam J, Tikhonov C, Feely M, Armstrong B, Ayotte P et al. Environmental contaminants and human health in the Canadian Arctic. Sci Total Environ. 2010;408:5156–234.
13. Clarkson TW, Strain JJ. Methylmercury: Loaves versus fishes. Reprod Neurotoxicol. 2020;81:282–7.
14. Boucher O, Jacobson SW, Plusquellec P, Dewailly E, Ayotte P, Forget-Dubois N et al. Prenatal methylmercury, postnatal lead exposure, and evidence of attention deficit/hyperactivity disorder among Inuit children in Arctic Quebec. Environ Health Perspect. 2012;120:1456–61.

15. Kodros JK, Bell ML, Dominici F, L'Orange C, Godri Pollitt KJ, Weichenthal S et al. Unequal airborne exposure to toxic metals associated with race, ethnicity, and segregation in the USA. Nat Commun. 2022;13:6329.
16. Gardener H, Bowen J, Callan SP. Lead and cadmium contamination in a large sample of United States infant formulas and baby foods. Sci Total Environ. 2018;651:822–7.
17. Upadhyay MK, Shukla A, Yadav P, Srivastava S. A review of arsenic in crops, vegetables, animals and food products. Food Chem. 2019;276:608–18.
18. Sadiq NW, Beauchemin D. Simultaneous speciation analysis of arsenic, chromium and selenium in the bioaccessible fraction for realistic assessment of food safety. Anal Chem. 2018;257:230–6.
19. Centers of Disease Control and Prevention A, GA, USA. Fourth national report on human exposure to environmental chemicals. 2009. https://stacks.cdc.gov/view/cdc/21813
20. Schober SE, Sinks TH, Jones TH, Bolger PM, McDowell M, Osterloh J et al. Blood mercury levels of US children and women of childbearing age, 1999–2000. J Am Med Assoc. 2003;289:1667–74.
21. Health Canada. Cadmium in Canadians. Ottawa: Health Canada, 2021 December 2021. https://www.canada.ca/en/health-canada/services/environmental-workplace-health/reports-publications/environmental-contaminants/human-biomonitoring-resources/cadmium-canadians.html.
22. Landrigan PJ, Fuller R, Acosta NJR, Adeyi O, Arnold R, Basu N et al. The Lancet commission on pollution and health. Lancet. 2018;391:462–512.
23. Fuller R, Landrigan PJ, Balakrishnan K, Bathan G, Bose-O'Rielly S, Brauer M et al. Pollution and health: A progress update. Lancet Planet Health. 2022;6:e535–47.
24. Doroudian M, Pourzadi N, Gautam A, Gailer J. Translational toxicology of metal(loid) species: Linking their bioinorganic chemistry in the bloodstream to organ damage onset. Biometals. 2024;37:739–53.
25. Sturla SJ, Boobis AR, Fitzgerald RE, Hoeng J, Kavlock RJ, Schirmer K et al. Systems toxicology: From basic research to risk assessment. Chem Res Toxicol. 2014;27:314–29.
26. Gailer J, Turner RJ. Environmental and Biochemical Toxicology. Concepts, Case Studies and Challenges. Munich: De Gruyter; 2022. 335 p.
27. Sarpong-Kumankomah S, Gibson MA, Gailer J. Organ damage by toxic metals is critically determined by the bloodstream. Coord Chem Rev. 2018;374:376–86.
28. Percy AJ, Korbas M, George GN, Gailer J. Reversed-phase high performance liquid chromatographic separation of inorganic mercury and methylmercury driven by their different coordination chemistry towards thiols. J Chromatogr A. 2007;1156:331–9.
29. Jahromi EZ, White W, Wu Q, Yamdagni R, Gailer J. Remarkable effect of mobile phase buffer on the SEC-ICP-AES derived Cu, Fe and Zn-metalloproteome pattern of rabbit blood plasma. Metallomics. 2010;2:460–8.
30. Izatt RM, Izatt SR, Bruening RL, Izatt NE, Moyer BA. Challenges to achievement of metal sustainability in our high-tech society. Chem Soc Rev. 2014;43:2451–75.
31. Khan S, Cao Q, Zheng YM, Huang YZ, Zhu YG. Health risks of heavy metals in contaminated soils and food crops irrigated with wastewater in Beijing, China. Environ Pollut. 2008;152:686–92.
32. Nordberg GF. Historical perspectives on cadmium toxicity. Toxicol Appl Pharmacol. 2009;238:192–200.
33. Kusak R, Nasiadek M, Stragierowicz J, Hanke W, Kilanowicz A. Changes in endogenous essential metal homeostasis in the liver and kidenys during a six-month follow-up period after subchronic cadmium exposure. Int J Mol Sci. 2024;25:3829.
34. Hill A, Gailer J. Linking molecular targets of Cd in the bloodstream to organ-based adverse health effects. J Inorg Biochem. 2021;216:111279.

35. Guan T-Q, Qiu B-H, Nurmamedov H, Talukder M, Lyv M-W, Li J-L. Cadmium-induced splenic lymphocytes anoikis is not mitigated by activating Nrf2-mediated antioxidative defense response. J Inorg Biochem. 2022;234:111882.
36. Thevenod F, Lee W-K. Toxicology of cadmium and its damage to mammalian organs. In: A. Sigel, H. Sigel, R.K.O. Sigel, editors. Cadmium: From Toxicity to Essentiality. Vol. 11. New York: Springer Science + Business Media Dordrecht; 2013. p. 415–90.
37. Thevenod F, Lee W-K. Cadmium transport by mammalian ATP-binding cassette transporters. Biometals. 2024;37:697–719.
38. Sagmeister P, Gibson MA, McDade KH, Gailer J. Physiologically relevant plasma D, L-homocysteine concentrations mobilize Cd from human serum albumin. J Chromatogr B. 2016;1027:181–6.
39. Bridges CC, Zalups RK. Mechanisms involved in the transport of mercuric ions in target tissues. Arch Toxicol. 2017;91:63–81.
40. Isokawa M, Kanamori T, Funatsu T, Tsunoda M. Analytical methods involving separation techniques for determination of low-molecular weight biothiols in human plasma and blood. J Chromatogr B. 2014;964:103–15.
41. Gautam A, Gomez A, Mendoza Rengifo E, George GN, Pickering IJ, Gailer J. Structural characterization of toxicologically relevant $Cd^{2+}$-L-Cysteine complexes. Toxics. 2023;11:294.
42. Holleman-Wiberg. Das Cadmium. In: Lehrbuch der Anorganischen Chemie. Berlin: Walter de Gruyter; 1976. p. 829.
43. Bridle TG, Doroudian M, White W, Gailer J. Physiologically relevant hCys concentrations mobilize MeHg from rabbit serum albumin to form MeHg-hCys complexes. Metallomics. 2022;14:mfac010.
44. Kurland LT, Faro SN, Siedler H. Minamata disease. The outbreak of a neurologic disorder in Minamata, Japan, and its relationship to the ingestion of seafood contaminated by mercury compounds. World Neurology. 1960;1:370–90.
45. James AK, Nehzati S, Dolgova NV, Sokaras D, Kroll T, Eto K et al. Rethinking the Minamata tragedy: What mercury species was really reponsible. Environ Sci Technol. 2020;54:2726–33.
46. Yorifuji T, Tsuda T, Inoue S, Takao S, Harada M. Long-term exposure to methylmercury and psychiatric symptoms in residents of Minamata. Environ Int. 2011;37:907–13.
47. Nogara PA, Oliveira CS, Schmitz GL, Piquini PC, Farina M, Aschner M et al. Methylmercury's chemistry: From the environment to the mammalian brain. Biochim Biophys Acta, General Subjects. 2019;1863:129284.
48. Simmons-Willis TA, Koh AS, Clarkson TW, Ballatori N. Transport of a neurotoxicant by molecular mimicry: The methylmercury-L-cysteine complex is a substrate for human L-type large neutral amino acid transporter (LAT) 1 and LAT 2. Biochem J. 2002;367:239–46.
49. Farina M, Rocha JBT, Aschner M. Mechanisms of methylmercury-induced neurotoxicity: Evidence from experimental studies. Life Sci. 2011;89:555–63.
50. Gautam A, Gailer J. More effective mobilization of $Hg^{2+}$ from human serum albumin compared to $Cd^{2+}$ by L-cysteine at near physiological conditions. Toxics. 2023;11:599.
51. Pourzadi N, Gailer J. L-cysteine mediated rapid baseline separation of $Hg^{2+}$ and MeHg by anion-exchange HPLC. J Chromatogr A 2024;1735:465309.
52. Cheesman BV, Arnold AP, Rabenstein DL. Nuclear magnetic resonance studies on the solution chemistry of metal complexes. 25. $Hg(thiol)_3$ complexes and Hg(II)-thiol ligand exchange kinetics. J Am Chem Soc. 1988;110:6359–64.
53. Watton SP, Wright JG, MacDonnel FM, Bryson JW, Sabat M, O'Halloran TV. Trigonal mercuric complex of an aliphatic thiolate: A spectroscopic and structural model for the receptor site in the Hg(II) biosensor MerR. J Am Chem Soc. 1990;112:2824–6.

54. Jalilehvand F, Parmar K, Zielke S. Mercury(II) complex formation with N-acetylcysteine. Metallomics. 2013;5:1368–76.
55. Ballatori N, Lieberman MW, Wang W. *N*-acetylcysteine as an antidote in methylmercury poisoning. Environ Health Perspect. 1998;106:267–71.
56. Koh AS, Simmons-Willis TA, Pritchard JB, Grassl SM, Ballatori N. Identification of a mechanism by which the methylmercury antidotes N-acetylcysteine and dimercaptopropanesulfonate enhance urinary metal excretion: Transport by the renal organic anion transporter-1. Mol Pharmacol. 2002;62:921–6.
57. Madejczyk MS, Aremu DA, Simmons-Willis TA, Clarkson TW, Ballatori N. Accelerated urinary excretion of methylmercury following administration of its antidote N-acetylcysteine requires Mrp2/Abcc2, the apical multidrug resistance-associated protein. J Pharmacol Exp Ther. 2007;322:378–84.
58. Pei KL, Sooriyaarachchi M, Sherrell DA, George GN, Gailer J. Probing the coordination behavior of $Hg^{2+}$, $CH_3Hg^+$, and $Cd^{2+}$ towards mixtures of two biological thiols by HPLC-ICP-AES. J Inorg Biochem. 2011;105:375–81.
59. Gailer J, Lindner W. On-column formation of arsenic-glutathione species detected by size-exclusion chromatography in conjunction with arsenic-specific detectors. J Chromatogr B. 1998;716:83–93.
60. Doroudian M, Thibault ME, Gailer J. N-acetylcysteine displaces glutathionyl-moieties from $Hg^{2+}$ and $MeHg^+$ to form more hydrophobic complexes at near physiologoical conditions. Molecules. 2023;28:6762.
61. Glennon JD, Sarkar B. Nickel(II) transport in human blood serum. Biochem J. 1982;203:15–23.
62. Aschner M, Guilarte TR, Schneider JS, Zheng W. Manganese: Recent advances in understanding its transport and neurotoxicity. Toxicol Appl Pharmacol. 2007;221:131–47.
63. Gibson MA, Sarpong-Kumankomah S, Nehzati S, George GN, Gailer J. Remarkable differences in the biochemical fate of $Cd^{2+}$, $Hg^{2+}$, $CH_3Hg^+$ and thimerosal in red blood cell lysate. Metallomics. 2017;9:1060–72.
64. Weng M, Dolgova NV, Vogt LI, Quereshi M, Sokaras D, Kroll T, et al. Synchrotron speciation of umbilical cord mercury and selenium after environmental exposure at Niigata. Neurotoxicology. 2024;100:117–23.
65. Zhou Y, Li H, Sun H. Metalloproteomics for biomedical research: Methodology and applications. Annu Rev Biochem. 2022;91:449–73.
66. Sanchez-Alarcon J, Milic M, Bustamante-Montes LP, Isaac-Olive K, Valencia-Quintana R, Ramirez-Duran N. Genotoxicity of mercury and its derivatives demonstrated in vitro and in vivo in human populations studies. Systematic review. Toxics. 2021;9:326.
67. Ke T, Rajoo A, Tinkov AA, Skalny AV, Tizabi Y, Rocha JBT et al. Intestinal microbiota protects against methylmercury-induced neurotoxicity. Biometals. 2024;37:561–576.
68. Bradley MA, Barst BD, Basu N. A review of mercury bioavailability in humans and fish. Int J Environ Res Publ Health. 2017;14:169.
69. Deng X, Guo Y, Jin X, Si H, Dai K, Deng M et al. Manganese accumulation in red blood cells as a biomarker of manganese exposure and neurotoxicity. Neurotoxicology. 2024;102:1–11.
70. Yokel RA. Blood-brain barrier flux of aluminium, manganese, iron and other metals suspected to contribute to metal-induced neurodegeneration. J Alzheimer's Dis. 2006;10:223–33.
71. Sussulini A, Hauser-Davis RA. Metallomics applied to the study of neurodegenerative and mental diseases. Adv Exp Med Biol. 2018;1055:21–37.

# 3 Minor To Whom? The Interaction Between Minor Metals and Plants

*Davide Marzi, Maria Luisa Antenozio, Massimo Zacchini, and Patrizia Brunetti*

## 3.1 INTRODUCTION

During human evolution, the exploitation of natural resources marked strategic turning points in the development of advanced societies. Metals such as copper (Cu), iron (Fe), and gold (Au) have long shaped cultures worldwide, and today a growing suite of so-called minor metals is gaining importance on a global scale. The term "minor metals" might suggest limited importance, yet these elements play a pivotal role in driving modern technologies. Unlike noble metals, which are highly resistant to corrosion, oxidation, and acids, or base metals that are common and readily oxidize or corrode, minor metals are not a chemically defined group but a socioeconomically constructed one. The expression "minor metals" is an industry term coined to describe metals produced in relatively low quantities, predominantly extracted as by-products of base metals that rarely receive the spotlight in main mining operation. These elements are generally subject to limited market activity and are excluded from the major public exchanges such as the London Metal Exchange (LME). Thus, they are not "minor" in function but in market visibility. In this context, within the Critical Raw Materials (CRMs) framework [1], the European Commission identifies substances that are of great importance to the economy and at risk of supply disruption [2]. The overlap between CRMs and minor metals is striking as many elements, such as bismuth (Bi), gallium (Ga), germanium (Ge), lithium (Li), rare earth elements (REEs), and tungsten (W), appear in both lists (Figure 3.1).

Many of these metals are essential for technologies ranging from semiconductors and permanent magnets to catalysts and superalloys, enabling the production of smartphones, wind turbines, advanced alloys, and life-saving medical equipment [3]. Thus, the European Commission has derived an additional list of Strategic CRMs (SCRMs) from the CRMs framework, and the most up-to-date list includes minor metals such as Bi, Ga, Ge, Li, REEs, and W. These elements are the paradigmatic "technology metals" of the green digital transition and they best illustrate how small-volume, high-function metals underpin technological development and international economic strategies, sustaining the backbone of the modern world [1, 2]. This chapter delves into the journey of these minor metals from their hidden ores to their interaction with vegetation.


DOI: 10.1201/9781003632092-4

| Minor metal list (MMTA) | Metals reported in both lists | | | EU Critical Raw Material list (2023) |
|---|---|---|---|---|
| Calcium | Arsenic | Lanthanum | Samarium | Boron |
| Cadmium | Beryllium | Lithium | Strontium | Baryte |
| Chromium | Bismuth | Lutetium | Tantalum | Bauxite |
| Mercury | Cerium | Magnesium | Terbium | Coking coal |
| Indium | Cobalt | Manganese | Titanium | Copper |
| Molybdenum | Dysprosium | Niobium | Thulium | Feldspar |
| Rhenium | Erbium | Neodymium | Vanadium | Fluorspar |
| Tellurium | Europium | Osmium | Tungsten | Helium |
| Thallium | Gallium | Praseodymium | Yttrium | Graphite |
| Zirconium | Gadolinium | Rhodium | Ytterbium | Nickel |
| | Germanium | Ruthenium | | Phosphorus |
| | Hafnium | Antimony | | Palladium |
| | Holmium | Scandium | | Phosphate |
| | Iridium | Silicon | | Platinum |

**FIGURE 3.1** Comparison between the list of minor metals proposed by the Minor Metals Trade Association (MMTA; https://mmta.co.uk) and the European Critical Raw Materials (CRMs) list (last update 2025). The central panel shows metals appearing in both lists, while the left and right panels report metals unique to the MMTA list and to the EU CRMs list, respectively.

## 3.2 ENVIRONMENTAL OCCURRENCE

Minor metals are heterogeneously distributed throughout the Earth's crust and typically dispersed in low concentrations within mineral lattices characterized by more abundant elements. In general, they are recovered as by-products of mining activities for those more abundant elements, meaning that their availability is tied to the production of host metals with which they occur [4]. The increasing worldwide demand for minor metals has prompted in-depth investigation into their occurrence in the environment, leading to the characterization of recurring associations with host metals that could be used to support more efficient mining (Figure 3.2).

This is the case of Bi, which is most often produced as a by-product of lead (Pb) processing. Abundance of Bi on Earth is estimated at 0.025 ppm and its average concentrations in natural soils are estimated to range between 0.13 and 1 μg/g. The highest amounts of Bi have been detected at mining sites such as Cerro de Choroloque and Tazna in Bolivia, as well as in New South Wales, Tasmania, and Queensland [5]. Ga is obtained as a by-product of processing bauxite rich in aluminum (Al) and from zinc (Zn) processing residues; however, the main routes of Ga environmental contamination derive from industry [6]. In recent years, the rapid development of high-technology industries has led to the production of large amounts of

| Al | Au | Cu | Fe | Ni | Pb | Pt | Sn | Ti | Zn |
|---|---|---|---|---|---|---|---|---|---|
| Ga | Ag | Se | Ce | Pd | Bi | Ir | Nb | Hf | Cd |
| V | Zn | Te | Nd | Co | Sb | Ru | W | Zr | In |
| | Cu | Mo | Pr | Os | Te | Os | In | Lu | Ge |
| | Sb | Co | La | Rh | Cu | Rh | Bi | Yb | Ag |
| | U | Re | Eu | Pt | Zn | Pd | Ta | Er | Ga |
| | | Ag | V | Ir | | Au | Er | Dy | Sn |
| | | Bi | Sm | Cu | | Co | Ho | Ho | Cu |
| | | U | Gd | Au | | Ni | Dy | Tb | Au |
| | | Au | Tb | Ru | | Cr | Tb | Gd | |
| | | In | Mn | | | Cu | Sb | Eu | |
| | | Sn | Dy | | | Ag | Gd | Sm | |
| | | Zn | Y | | | | Eu | Nd | |
| | | | | | | | Sm | Ce | |
| | | | | | | | Nd | Pr | |
| | | | | | | | Pr | La | |
| | | | | | | | Ce | Y | |
| | | | | | | | La | | |
| | | | | | | | Y | | |

**FIGURE 3.2 Table of host and companion metals.** Host metals are indicated in the upper panels, whereas their companion metals are listed in the lower panels. Aluminum (Al), antimony (Sb), bismuth (Bi), cerium (Ce), cobalt (Co), chromium (Cr), copper (Cu), dysprosium (Dy), erbium (Er), europium (Eu), gadolinium (Gd), gallium (Ga), germanium (Ge), gold (Au), hafnium (Hf), holmium (Ho), indium (In), iridium (Ir), iron (Fe), lanthanum (La), lead (Pb), lutetium (Lu), manganese (Mn), mercury (Hg), molybdenum (Mo), neodymium (Nd), nickel (Ni), niobium (Nb), osmium (Os), palladium (Pd), platinum (Pt), praseodymium (Pr), rhenium (Re), rhodium (Rh), ruthenium (Ru), samarium (Sm), selenium (Se), silver (Ag), tantalum (Ta), terbium (Tb), tellurium (Te), tin (Sn), titanium (Ti), tungsten (W), uranium (U), vanadium (V), ytterbium (Yb), yttrium (Y), zinc (Zn), and zirconium (Zr).

wastewater and waste containing Ga, as reported for Taiwan area [7]. Ge has an estimated abundance of 1.3–1.6 µg/g in Earth's crust; similar to other minor metals, Ge is mainly obtained as by-product and is associated with Pb, Zn, and Cu sulfide ores [8, 9]. Unlike most other minor metals, Li and W are primarily targeted for primary mining; however, their high demand places them among strategic materials [8]. As shown in Figure 3.3, China is both the largest global supplier and the EU's main supplier for the majority of the CRMs, including Bi, Ga, Ge, REEs, and W. Other relevant suppliers include Russia, Brazil, the Democratic Republic of Congo (DRC), and Australia. In addition, recent discoveries of new Li resources have increased the global availability of this metal, with Bolivia, Argentina, and the United States as the main producers [2, 8].

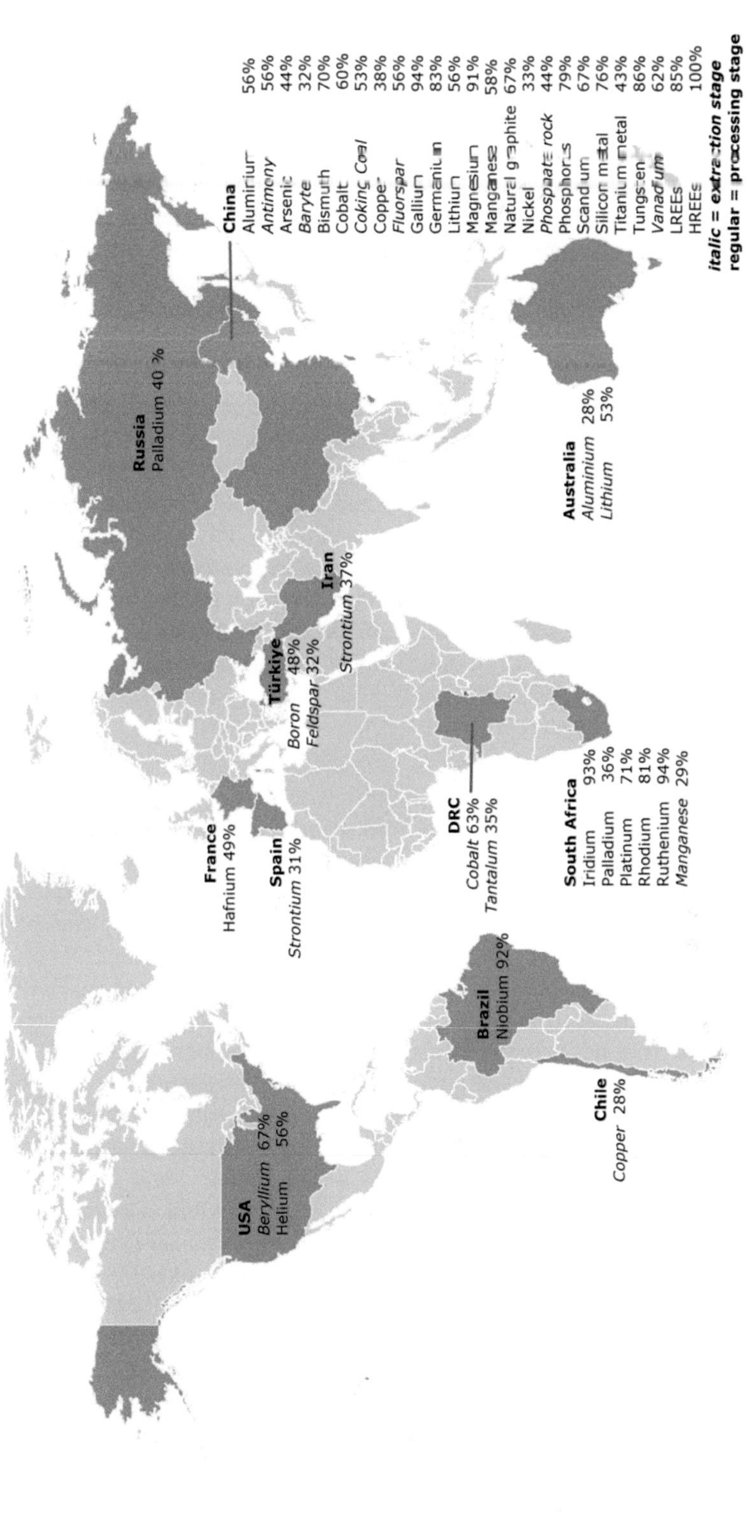


**FIGURE 3.3** Main global suppliers of individual CRMs. (Image adapted from [2]).

## 3.3 UPTAKE, ACCUMULATION, AND TOXICITY OF MINOR METALS IN PLANTS

The environmental risks associated with minor metals have so far received little attention, overshadowed by economic and geopolitical considerations. Their uptake by vegetation can trigger metal accumulation in different plant organs, posing risks to food safety and human health, yet also paving the way for technologies that exploit this characteristic, such as phytomining (i.e., the use of plants to extract valuable metals from environmental matrices). The molecular bases regulating minor metals uptake in plants have been little investigated; however, recent studies have provided new insights into these mechanisms. Minor metals with chemical similarities to essential nutrients can enter roots through the same membrane transporters used for micro- and macronutrient uptake, via non-selective cation channels (NSCCs) that allow the entry of various cations into cells, and through aquaporins, which are passive water membrane channels that may also act as NSCCs. Depending on the plant species, minor metals can be absorbed and accumulated in roots, or they can be translocated to aerial organs such as stems and leaves. Physiological parameters including seed germination, root elongation, and photosynthetic activity are used to assess the toxicity of these metals in plants, often corroborated by the analyses of oxidative stress responses. The tolerance and the ability to translocate and accumulate high metal amounts in aerial organs rather than in the roots, define plants as hyperaccumulators.

### 3.3.1 Bismuth

Bi is considered a "green metal" and has been suggested as a replacement for Pb in many industrial processes; however, its increasing use raises concerns about environmental safety. In the model plant species *Arabidopsis thaliana* L., Bi affects the expression of several genes involved in Fe uptake, including the $Fe^{2+}$ transporter *Iron-Regulated Transporter 1 (IRT1)* and the *Ferritin 1* (FER1) and *Ferritin 4* (FER4), suggesting that Bi may be absorbed by roots and translocated via Fe transport mechanisms [10]. Bi is mainly accumulated in roots, where it affects nutrient homeostasis and root development, whereas only small amounts are translocated to shoots, impairing photosynthetic performance as Bi concentration in the growth medium increases [5, 11–13]. For instance, *Arabidopsis* seedlings grown on 2-μM Bi accumulate up to 14 mg/kg Bi in roots and 2 mg/kg Bi in shoots, whereas garden cress (*Lepidium sativum* L.) grown on 1-mM Bi accumulate about 100 mg/kg Bi in roots and 5 mg/kg Bi in shoots [10, 13]. Root growth inhibition by Bi has been reported across species, including *Arabidopsis* and tomato (*Solanum lycopersicum* L.) seedlings treated with 2 μM and 3 μM Bi, respectively, and garden cress and perennial ryegrass (*Lolium perenne* L.) plants treated with about 60 μM Bi [10, 12, 14, 15]. Bi toxic effects also include reduced seed germination, although the magnitude of the effect varies among plant species. Germination is reduced by about 35% in *Arabidopsis* seeds exposed to 4 μM Bi, by 66% in tomato seeds exposed to 12 μM Bi, and by 40% in ryegrass seeds exposed to 1 mM Bi, respectively [10, 12, 14]. In addition, Bi exerts different detrimental effects in leaves by impairing the photosynthetic machinery, reducing the total chlorophyll index, chlorophyll *a*, chlorophyll *b*, and the carotenoid pigment-specific simple ratio (PSSR), as well as decreasing transpiration rate and stomatal conductance in garden cress, horse gram (*Macrotyloma uniflorum* Lam. Verdc), wheat (*Triticum aestivum* L.), and bean (*Phaseolus vulgaris* L.) plants [11, 13, 16]. Like other metals, high Bi concentrations induce oxidative stress in plants, causing increased production of reactive oxygen species (ROS), such as hydrogen peroxide ($H_2O_2$), and malondialdehyde (MDA) levels, along with increased activity or abundance of detoxification-related glutathione-S-transferase and antioxidative ROS-scavenging enzymes, including superoxide dismutase (SOD), catalase (CAT), peroxidase (POX), ascorbate peroxidase (APX), glutathione peroxidase (GPX), glutathione reductase (GR), dehydroascorbate reductase (DHAR), and monodehydroascorbate reductase (MDHAR), in both roots and shoots (10–12, 14, 16].

### 3.3.2 Gallium

Gallium (Ga) is an emerging pollutant mostly associated with high-tech industries, particularly semiconductors; however, Ga mobility in the soil–plant system is low compared with other common contaminants such as cadmium (Cd), Cu, and Zn [17]. Studies on the interaction between Ga and plants are still scarce, and no clear uptake mechanisms have been described so far. Its chemical similarity to Al suggests that Ga may enter the roots by exploiting the same transporters, such as the Nramp aluminum transporter 1 (Nrat1), which belongs to the natural resistance-associated macrophage protein (Nramp) family, and the nodulin 26-like intrinsic protein 1;2 (NIP2;1), which is part of the aquaporin family [18]. This hypothesis is further supported by the finding that the Al hyperaccumulator plant *Camellia sinensis* L. Kuntze accumulates high levels of Ga, up to 150 mg/kg, suggesting that other plants able to hyperaccumulate Al may also take up and store Ga [19]. As indicated by studies in *Arabidopsis* and rice, Ga accumulation mainly occurs in roots, with little translocation to shoots, likely mediated by complexation with organic acids and subsequent sequestration in vacuoles, as occurs for Al [20–22]. *Arabidopsis* seedlings grown on medium containing 500 μM Ga accumulate about 950 and 175 mg/kg Ga in roots and shoots, whereas rice (*Oryza sativa* L.) grown on 85 μM Ga accumulate about 800 mg/kg Ga in roots and tenfold less in shoots, respectively [20, 21]. Root elongation and shoot growth are reduced in Arabidopsis treated with 6 and 500 μM Ga, respectively [20]. In contrast, these parameters are significantly promoted in rice treated with up to 85 μM Ga; moreover, at this concentration, Ga increases chlorophyll SPAD values, confirming stimulatory effect on growth and photosynthetic activity in rice [21]. However, excessive Ga accumulation in plant tissues may lead to inhibition of relative growth rate, decrease of water use efficiency, and induction of oxidative stress, in both Arabidopsis and rice [20–23]. Interestingly, Ga triggers stress responses highly similar to those induced by Al in *Arabidopsis*, suggesting that tolerance to both metals may rely on shared mechanisms in plants, including the production and complexation of organic acids, such as malate, citrate, and oxalate, secreted into the rhizosphere to immobilize reactive cations [18, 20].

### 3.3.3 Germanium

Ge is a semiconductor metalloid recently included in the list of the CRMs due to its central role in the development of high-tech electronics and infrared optics. The increasing demand for Ge poses new environmental risks related to the mining and processing of this minor metal. The chemical similarities between Ge and silicon (Si) enable the uptake of these metals with almost the same kinetics in plants such as rice, barley (*Hordeum vulgare* L.), cucumber (*Cucumis sativus* L.), and tomato, indicating little or no discrimination during their uptake [24]. In rice, Ge uptake occurs through Si transporters such as Low silicon 1 (Lsi1), which belongs to a NIP subfamily of aquaporins, and Low silicon 2 (Lsi2), which mediates the active transport of $Si(OH)_4$ in antiport with hydrogen ion ($H^+$) across the endodermis, through the Casparian strip toward the sap streams [25–27]. In mung bean (*Vigna radiata* L.), the aquaporin NIP2-1 regulates Si uptake; heterologous expression of NIP2-1 in yeast showed that this channel also mediates Ge transport [28]. Since Ge has been used as a tracer for Si in biogeochemical investigations, it is generally assumed that Ge is absorbed through the same mechanisms as Si in different plant species [9]. However, data on Ge and Si accumulation and distribution show high variability across plant species and experimental conditions. A survey of maize (*Zea mays* L.), oat (*Avena sativa* L.), and canary grass (*Phalaris arundinacea* L.) grown on soil containing about 1.6 μg/g Ge and 257 mg/g Si (values comparable to average Earth's crust), showed that Ge is mainly retained in roots rather than in leaves, reaching up to 0.5–0.6 μg/g Ge and 0.07–0.3 μg/g Ge, respectively. Likewise, Si also accumulated predominantly in roots compared with leaves, with concentrations of 8.4–30.3 mg/g and 3.0–4.9 mg/g Si, respectively [29]. However, studies on blue grass (*Poa annua* L.) and orchard grass (*Dactylis glomerata* L.) grown in hydroponics suggest that selectivity for these metals increases during translocation, as Ge is largely retained in roots whereas Si is translocated and accumulates in leaves [30]. Although not recognized as an essential nutrient, Si promotes plant growth, productivity, and enhances tolerance to abiotic stress and diseases. Seeds treatments with Ge significantly improve growth in spinach (*Spinacia oleracea* L.), enhance tolerance to Bi stress in oat, alleviate sodium chloride (NaCl)-induced toxicity on seed germination and seedlings growth in black goji berry (*Lycium ruthenicum* Murr), improving the oxidative status and the activity of ROS-scavenging enzymes [31–33]. Ge toxic effects have been reported in lettuce (*Lactuca sativa* L.) treated with inorganic ($GeO_2$) and organic (Ge-132) Ge at concentrations equal or above 5 mg/L and 25 mg/L, respectively, which caused the reduction of root development, plant growth and biomass production [34]. Treatments with $GeO_2$ at lower concentrations, below 5 mg/L, enhance photosynthetic efficiency, increasing the maximum photochemical quantum yield and actual quantum yield of photosystem II (PSII), while reducing ROS accumulation in dwarf apple (*Malus domestica* Borkh. cv. *Red Fuji*/M26 *Malus pumila* Mill.) trees [35]. Recently, the beneficial effects of Ge on plants have been confirmed using ZIF-8@Ge-132, a Ge-containing compound sprayed onto plants that reduces arsenic (As) and Cd accumulation in rice [36]. However, increasing $GeO_2$ concentration results in phytotoxicity [35].

### 3.3.4 Lithium

The increasing economic interest in Li, driven by its use in lithium-ion batteries for electronics and electric vehicles, has raised concerns in its environmental cycling, toxicity, and accumulation in soils, waters, and plants [37]. Several studies on nutrient uptake in plants have used Li as an analog of sodium (Na), showing that these elements may exploit the same membrane transporters to enter roots, such as the non-selective $Na^+$ importer NRT1.2/NPF4.6/AIT1, which is regulated by the salt overly sensitive (SOS) pathway in *Arabidopsis* [38, 39]. The SOS pathway regulates responses and tolerance to Na in plants, and evidence suggests that SOS1, a $Na^+/H^+$ antiporter, may regulate Li transport as well [38]. Chemical similarity to potassium (K) enables Li uptake also via $K^+$ channels, such as the plasma membrane intrinsic protein PIP2;1, which is included in the aquaporin family, and through the high affinity $K^+$ transporter (HKT1), both of which can act as NSCCs in Arabidopsis [38, 40, 41]. Interestingly, a specific $Li^+/H^+$ antiporter involved in $Li^+$ detoxification in Arabidopsis roots, namely sodium/hydrogen exchanger 8 (NHX8), has been identified within the $Na^+/H^+$ antiporter family [42]. Interference of Li with nutrients uptake has been reported, possibly due to competition for binding sites in channels and transporters, while also affecting the functionality of the enzymes requiring cations as co-factors [43, 44]. Treatments with Li at concentrations up to 10 and 20 mg/L enhance seed germination and root development in garden cress and increased biomass growth in spinach, respectively, whereas further increases in Li concentration trigger detrimental effects in both plant species [45, 46]. Similar dual responses, with enhanced growth at low doses and toxic responses to high concentrations of Li, have been observed also in sunflower (*Helianthus annuus* L.), maize and hemp (*Cannabis sativa* L.) plants [47, 48]. In these species, Li mainly accumulates in shoots (i.e., stems and leaves), and uptake increases as Li concentration in the medium raises [46–48]. Recent studies have highlighted the potential of different plant species to accumulate high Li levels, exploring their use for phytomining strategies; for example, dogbane (*Apocynum venetum* L.) can accumulate up to 1800 mg/kg Li in leaf tissues while tolerating up to 400 mg/kg Li in the growth medium [49]. The halophytes plant species river saltbush (*Atriplex amnicola* Paul G.Wilson) and *Salsola australis* R.Br., which tolerate high Na levels, show enhanced Li uptake, translocating up to 10 g/kg Li to leaves; these results support the hypothesis that halophytes could be Li hyperaccumulators [19]. Further strategies to enhance Li uptake by plants rely on the addition of disodium ethylenediaminetetraacetate dihydrate (EDTA) or ethylenediamine-N,N′-disuccinic acid (EDDS), two chelating agents that have been shown to dramatically increase Li uptake in rapeseed (*Brassica napus* L.) and sunflower [50]. The symptoms of Li toxicity are species-specific and generally include reduced root length, plant height, and biomass, as well as decreased photosynthetic efficiency and impaired nutrients homeostasis, as reported for hemp [47]. In addition, Li stress responses can include oxidative damage through the induction of ROS, high levels of MDA, and increased activity of ROS-scavenging enzymes in spinach [45]. Impaired nutrients homeostasis and reduced maximum quantum yield of PSII have been reported in Li-treated hemp microshoots, providing evidence that photosynthetic efficiency and pigment metabolism are highly sensitive to Li [44]. Indeed, Li may limit or inhibit carbon fixation by reducing nicotinamide adenine dinucleotide phosphate ($NADP^+$) regeneration during the Calvin cycle or by inducing oxidative damage to the electron transport chain during photosynthesis [43]. Indirect effects of Li on photosynthesis include the formation of necrotic lesions, spots and regions extending from the margins to the entire leaf blade, primarily on older leaves, thereby decreasing photosynthetically active leaf area [47]. It has been suggested that the negative effect of Li on chlorophyll content may be due to increased chlorophyllase catalytic activity and other enzymes responsible for chlorophyll bleaching [51].

### 3.3.5 Rare Earth Elements

Over the past decades, REEs have played a pivotal role in technological development, as these elements underpin high-tech applications and the green transition; however, their mining, processing, and disposal increase environmental release, raising ecotoxicological considerations. Based on their chemical properties, REEs are classified as light, from lanthanum (La) to gadolinium (Gd), and heavy, from terbium (Tb) to lutetium (Lu). In the form of trivalent cations, REEs such as $La^{3+}$ and europium ($Eu^{3+}$) have ionic radii and physicochemical properties similar to calcium ($Ca^{2+}$), and their entry into roots is mainly mediated by $Ca^{2+}$ channels in plants, as demonstrated in barley and *Amaranthus caudatus* L. [52, 53]. These REEs may also be taken up through NSCCs, $K^+$ and $Na^+$ channels, although to a lesser extent [53]. In most plants, REEs accumulation mainly occurs in roots, except for hyperaccumulator plants such as the fern *Dicranopteris linearis* (Burm.f.) Underw., which can translocate high REEs quantities, up to 3830 mg/kg, to the fronds [54, 55]. Recent research demonstrated that the remarkable uptake of REEs by *Dicranopteris* is mediated in roots by the Nramp REE Transporter 1 (NREET1), a specific active transporter for REEs with high affinity for light REEs [56]. Driven by growing economic interest and prospects for biotechnological applications, identifying plant species able to hyperaccumulate REEs is an active research field. Interestingly, the application of non-destructive technologies to monitor metal uptake in plants, such as X-ray fluorescence spectroscopy (XRF), has accelerated the identification of candidate hyperaccumulator species [57]. In a survey conducted at three legacy mine sites in Portugal characterized by high REEs contamination, several plants were identified as putative REEs accumulators. Among them, *Pteridium aquilinum* (L.) Kuhn showed the highest REEs accumulation in shoots, whereas *Juncus acutus* L. showed the highest REEs accumulation in roots, reaching up to 27.3 mg/kg and 112 mg/kg of REEs, respectively. Across these mine sites, La, cerium (Ce) and neodymium (Nd) were the most abundant REEs; accordingly, they were also the main REEs detected in both *Pteridium* and *Juncus* [58]. Beneficial effects of REEs on plant growth and crop yield have been reported, as they can enhance root growth, biomass development, and photosynthetic activity, increasing chlorophyll and carotenoids contents and accelerating photosynthetic reactions, as observed in tomato [59, 60]. Although it is generally accepted that REEs exert positive effects on plants at low concentrations, toxicity of high doses of REEs has been reported in several plant species, including Arabidopsis, maize, tomato and rice, causing repression of primary root elongation, decreased biomass production, increased lipid peroxidation, reduced photosynthetic rate and chlorophyll index, and dysregulation of ROS and antioxidant responses [60].

### 3.3.6 Tungsten

The recent use of tungsten (W) to produce materials with increased hardness and high-temperature components has sharply increased interest in this transition metal, raising questions about the management of environmental contamination. Indeed, because of its similarity to molybdenum (Mo), an essential nutrient, W can be readily absorbed by plants and microorganisms; however, few studies have investigated its impacts on vegetation. The structural and chemical similarity between tungstate ($WO_4^{2-}$) and molybdate ($MoO_4^{2-}$) supports the hypothesis that these oxyanions are taken up through the same channels in plants, likely via the Mo transporters MOT1 and MOT2 [61]. Furthermore, $WO_4^{2-}$ chemically mimics the oxyanion sulfate ($SO_4^{2-}$) and can be absorbed through the high-affinity sulfate transporter 1 (SHST1) in *Stylosanthes hamata* L. [62]. In general, plant tolerance to W can be high, as shown by healthy ryegrass, sunflowers, and soybean (*Glycine max* cv. Primus) grown on soil amended with high concentrations of metallic W, ranging from 1 to 5 g/kg W, whereas higher concentrations induced toxic responses. Under these conditions, plants accumulated up to 1 g/kg W in roots and 201 mg/kg W in leaves [63–65]. In contrast, inhibition of root length, shoot height, and fresh weight has been observed in celery (*Apium graveolens* L.) and pepper (*Capsicum* spp.) grown in hydroponic solution containing 60 mg/kg W. The high sensitivity shown by celery and pepper suggests that these species could be promising biomarkers for W [66]. In addition, W toxicity can include membrane lipid peroxidation and repression of photosynthesis, as well as increased activities ROS-scavenging enzymes in celery, peppers, and ryegrass [66, 67].

## 3.4 CONCLUSIONS AND FUTURE PERSPECTIVES

Minor metals are gaining pivotal importance for the development of high-tech technologies and economic strategies, driving a rapid increase in their exploitation. Therefore, environmental impacts of the extraction, processing, and disposal of these metals must be carefully assessed. The mechanisms regulating the uptake of minor metals by plants remain unclear; however, recent findings on the regulation of cation uptake through NSCCs, as well as newly recognized roles of aquaporin family members in metal uptake and translocation, are helping to fill key knowledge gaps (Table 3.1).

Evidence suggests that aquaporins may play an underestimated role and further studies are required to deepen our understanding of their involvement in plant–minor metals interactions. Several convergent mechanistic patterns emerge despite substantial metal-specific variability. First, plant uptake is often governed by transporter promiscuity, whereby chemically similar non-essential ions "hitchhike" on pathways evolved for essential nutrients: Bi can interfere with Fe homeostasis and likely

**TABLE 3.1**
**Summary of the Effective and Putative Uptake Mechanisms of Minor Metals in Plants**

| Minor Metal | Transporters and Channels | Pathways Involved in Plants Uptake | Effects on Plant Growth | Reference |
|---|---|---|---|---|
| Bi | IRT1 | $Fe^{2+}$ uptake Strategy I<br>NSCCs | At low concentration: hormetic effect, promotion of plant growth<br>At medium/high concentration: repression of seed germination and root elongation, impairment of photosynthetic machinery and stomatal conductance, impairment of nutrients homeostasis | [10–16] |
| Ga | Nrat1<br>NIP2;1 | Al uptake mechanisms<br>Aquaporins<br>NSCCs | At low concentration: increase of root length, shoot height, biomass, chlorophyll SPAD values<br>At high concentration: repression of seed germination, root elongation and shoot biomass | [18, 20–22] |
| Ge | Lsi1<br>Lsi2<br>NIP2-1 | Si uptake mechanisms<br>Aquaporins<br>NSCCs | At low/medium concentration: increase of biomass and photosynthetic efficiency, reduction of ROS accumulation, attenuation of abiotic stress and metal-induced stress<br>At high concentration: reduction of root development, plant growth, and biomass production | [25–30]<br>[31, 34] |
| Li | NRT1.2/ NPF4.6/AIT<br>PIP2;1<br>HKT1<br>NHX8 | $Na^+$ uptake and tolerance mechanisms<br>Specific Li transporters<br>Aquaporins<br>NSCCs | At low concentration: increase in seed germination, root development and biomass<br>At high concentration: reduction of root length, plant height, biomass, photosynthetic efficiency; impairment of nutrients homeostasis | [38–41]<br>[42, 45–49] |
| REEs | NREET1 | $Na^+$ uptake mechanisms<br>$K^+$ uptake mechanisms<br>$Ca^{2+}$ uptake mechanisms<br>Specific REEs transporters | At low concentration: promotion of growth, development and photosynthetic efficiency<br>At high concentration: repression of primary root elongation, biomass and photosynthetic efficiency; increased lipid peroxidation | [53–56, 58–60] |
| W | SHST1 | $SO_4^{2-}$ uptake mechanisms<br>Mo uptake mechanisms | At very high concentration: decrease in total weight, root length and shoot length | [61–64, 66] |

exploits Fe-related transport processes; Ga largely follows Al-associated uptake and transport routes; Ge is taken up via Si transporters with limited discrimination from Si; REEs frequently enter through $Ca^{2+}$ channels; W can use molybdate and sulfate transport systems (MOTs, SHST1), reflecting strong anion mimicry. Second, several metals show dose dependent responses, with growth stimulation at low exposure (hormesis) and toxicity at higher doses. Third, when toxic thresholds are exceeded, responses converge on oxidative stress and redox imbalance, typically evidenced by increased lipid peroxidation, altered ROS status, and induction of ROS-scavenging enzymes, accompanied by impairments in photosynthesis (chlorophyll loss, reduced PSII performance) and growth inhibition, particularly in roots.

The combined use of omics approaches, including transcriptomics, proteomics, ionomics, and metabolomics, could accelerate the identification mechanisms underlying uptake, translocation, and storage of these elements. Moreover, novel nondestructive technologies to monitor metal accumulation in plants, such as XRF and hyperspectral imaging, are accelerating the characterization of accumulation patterns. Recent research has led to the identification of new REEs hyperaccumulator plant species, shedding light on the evolution of channels and transporters that regulate metal uptake, translocation, and storage. The discovery of REEs-specific transporters highlights the remarkable plasticity and adaptability of plants to diverse environments and suggests that additional, functionally analogous transporters remain to be uncovered. Expanding knowledge of the dynamics regulating minor metals absorption in plants could pave the way for sustainable biotechnologies to recover these valuable materials from contaminated matrices, further supported by non-destructive technologies that can improve the management of field phytoremediation and phytomining by informing optimal harvest timing. Future investigations will be pivotal for designing sustainable frameworks to manage minor metals, converting potential environmental risks into strategic resources, and consolidating the basic and applied research needed to sustain the green technology transition.

## REFERENCES


1. European Commission: Regulation (EU) 2024/1252 of the European Parliament and of the Council of 11 April 2024. (2024). *Establishing a framework for ensuring a secure and sustainable supply of critical raw materials and amending Regulations (EU) No 168/2013, (EU) 2018/858, (EU) 2018/1724 and (EU) 2019/1020.* Publications Office of the European Union. http://data.europa.eu/eli/reg/2024/1252/oj.
2. European Commission: Directorate-General for Internal Market, Industry, Entrepreneurship and SMEs, Grohol, M. and Veeh, C. (2023). *Study on the critical raw materials for the EU 2023 – Final report.* Publications Office of the European Union. https://data.europa.eu/doi/10.2873/725585.
3. Moss, Raymond., Tzimas, Evangelos., Willis, Peter., Arendorf, Josephine., Thompson, Paul., Chapman, Adrian., Morley, Nick., Sims, Edward., Bryson, Ruth., Pearson, James., Tercero, E. Luis., Marscheider-Weidemann, Frank., Soulier, Marcel., Lüllmann, Arne., Sartorius, Christian., & Ostertag, Katrin. (2013). *Critical Metals in the Path Towards the Decarbonisation of the EU Energy Sector: Assessing Rare Metals as Supply-Chain Bottlenecks in Low-Carbon Energy Technologies.* Publications Office of the European Union.

4. Nassar, N. T., Graedel, T. E., & Harper, E. M. (2015). By-product metals are technologically essential but have problematic supply. *Science Advances*, *1*(3). https://doi.org/10.1126/sciadv.1400180.
5. Zacchini, M. (2024). Bismuth interaction with plants: Uptake and transport, toxic effects, tolerance mechanisms: A review. *Chemosphere*, 360). https://doi.org/10.1016/j.chemosphere.2024.142414.
6. Yu, H.-S., & Liao, W.-T. (2011). "Gallium: Environmental pollution and health effects." In *Encyclopedia of Environmental Health* (pp. 829–833). Elsevier. https://doi.org/10.1016/B978-0-444-52272-6.00474-8.
7. Syu, C. H., Chen, P. W., Huang, C. C., & Lee, D. Y. (2020). Accumulation of gallium (Ga) and indium (In) in rice grains in Ga- and In-contaminated paddy soils. *Environmental Pollution*, *261*. https://doi.org/10.1016/j.envpol.2020.114189.
8. U.S. Geological Survey. (2023). *Mineral commodity summaries 2023: U.S. Geological Survey*. 210. https://doi.org/10.3133/mcs2023.
9. Wiche, O., Székely, B., Moschner, C., & Heilmeier, H. (2018). Germanium in the soil-plant system—A review. *Environmental Science and Pollution Research*, *25*(32), 31938–31956. https://doi.org/10.1007/s11356-018-3172-y.
10. Nagata, T. (2015). Growth inhibition and IRT1 induction of *Arabidopsis thaliana* in response to bismuth. *Journal of Plant Biology*, *58*(5), 311–317. https://doi.org/10.1007/s12374-015-0256-9.
11. Mohammed, A. E., Pawelzik, E., Nour, M. M., Alotaibi, M. O., Abdelgawad, H., & Saleh, A. M. (2023). Mycorrhized wheat and bean plants tolerate bismuth contaminated soil via improved metal detoxification and antioxidant defense systems. *Plant Physiology and Biochemistry*, *205*. https://doi.org/10.1016/j.plaphy.2023.108148.
12. Nagata, T., & Kimoto, S. (2020). *Growth inhibition and root damage of bismuth in Solanum lycopersicum*. http://www.scirea.org/journal/Biology.
13. Pietrini, F., Passatore, L., Carloni, S., Massimi, L., Astolfi, M. L., Giusto, C., & Zacchini, M. (2023). Bismuth exposure affects morpho-physiological performances and the ionomic profile in garden cress (*Lepidium sativum* L.) plants. *Frontiers in Environmental Science*, *11*. https://doi.org/10.3389/fenvs.2023.1221573.
14. Omouri, Z., Hawari, J., Fournier, M., & Robidoux, P. Y. (2019). Phytotoxicity of bismuth nitrate and bismuth citrate on germination and growth of *Lolium perenne* exposed on filter paper and on artificially contaminated soil. *Journal of Soil Science and Plant Health*, 3(1).
15. Passatore, L., Pietrini, F., Carloni, S., Massimi, L., Giusto, C., Zacchini, M., & Iannilli, V. (2022). Morpho-physiological and molecular responses of *Lepidium sativum* L. seeds induced by bismuth exposure. *Science of the Total Environment*, *831*. https://doi.org/10.1016/j.scitotenv.2022.154896.
16. Prabhavati, E., Vishnuvardhan, Z., & Mallaiah, K. V. (2017). Photosynthetic ability of Horse gram [*Macrotyloma uniflorum* (Lam.) Verdc] plants in response to metal stress. *Russian Agricultural Sciences*, *43*(2), 144–148. https://doi.org/10.3103/s1068367417020148.
17. Jensen, H., Gaw, S., Lehto, N. J., Hassall, L., & Robinson, B. H. (2018). The mobility and plant uptake of gallium and indium, two emerging contaminants associated with electronic waste and other sources. *Chemosphere*, *209*, 675–684. https://doi.org/10.1016/j.chemosphere.2018.06.111.
18. Kocjan, A., Kwasniewska, J., & Szurman-Zubrzycka, M. (2025). Understanding plant tolerance to aluminum: Exploring mechanisms and perspectives. *Plant and Soil*, 507(1), 195–219). https://doi.org/10.1007/s11104-024-06745-0.
19. Nkrumah, P. N., & van der Ent, A. (2023). Possible accumulation of critical metals in plants that hyperaccumulate their chemical analogues? *Science of the Total Environment*, *878*. https://doi.org/10.1016/j.scitotenv.2023.162791.

20. Chang, H. F., Wang, S. L., & Yeh, K. C. (2017). Effect of gallium exposure in *Arabidopsis thaliana* is similar to aluminum stress. *Environmental Science and Technology, 51*(3), 1241–1248. https://doi.org/10.1021/acs.est.6b05760.
21. Syu, C. H., Chien, P. H., Huang, C. C., Jiang, P. Y., Juang, K. W., & Lee, D. Y. (2017). The growth and uptake of Ga and In of rice (*Oryza sativa* L.) seedlings as affected by Ga and In concentrations in hydroponic cultures. *Ecotoxicology and Environmental Safety, 135*, 32–39. https://doi.org/10.1016/j.ecoenv.2016.09.016.
22. Yu, X. Z., Feng, X. H., & Feng, Y. X. (2015). Phytotoxicity and transport of gallium (Ga) in rice seedlings for 2-day of exposure. *Bulletin of Environmental Contamination and Toxicology, 95*(1), 122–125. https://doi.org/10.1007/s00128-015-1471-2.
23. Zhang, D. C., Ullah, A., Tian, P., & Yu, X. Z. (2023). Response to gallium (Ga) exposure and its distribution in rice plants. *Environmental Science and Pollution Research International, 30*(58), 121908–121914. https://doi.org/10.1007/s11356-023-30975-x.
24. Nikolic, M., Nikolic, N., Liang, Y., Kirkby, E. A., & Römheld, V. (2007). Germanium-68 as an adequate tracer for silicon transport in plants. Characterization of silicon uptake in different crop species. *Plant Physiology, 143*(1), 495–503. https://doi.org/10.1104/pp.106.090845.
25. Konishi, N., Mitani-Ueno, N., & Ma, J. F. (2025). Role of polar localization of the silicon transporter OsLsi1 in metalloid uptake by rice roots. *Plant Physiology, 198*(1). https://doi.org/10.1093/plphys/kiaf196.
26. Ma, J. F., Tamai, K., Yamaji, N., Mitani, N., Konishi, S., Katsuhara, M., Ishiguro, M., Murata, Y., & Yano, M. (2006). A silicon transporter in rice. *Nature, 440*(7084), 688–691. https://doi.org/10.1038/nature04590.
27. Ma, J. F., Yamaji, N., Mitani, N., Tamai, K., Konishi, S., Fujiwara, T., Katsuhara, M., & Yano, M. (2007). An efflux transporter of silicon in rice. *Nature, 448*(7150), 209–212. https://doi.org/10.1038/nature05964.
28. Thakral, V., Sharma, Y., Mandlik, R., Kumawat, S., Patil, G., Sonah, H., Isenring, P., Bélanger, R., Sharma, T. R., & Deshmukh, R. (2023). Identification of VrNIP2-1 aquaporin with novel selective filter regulating the transport of beneficial as well as hazardous metalloids in mungbean (*Vigna radiata* L.). *Plant Physiology and Biochemistry, 203*. https://doi.org/10.1016/j.plaphy.2023.108057.
29. Kaiser, S., Wagner, S., Moschner, C., Funke, C., & Wiche, O. (2020). Accumulation of germanium (Ge) in plant tissues of grasses is not solely driven by its incorporation in phytoliths. *Biogeochemistry, 148*(1), 49–68. https://doi.org/10.1007/s10533-020-00646-x.
30. Sparks, J. P., Chandra, S., Derry, L. A., Parthasarathy, M. V., Daugherty, C. S., & Griffin, R. (2011). Subcellular localization of silicon and germanium in grass root and leaf tissues by SIMS: Evidence for differential and active transport. *Biogeochemistry, 104*(1–3), 237–249. https://doi.org/10.1007/s10533-010-9498-2.
31. Liu, Y., Hou, L. Y., Li, Q. M., Jiang, Z. P., Liu, D., & Zhu, Y. (2016). The effects of exogenous antioxidant germanium (Ge) on seed germination and growth of *Lycium ruthenicum* Murr subjected to NaCl stress. *Environmental Technology (United Kingdom), 37*(8), 909–919. https://doi.org/10.1080/09593330.2015.1091512.
32. Najar, B., Zrig, A., Alsherif, E. A., Selim, S., Aloufi, A. S., Korany, S. M., Nhs, M., Aldilam, M., & Bouqellah, N. A. (2024). Synergistic effect of arbuscular mycorrhizal fungi and germanium on the growth, nutritional quality, and health-promoting activities of *Spinacia oleracea* L. *Plants, 13*(20). https://doi.org/10.3390/plants13202869.
33. Yang, X., Alsherif, E. A., El-Shafey, N. M., Korany, S. M., Alhaj Hamoud, Y., Shaghaleh, H., Sheteiwy, M. S., Suliemani, S., & Madany, M. M. Y. (2025). Arbuscular mycorrhizal Fungi and germanium: A dual approach to enhancing oat growth and mitigating oxidative damage due to bismuth contamination. *BMC Plant Biology, 25*(1). https://doi.org/10.1186/s12870-025-06989-7.

34. Cheong, Y. H., Kim, S. U., Seo, D. C., Chang, N. I., Lee, J. B., Park, J. H., Kim, K. S., Kim, S. D., Kim, H. T., Heo, J.-S., & Cho, J.-S. (2009). Effect of inorganic and organic germanium treatments on the growth of lettuce (*Lactuca sativa*). *Journal of the Korean Society for Applied Biological Chemistry*, *52*(4), 389–396. https://doi.org/10.3839/jksabc.2009.069.
35. Wang, Z. B., Wang, Y. F., Zhao, J. J., Ma, L., Wang, Y. J., Zhang, X., Nie, Y. T., Guo, Y. P., Mei, L. X., & Zhao, Z. Y. (2018). Effects of $GeO_2$ on chlorophyll fluorescence and antioxidant enzymes in apple leaves under strong light. *Photosynthetica*, *56*(4), 1081–1092. https://doi.org/10.1007/s11099-018-0807-7.
36. Cui, H., Tang, S., Huang, S., Lei, L., Jiang, Z., Li, L., & Wei, S. (2023). Simultaneous mitigation of arsenic and cadmium accumulation in rice grains by foliar inhibitor with ZIF-8@Ge-132. *Science of the Total Environment*, *860*. https://doi.org/10.1016/j.scitotenv.2022.160307.
37. Shakoor, N., Adeel, M., Azeem, I., Ahmad, M. A., Zain, M., Abbas, A., Hussain, M., Jiang, Y., Zhou, P., Li, Y., Xu, M., Rui, Y. (2023). Interplay of higher plants with lithium pollution: Global trends, meta-analysis, and perspectives. *Chemosphere*, *310*, 136663. https://doi.org/10.1016/j.chemosphere.2022.136663.
38. Garcia-Daga, S., Fischer, S., & Gilliham, M. (2025). Lithium in plants. *New Phytologist*, 248(4), pp. 1639–1654. https://doi.org/10.1111/nph.70523.
39. Liu, X., Zhang, L., Zhao, Z., Zheng, Y., Ren, Y., Zhao, X., Zhang, S., Yang, G., Huang, J., Yan, K., Li, C., Zheng, C., & Wu, C. (2025). Regulation of the non-selective $Na^+$ importer capacity of NRT1.2/NPF4.6/AIT1 by SOS2-mediated phosphorylation in *Arabidopsis*. *Cell Reports*, *44*(6). https://doi.org/10.1016/j.celrep.2025.115729.
40. Qiu, J., McGaughey, S. A., Byrt, C. S., & Tyerman, S. D. (2025). Post-translational modification acts as a digital like switch influencing AtPIP2;1 water and cation permeability. *Scientific Reports*, *15*(1). https://doi.org/10.1038/s41598-025-06200-9.
41. Rus, A., Yokoi, S., Sharkhuu, A., Reddy, M., Lee, B., Matsumoto, T. K., Koiwa, H., Zhu, J.-K., Bressan, R. A., & Hasegawa, P. M. (2001). AtHKT1 is a salt tolerance determinant that controls $Na^+$ entry into plant roots. *Proceedings of the National Academy of Sciences*, *98*(24), 14150–14155. https://doi.org/10.1073/pnas.241501798.
42. An, R., Chen, Q. J., Chai, M. F., Lu, P. L., Su, Z., Qin, Z. X., Chen, J., & Wang, X. C. (2007). AtNHX8, a member of the monovalent cation:proton antiporter-1 family in *Arabidopsis thaliana*, encodes a putative Li+/H+ antiporter. *Plant Journal*, *49*(4), 718–728. https://doi.org/10.1111/j.1365-313X.2006.02990.x.
43. Shahzad, B., Tanveer, M., Hassan, W., Shah, A. N., Anjum, S. A., Cheema, S. A., & Ali, I. (2016). Lithium toxicity in plants: Reasons, mechanisms and remediation possibilities: A review. *Plant Physiology and Biochemistry*, *107*, 104–115. https://doi.org/10.1016/j.plaphy.2016.05.034.
44. Zacchini, M., Gullotta, G., D'Onofrio, G., Bertolotto, P., Massimi, L., & Pietrini, F. (2023). Effects of lithium on morpho-physiological and ionomic traits in *Cannabis sativa* L. microshoots under in vitro conditions. *Plant Cell, Tissue and Organ Culture*, *155*(3), 873–882. https://doi.org/10.1007/s11240-023-02606-3.
45. Bakhat, H. F., Rasul, K., Farooq, A. B. U., Zia, Z., Natasha, Fahad, S., Abbas, S., Shah, G. M., Rabbani, F., & Hammad, H. M. (2020). Growth and physiological response of spinach to various lithium concentrations in soil. *Environmental Science and Pollution Research*, *27*(32), 39717–39725. https://doi.org/10.1007/s11356-019-06877-2.
46. Iannilli, V., D'Onofrio, G., Marzi, D., Passatore, L., Pietrini, F., Massimi, L., & Zacchini, M. (2024). Lithium toxicity in *Lepidium sativum* L. seedlings: Exploring Li accumulation's impact on germination, root growth, and DNA integrity. *Environments MDPI*, *11*(5). https://doi.org/10.3390/environments11050093.
47. D'Onofrio, G., Marzi, D., Passatore, L., Massimi, L., Astolfi, M. L., Zacchini, M., & Pietrini, F. (2025). Lithium (Li) accumulation and toxicity assessment in hemp (*Cannabis sativa* L.) plants by biometric, physiological and ionomic analyses under hydroponics. *Environments*, *12*(12), 461. https://doi.org/10.3390/environments12120461.

48. Hawrylak-Nowak, B., Kalinowska, M., & Szymańska, M. (2012). A study on selected physiological parameters of plants grown under lithium supplementation. *Biological Trace Element Research*, *149*(3), 425–430. https://doi.org/10.1007/s12011-012-9435-4.
49. Jiang, L., Wang, L., Mu, S. Y., & Tian, C. Y. (2014). *Apocynum venetum*: A newly found lithium accumulator. *Flora: Morphology, Distribution, Functional Ecology of Plants*, *209*(5–6), 285–289. https://doi.org/10.1016/j.flora.2014.03.007.
50. Kavanagh, L., Keohane, J., Cabellos, G. G., Lloyd, A., & Cleary, J. (2018). Induced plant accumulation of lithium. *Geosciences (Switzerland)*, *8*(2). https://doi.org/10.3390/geosciences8020056.
51. Kastori, R., Maksimović, I., & Putnikdelić, M. (2022). Lithium in the environment and its effects on higher plants. *Contemporary Agriculture*, *71*(3–4), 226–239. https://doi.org/10.2478/contagri-2022-0030.
52. Gao, Y., Zeng, F., Yi, A. N., Ping, S., & Jing, L. (2003). Research of the entry of rare earth elements $Eu^{3+}$ and $La^{3+}$ into plant cell. *Biological Trace Element Research*, *91*, 253–265.
53. Han, F., Shan, X. Q., Zhang, J., Xie, Y. N., Pei, Z. G., Zhang, S. Z., Zhu, Y. G., & Wen, B. (2005). Organic acids promote the uptake of lanthanum by barley roots. *New Phytologist*, *165*(2), 481–492. https://doi.org/10.1111/j.1469-8137.2004.01256.x.
54. Liu, W. S., Zheng, H. X., Guo, M. N., Liu, C., Huot, H., Morel, J. L., van der Ent, A., Tang, Y. T., & Qiu, R. L. (2019). Co-deposition of silicon with rare earth elements (REEs) and aluminium in the fern *Dicranopteris linearis* from China. *Plant and Soil*, *437*(1–2), 427–437. https://doi.org/10.1007/s11104-019-04005-0.
55. Shan, X., Wang, H., Zhang, S., Zhou, H., Zheng, Y., Yu, H., & Wen, B. (2003). Accumulation and uptake of light rare earth elements in a hyperaccumulator *Dicropteris dichotoma*. *Plant Science*, *165*(6), 1343–1353. https://doi.org/10.1016/S0168-9452(03)00361-3.
56. Zheng, H. X., Liu, W. S., Sun, D., Zhu, S. C., Li, Y., Yang, Y. L., Liu, R. R., Feng, H. Y., Cai, X., Cao, Y., Xu, G. H., Morel, J. L., van der Ent, A., Ma, L. Q., Liu, Y. G., Rylott, E. L., Qiu, R. L., & Tang, Y. T. (2023). Plasma-membrane-localized transporter NREET1 is responsible for rare earth element uptake in hyperaccumulator *Dicranopteris linearis*. *Environmental Science and Technology*, *57*(17), 6922–6933. https://doi.org/10.1021/acs.est.2c09320.
57. Goudard, L., Blaudez, D., Sirguey, C., Purwadi, I., Invernon, V., Rouhan, G., & van der Ent, A. (2024). Prospecting for rare earth element (hyper)accumulators in the Paris Herbarium using X-ray fluorescence spectroscopy reveals new distributional and taxon discoveries. *Annals of Botany*, *133*(4), 573–583. https://doi.org/10.1093/aob/mcae011.
58. Forsyth, K., Dia, A., Marques, R., Prudêncio, M. I., Diamantino, C., Carvalho, E., Russo, D., Dionisio, I., Davranche, M., Bouhnik-Le-Coz, M., & Pédrot, M. (2023). Bioconcentration and translocation of rare earth elements in plants collected from three legacy mine sites in Portugal. *Frontiers in Environmental Science*, *11*. https://doi.org/10.3389/fenvs.2023.1191909.
59. Dang, D. H., Kernaghan, A., Emery, R. J. N., Thompson, K. A., Kisiala, A., & Wang, W. (2024). The mixed blessings of rare earth element supplements for tomatoes and ferns. *Science of the Total Environment*, *906*. https://doi.org/10.1016/j.scitotenv.2023.167822.
60. Ozturk, M., Metin, M., Altay, V., Prasad, M. N. V., Gul, A., Bhat, R. A., Darvash, M. A., Hasanuzzaman, M., Nahar, K., Unal, D., Unal, B. T., García-Caparrós, P., Kawano, T., Toderich, K., & Shahzadi, A. (2023). Role of rare earth elements in plants. *Plant Molecular Biology Reporter*, *41*(3), 345–368. https://doi.org/10.1007/s11105-023-01369-7.
61. Adamakis, I. D. S., Panteris, E., & Eleftheriou, E. P. (2012). Tungsten toxicity in plants. *Plants*, *1*(2), 82–99. https://doi.org/10.3390/plants1020082.
62. Lindblom, S. D., Abdel-Ghany, S., Hanson, B. R., Hwang, S., Terry, N., & Pilon-Smits, E. A. H. (2006). Constitutive expression of a high-affinity sulfate transporter in Indian mustard affects metal tolerance and accumulation. *Journal of Environmental Quality*, *35*(3), 726–733. https://doi.org/10.2134/jeq2005.0119.

63. Johnson, D. R., Inouye, L. S., Bednar, A. J., Clarke, J. U., Winfield, L. E., Boyd, R. E., Ang, C. Y., & Goss, J. (2009). Tungsten bioavailability and toxicity in sunflowers (*Helianthus annuus* L.). *Land Contamination and Reclamation*, *17*(1), 141–151. https://doi.org/10.2462/09670513.939.
64. Oburger, E., Vergara Cid, C., Preiner, J., Hu, J., Hann, S., Wanek, W., & Richter, A. (2018). PH-dependent bioavailability, speciation, and phytotoxicity of tungsten (W) in soil affect growth and molybdoenzyme activity of nodulated soybeans. *Environmental Science and Technology*, *52*(11), 6146–6156. https://doi.org/10.1021/acs.est.7b06500.
65. Strigul, N., Koutsospyros, A., Arienti, P., Christodoulatos, C., Dermatas, D., & Braida, W. (2005). Effects of tungsten on environmental systems. *Chemosphere*, *61*(2), 248–258. https://doi.org/10.1016/j.chemosphere.2005.01.083.
66. Li, Q., Zheng, X., & Chen, M. (2024). Ecotoxicological effects of tungsten on celery (*Apium graveolens* L) and pepper (*Capsicum* spp.). *PeerJ*, *12*(6). https://doi.org/10.7717/peerj.17601.
67. Alsherif, E. A., Hajjar, D., & AbdElgawad, H. (2023). Future climate $CO_2$ reduces the tungsten effect in rye plants: A growth and biochemical study. *Plants*, *12*(10):1924. https://doi.org/10.3390/plants12101924.

# 4 The Role of Chemical Metals in SARS-CoV-2 Infection

*Ketolly Natanne da Silva Leal, Isabela Rocha, Alan J. Stewart, Carlos Henrique Inácio Ramos, and Marco Aurélio Zezzi Arruda*

## 4.1 INTRODUCTION

The outbreak of severe acute respiratory syndrome coronavirus 2 (SARS-CoV-2) in 2019 underscored the urgency of deciphering the molecular determinants of viral infection and host response [1]. Considerable progress has been made in characterizing viral proteins, genomic organization, and immune evasion strategies [2]. However, the role of chemical elements, particularly trace and transition metals, remains comparatively underexplored, despite their critical participation in viral replication, host cellular processes, and immune regulation [3].

Essential metals such as zinc, copper, iron, selenium, magnesium, and calcium function as cofactors in enzymes involved in nucleic acid metabolism, redox homeostasis, and immunomodulation [4]. Their availability and regulation directly influence viral entry, replication efficiency, and the host's ability to mount an effective defense. Conversely, exposure to non-essential or toxic metals, including cadmium, lead, mercury, and arsenic, may aggravate oxidative stress, impair immune function, and potentiate severe outcomes. This duality illustrates the complexity of metal–virus–host interactions and highlights their relevance in understanding COVID-19 pathogenesis.

The present chapter is structured to provide a comprehensive overview of these aspects. Section 4.2 describes the structural organization of SARS-CoV-2, with emphasis on the involvement of chemical elements in viral proteins and stability. Section 4.3 reviews the infection mechanism, focusing on the role of metals in viral entry and replication. Section 4.4 examines the effects of COVID-19 on the immune system, distinguishing the contributions of essential and non-essential elements, while Section 4.5 discusses current challenges and perspectives, emphasizing the implications of metal dysregulation in long COVID.

By integrating concepts from virology, immunology, and bioinorganic chemistry, this book aims to advance the understanding of the interplay between SARS-CoV-2 and chemical elements, thereby providing new perspectives for the development of diagnostic tools, therapeutic interventions, and preventive strategies.

DOI: 10.1201/9781003632092-5

## 4.2 UTILIZATION OF METAL IONS FOR SARS-CoV-2 SURVIVAL AND REPLICATION

Metal ions such as $Zn^{2+}$, $Cu^{2+}$, and $Mg^{2+}$ are essential for the maintenance of SARS-CoV-2 protein stability and function. Specifically, these metal ions interact with viral proteins playing both structural and catalytic roles to regulate the survival, pathogenesis, and replication of the virus [5]. As such, understanding these metal ion-dependent mechanisms in SARS-CoV-2, and indeed other viruses, may aid the development of novel therapies for this and other infections. In this section, the metal-binding characteristics of SARS-CoV-2 proteins and their functional significance will be examined.

### 4.2.1 ZINC

$Zn^{2+}$ is believed to be the most common metal ion bound to SARS-CoV-2 proteins [6]. To better characterize the zinc proteome of the virus, Andreini et al. adopted an *in silico* approach identifying 15 experimental $Zn^{2+}$-binding sites and predicting further 22 sites across various proteins [6]. SARS-CoV-2 proteins containing structurally characterized sites included the replicase polyproteins 1a and 1b which are processed into non-structural proteins (NSPs). Additional $Zn^{2+}$-binding sites were predicted in these proteins as well as various NSPs, spike glycoprotein and open reading frame (ORF) proteins, such as ORF3a, ORF3d, and ORF8. The protein with the largest number of potential sites was identified as nsp3 (six sites plus a bridging site).

Nsp3 is the largest protein encoded by the SARS-CoV-2 genome, and it is an essential component of the replication–transcription complex. An experimentally characterized site is present in the papain-like protease 2 (PLpro) domain and appears to serve a structural role in this domain [7]. This site is composed of four cysteines, as commonly observed in zinc finger proteins, and is known to be essential for activity.

Nsp13 is another SARS-CoV-2 protein known to bind $Zn^{2+}$ [8]. Nsp13 is an essential helicase required for viral replication. The zinc-binding domain (ZBD) of nsp13 contains three $Zn^{2+}$-binding sites which are critical for its structure and activity [9]. In addition, the ZBD is involved in the interaction between nsp13 and other viral proteins, particularly nsp12 which is part of the replication–transcription complex. It has also been revealed that the ZBD of nsp13 also contains an iron–sulfur (Fe–S) cluster, specifically an $Fe_4S_4$ cluster that modulates both its binding to template RNA and its unwinding activity [10]. The importance of zinc binding to nsp13 has made it a potential target for antiviral strategies. For example, bismuth-based compounds have been found to inhibit nsp13 activity by replacing $Zn^{2+}$ ions with $Bi^{3+}$ ions, disrupting viral replication [8].

The spike glycoprotein of SARS-CoV-2 plays a crucial role in cell entry and infection of host cells, and it is thought to possess several $Zn^{2+}$-binding sites that play important roles in its structure and function. Among the sites predicted by Andreini et al., the C-terminal cytoplasmic region of the spike protein contains ten potential zinc-binding cysteines that are highly conserved among coronaviruses [5, 6]. These cysteines have a distribution similar to that observed in metallothioneins. It has been suggested that $Zn^{2+}$ ions may interact with the recognition interface between the spike protein and angiotensin-converting enzyme 2 (ACE2), the receptor on cell surfaces through which SARS-CoV-2 enters the cells [11]. Thus, $Zn^{2+}$-binding to the spike glycoprotein may potentially influence the virus's ability to enter cells.

In addition to the potential $Zn^{2+}$-binding sites identified by Andreini et al. in ORF3a/d and 8, another ORF protein, ORF7a contains two potential zinc-binding motifs (a CX3HX3C motif at residues 15–23 and a CX8CX5H motif at residues 56–73). These are similar to sites in the zinc finger domains of other viral proteins such as the HIV nucleocapsid protein. $Zn^{2+}$ binding is thought to play structural role in these proteins to facilitating their respective functions.

### 4.2.2 Copper

Significantly higher plasma copper concentrations have been found in plasma of COVID-19 patients compared to healthy controls [12], and reduced urinary excretion of copper has been reported during COVID-19 infection [13]. Also, copper death-related genes show differential expression in COVID-19 patients, with potential implications for disease pathogenesis [14]. These findings suggest that the virus significantly affects copper homeostasis in the host during infection.

The binding of copper to SARS-CoV-2 proteins has the potential to play a dual role by not only acting to stabilize viral proteins but also generating oxidative stress that may damage the host. $Cu^{2+}$ ions bind to the intrinsically disordered C-terminal region of Nsp1 with micromolar affinity. This protein inhibits host protein synthesis by blocking messenger RNA (mRNA) access to the entrance tunnel of the 40S ribosome. As such, it is possible that the $Cu^{2+}$–Nsp1 interaction may inhibit host protein synthesis. It has also been suggested that $Cu^{2+}$ ions, like $Zn^{2+}$ ions, can interact with the recognition interface between ACE2 receptor and the spike glycoprotein to modulate virus entry into cells [11].

$Cu^{+}$ ions are likely to play a significant role in combating SARS-CoV-2 and other viruses. $Cu^{+}$ specifically targets the viral genome inhibiting its ability to replicate [15]. It is important to note that while copper, including $Cu^{+}$, shows promising antiviral properties, current evidence does not support using copper supplements as a treatment or preventive measure for COVID-19. The primary benefits of copper against SARS-CoV-2 appear to be in surface disinfection. Studies have shown that SARS-CoV-2 survives for less than 4 hours on copper surfaces compared to up to 3 days on plastic and stainless-steel surfaces [16].

### 4.2.3 Magnesium (and Phosphorus)

Structural and biochemical studies demonstrate that $Mg^{2+}$ acts as an essential cofactor for the Nsp16/Nsp10 2′-O-methyltransferase enzyme, helping to orient the viral mRNA substrate to ensure methylation [17]. This is required by the virus to evade host immune detection. $Mg^{2+}$ ions also have the potential to modulate SARS-CoV-2 infection. One mechanism may be through the inhibition of host cell proteases, transmembrane protease serine 2 (TMPRSS2) and furin, which are crucial for the activation of the SARS-CoV-2 spike glycoprotein and viral entry into host cells. It is thought that $Mg^{2+}$ increases methylation of the TMPRSS2 promoter, hindering its transcription to reduce the expression of this enzyme [18]. It is also possible that $Mg^{2+}$ hampers furin activity, which is $Ca^{2+}$-dependent, through an antagonistic effect. Indeed, higher plasma levels of magnesium are associated with lower severity of COVID-19 symptoms, suggesting a protective effect against severe infection [19].

Similarly, deficiencies of magnesium and phosphate are commonly observed in severe COVID-19 cases and are associated with poor outcomes [20, 21]. Both magnesium and phosphate are essential for ATP production, which is important for energy metabolism and immune functioning. Deficiencies of these minerals in the host can lead to impaired ATP synthesis, leading to muscle weakness and respiratory failure. Upon infecting host cells, SARS-CoV-2 induces significant changes in cellular phosphorylation patterns. In infected cells, this alteration affects various signaling pathways, including the p38/MAPK pathway which is involved in inflammation and cell stress responses [22]. These modifications facilitate viral replication and the host cell ability to adapt to the infection. Interestingly, polyphosphate has been shown to inhibit binding of the spike protein to ACE2 receptor [23], suggesting it could serve as a potential therapeutic agent for prevention or treatment of COVID-19.

### 4.2.4 Iron

Iron dysregulation is associated with increased severity of COVID-19 infection. High iron levels can enhance ACE2 receptor expression in host cells, contributing to higher degree of SARS-CoV-2 infectivity [24]. Iron overload synergistically enhances the production of reactive oxygen species, reactive nitrogen species, and pro-inflammatory cytokines in response to spike protein stimulation [25, 26] which can exacerbate the host's inflammatory response to the virus.

Iron metabolism alterations may also lead to anemia of inflammation, also known as anemia of chronic disease. This condition is characterized by hypoferremia (low serum iron) and hyperferritinemia (high ferritin levels). The underlying mechanism involves body's innate immune response to infection and production of pro-inflammatory cytokines, particularly interleukin-6 (IL-6), which stimulate increased hepcidin production in the liver [27, 28]. Elevated hepcidin levels cause iron to be sequestered within cells (especially macrophages and hepatocytes) by degrading ferroportin, resulting in decreased circulating iron and increased intracellular iron storage, ultimately reducing iron availability for erythropoiesis and pathogen use [29].

A spike protein-derived peptide dubbed "Covidin" may be involved in host ferroportin binding, potentially mimicking the iron-regulating hormone hepcidin and contributing to iron dysregulation in severe COVID-19 cases [30]. Iron dysregulation and hypoxia may sustain a destructive cycle of impaired immune function and inflammation, contributing to both acute and long-term manifestations of COVID-19. These findings suggest that iron plays a complex and multifaceted role in SARS-CoV-2 infection, affecting both viral entry and the host's immune response.

### 4.2.5 Other Metals

Beyond zinc, copper, magnesium, and iron, there is emerging evidence that other metals may bind and regulate SARS-CoV-2 proteins. For example, $Mn^{2+}$ and $Ca^{2+}$

**TABLE 4.1**
**Metals and Their Roles in Binding to SARS-CoV-2 Viral Proteins**

| Metal | Notes |
|---|---|
| Zn | Key for viral proteins including nsp3, nsp13, the spike glycoprotein, and several ORF proteins |
| Cu | Can replace weaker binding metals, may disrupt viral proteins |
| Mg | Interactions with RNA–protein complexes, essential for methyltransferase activity |
| Fe | Important for general viral protein function |
| Mn/Ca | May substitute for $Mg^{2+}$ in some roles (e.g., for protein stabilization) |
| Au/Ag | Interactions observed on surfaces, but not within protein sites |
| Ni/Co | No direct experimental evidence of binding to SARS-CoV-2 proteins |

*Abbreviation*: Nsp, non-structural protein; ORF, open reading frame.

may (like $Mg^{2+}$) act to stabilize the SARS-CoV-2 mRNA capping process through binding to the nsp16/nsp10 protein complex. This is supported by clinical observations that report altered levels of manganese and calcium in COVID-19 patients [31, 32], suggesting a role in viral protein interactions or host response to infection. Another area of focus has been examination of SARS-CoV-2 spike protein adsorption onto gold, silver, and copper surfaces using computational methods [33]. These studies show that the spike protein can interact with such surfaces, although it is important to note that these interactions do not represent specific coordination of a metal within protein sites. While direct binding of $Ni^{2+}$ and $Co^{2+}$ to SARS-CoV-2 proteins has not been conclusively demonstrated, the Irving–Williams series suggests that proteins bind divalent metal ions in the order: $Mg^{2+} < Mn^{2+} < Fe^{2+} < Co^{2+} < Ni^{2+} < Cu^{2+} (Cu^{+}) > Zn^{2+}$. This implies that, under certain conditions, viral proteins can coordinate with $Co^{2+}$ or $Ni^{2+}$ ions, although experimental confirmation is lacking and this may be hindered in vivo by the limited bioavailability of these metal ions. A summary of the metals discussed and their interactions with SARS-CoV-2 proteins is provided in Table 4.1.

## 4.3 UNDERSTANDING SARS-CoV-2 INFECTION

This chapter provides a concise overview of SARS-CoV-2 infection mechanisms as a comprehensive treatment would require a dedicated volume. The focus is on the following key stages: viral entry, RNA delivery, transcription, translation, assembly, and release. Where relevant, the roles of metal ions and host responses are addressed.

### 4.3.1 Viral Entry

Understanding SARS-CoV-2 infection begins with viral entry into host cells. Like other coronaviruses, infection is initiated through binding of viral particles to host surface receptors. SARS-CoV-2 utilizes human ACE 2 (hACE2) for cell entry [34], similar to SARS-CoV. The virion comprises four structural proteins: nucleocapsid (N), membrane (M), envelope (E), and spike (S).

hACE2 (UniProt Q9BYF1) is an 805-residue metallopeptidase that converts angiotensin I to angiotensin-(1–9) (exerting antihypertrophic effects in cardiomyocytes) and angiotensin II to angiotensin-(1–7) (a vasodilator) [35]. Expressed in multiple human organs, its activity is modulated by chloride (enhancing angiotensin I cleavage while inhibiting angiotensin II cleavage) and fluoride [36]. Zinc is coordinated at its catalytic site via an H374E375XXH378 motif (with E402 completing the binding site), inducing conformational changes critical for enzymatic function [37]. Clinically, zinc deficiency correlates with increased COVID-19 severity [38]. Key spike-binding residues (Q24, D30, H34, E35, E37, D38, Y41, and Q42) include a metal-binding region (D30-E37) that interacts with both $Zn^{2+}$ and $Cu^{2+}$ [11].

The spike (S) glycoprotein mediates receptor binding and membrane fusion (for review see [39]). Comprising S1 (receptor-binding) and S2 (membrane-anchoring/fusion) subunits, it forms homotrimers decorated with ~66 N-linked glycans [40–42]. The S1 subunit contains an N-terminal domain (NTD), receptor-binding domain (RBD), and C-terminal domains (CTD1/CTD2); while S2 (587 residues) houses the fusion machinery [43]. The receptor-binding motif (RBM) within the RBD makes direct contact with hACE2 [44]. In prefusion conformation, RBMs are partially occluded ("down" state), transitioning to a fully exposed ("up") state upon receptor engagement [45]. Critical hACE2-interacting residues include N487, Q493, Y499, T500, Y505, and G446 [46]. Receptor binding triggers two essential cleavage events [47]: furin-mediated cleavage at the S1/S2 boundary (residues 685–686) primes the spike protein; and exposure and cleavage at the S2′ site (residues 816–817) by host proteases.

When TMPRSS2 [48] is present, it cleaves S2′ at the plasma membrane, enabling immediate fusion. In TMPRSS2-deficient cells, the virus–hACE2 complex undergoes clathrin-mediated endocytosis. Within acidified endolysosomes, cathepsin L activates and cleaves S2′. Both pathways expose the fusion peptide (FP), triggering S1 dissociation and FP insertion into the host membrane. S2′ cleavage releases conformational constraints, enabling FP-mediated membrane fusion through large-scale refolding events (helical bundle formation, coiled-coil transitions). This process resembles class I viral fusion mechanisms [49] and depends on lipid raft microdomains enriched with sphingolipids and cholesterol in both viral and host membranes [50]. Calcium ions ($Ca^{2+}$) facilitate fusion by stabilizing FP conformation and organizing membrane lipids-chelating intracellular $Ca^{2+}$ reduces SARS-CoV entry [51].

### 4.3.2 Release of Viral RNA

Following membrane fusion between the viral envelope and the host cell membrane, a transient fusion pore forms, enabling the release of the viral genome RNA (gRNA) into the cytoplasm. The SARS-CoV-2 genome is a positive-sense, single-stranded RNA (~29.9 kb), one of the largest known RNA genomes among viruses. It contains a transcriptional core sequence (TRS) that guides subgenomic mRNA (sgmRNA) synthesis [52], and it is protected from cellular exonuclease degradation by its 5′ cap structure and 3′ poly(A) tail [53]. The genome also encodes factors essential for maintaining RNA integrity and post-transcriptional processing, such as nsp8 which contributes to the polyadenylation and extension of viral RNAs [54].

Once inside the cytoplasm, the gRNA serves directly as mRNA, initiating translation by host ribosomes. Two large ORFs (ORF1a and ORF1b), which together comprise approximately two-thirds (~70%) of the genome, are translated into two overlapping polyproteins (pp1a and pp1ab). The production of pp1ab requires a ribosomal frameshift between ORF1a and ORF1b, allowing for extended translation of the downstream region.

These polyproteins are subsequently cleaved by virus-encoded proteases, specifically the papain-like protease (PLpro), part of nsp3, and the 3C-like protease (Mpro or 3CLpro), nsp5, generating a total of 16 NSPs (nsps 1–16). These NSPs assemble into the replication–transcription complex (RTC) which forms on double-membrane vesicles (DMVs) derived from the endoplasmic reticulum (ER). This complex performs several vital functions, including genome replication, transcription of sgmRNAs, proofreading (via nsp14 exonuclease activity), and evasion of host innate immune responses.

### 4.3.3 Transcription and Translation

The translation of both structural proteins (SPs) and NSPs depends on sgmRNAs which are synthesized alongside new genomic RNAs (gRNAs) within DMVs in the host cytoplasm (for review see [55]). These vesicles, formed under the guidance of the RTC, serve a dual purpose: They create a confined microenvironment for efficient viral RNA synthesis and shield viral RNA intermediates from host innate immune sensors such as RIG-I and MDA5.

DMV formation results from extensive membrane remodeling of the ER. NSPs nsp3 and nsp4 initiate this rearrangement by inducing membrane curvature and tethering apposed bilayers [56]. Nsp6 contributes further by forming bridges that connect DMVs to the ER network, facilitating substrate exchange and vesicle stability [57]. Interestingly, the lipid composition of DMVs differs from that of the ER, with DMVs enriched in cholesterol and sphingolipids which enhance membrane rigidity and replication compartment integrity [58].

Once established, viral positive-sense genomic RNA (+gRNA) acts directly as mRNA for the translation of ORF1a and ORF1b, producing two large polyproteins – pp1a and pp1ab, through a programmed -1 ribosomal frameshift between the two ORFs. These polyproteins are autoproteolytically processed by two virus-encoded cysteine proteases, PLpro, part of nsp3, and 3C-like protease (Mpro or 3CLpro, nsp5), to yield 16 mature NSPs (nsps 1–16) [7]. Translation occurs on cytoplasmic ribosomes and the nascent NSPs rapidly assemble to form the RTC on DMV membranes. Nsp1 is one of the first proteins released and rapidly expressed in the cytoplasm. It plays a central role in immune evasion by binding to the 40S ribosomal subunit, thereby blocking host mRNA translation and promoting its degradation [59]. This mechanism efficiently redirects the host translational machinery toward viral RNA synthesis.

The RTC encompasses a set of proteins with enzymatic functions required for RNA replication, transcription, modification, and proofreading [60]. The core polymerase complex includes nsp12 (the RNA-dependent RNA polymerase, RdRp), two copies of nsp8, and one copy of nsp7, which act together to drive RNA elongation [61]. Additional cofactors, nsp9, nsp10, nsp13, nsp14, and nsp16, modulate various steps of replication and RNA maturation [62]. Nsp13, the viral helicase, contains two zinc-finger motifs and interacts with nsp8 and nsp12 to unwind RNA secondary structures using ATP hydrolysis [63]. The RdRp (nsp12) catalyzes RNA synthesis, while nsp14 provides 3′→5′ exonuclease activity, essential for proofreading and maintaining replication fidelity, compensating for the intrinsically low fidelity of RdRp alone [64]. Nsp16, together with its cofactor nsp10, functions as a 2′-O-methyltransferase, adding a cap modification that renders viral mRNAs indistinguishable from host transcripts, thus preventing immune recognition.

Upon synthesis within DMVs, sgmRNAs are exported via a molecular pore complex spanning the DMV membranes. These sgmRNAs are then translated into the SPs S, M, E, and N, and accessory proteins. Translation of S, M, and E proteins occurs on the rough ER, from which they are co-translationally inserted into the ER membrane, while N protein is synthesized in the cytoplasm. Each gRNA associates with multiple N proteins to form nucleocapsids which are then trafficked to the ER–Golgi intermediate compartment (ERGIC). Assembly proceeds as nucleocapsids bud into membranes enriched with S, M, and E proteins, giving rise to progeny virions that later mature and are released through exocytosis.

Zinc appears to play crucial structural and catalytic roles in these processes. Both nsp13 (helicase) and nsp14 (exonuclease and methyltransferase) contain zinc-finger motifs that stabilize their tertiary structures and mediate protein–RNA interactions. Experimental studies have shown that zinc-ejector compounds can impair nsp13, nsp14, and nsp10 functions, effectively inhibiting viral replication [65]. Moreover, zinc interferes with RdRp (nsp12) activity in related coronaviruses, suggesting that targeting viral zinc-binding sites could represent a promising antiviral strategy [66].

### 4.3.4 Assembly, Maturation, and Release of New Virions

The newly synthesized gRNA associates with the nucleocapsid (N) protein to form ribonucleoprotein complexes (RNPs) – a process thought to be mediated by liquid–liquid phase separation (LLPS) between the highly positive N protein regions and the negatively charged RNA backbone [67]. This dynamic condensation ensures efficient RNA packaging and spatial organization within the cytoplasm.

Assembly of new virions occurs primarily in the ERGIC (for review see [68]). Within this region, the structural proteins S (spike), M (membrane), and E (envelope), which have been co-translationally inserted into the ER membrane, traffic and concentrate at specific ERGIC subdomains. Here, they coordinate with the RNP complexes to drive particle budding into the lumen of the ERGIC (for review see [69]).

The M protein, a triple-spanning transmembrane glycoprotein, plays a central organizational role in virion morphogenesis [70]. It forms homodimers and interacts with all major structural proteins (S, E, and N) acting as a scaffold that defines the shape, curvature, and particle size of the virion [71]. M binds directly to N, enabling encapsidation of gRNA and positioning of the N at budding sites [72]. It also interacts with E to guide assembly and recruit S to the correct membrane region, ensuring spatial coordination [73]. Because of its abundance, M is considered the most dominant structural component of the coronavirus envelope and a key determinant of virion integrity.

The E protein is a small, hydrophobic homopentameric protein that assembles into cation-selective ion channels (viroporins) located in ERGIC membranes [74]. These channels regulate ionic homeostasis and promote membrane curvature essential for virion budding and fission [75]. E protein also influences virus pathogenicity: its ion channel activity modulates host secretory pathways, while its ability to disrupt epithelial tight junctions contributes to increased permeability and viral dissemination in lung tissue [76]. Although present in small quantities within the virion, E plays a disproportionately large role in assembly, budding, and virulence.

The S protein, previously processed in the ER, undergoes additional cleavage and glycosylation within the Golgi and trans-Golgi network (TGN) (for review see [77]). These post-translational modifications, including furin-dependent cleavage at the S1/S2 site and complex N-linked glycan maturation, are essential for fusion competence and immune evasion.

Once the nucleocapsid and structural components coalesce at the ERGIC, budding produces enveloped progeny virions which accumulate in vesicular carriers (for review see [78]). These carriers migrate through the Golgi-TGN pathway, where final maturation occurs. Maturation processes likely include spike trimming, lipid remodeling, and E protein quality control. Unlike many enveloped viruses that utilize classical ER-Golgi exocytosis, SARS-CoV-2 appears to exploit the lysosomal exocytosis pathway for egress. Mature virions are trafficked to late endosomes/lysosomes and released into the extracellular space when these compartments fuse with the plasma membrane. This egress mechanism not only allows sustained viral release but may also interfere with normal lysosomal degradation, aiding immune evasion and persistent infection.

### 4.3.5 Host Cell Response

Several viral proteins and mechanisms described above are directly involved in modulating the host immune response and deserve to be mentioned shortly. First, the formation of DMVs shields nascent viral RNA and replication intermediates from detection by host innate immune sensors, such as RIG-I and MDA5, effectively delaying early antiviral signaling.

Viral proteins also strongly influence immune activity. The N protein, one of the most abundant viral proteins in infected cells [79], elicits a robust adaptive immune response while simultaneously inhibiting the host interferon (IFN) pathway [80]. Mutations in the S protein among variants of concern, such as Alpha, Beta, Gamma, Delta, and Epsilon (B.1.429), reduce susceptibility to neutralization by convalescent or vaccine-induced antibodies [81]. The M protein is highly immunogenic, capable of stimulating neutralizing antibody production and $CD8^+$ T cell responses [82].

Among nonstructural and accessory proteins, nsp1 suppresses host translation, hindering IFN signaling [83]; nsp15 degrades viral dsRNA intermediates to avoid immune detection [84]; and ORF3a and ORF7a proteins disrupt antigen presentation via major histocompatibility complex (MHC)-I [85]. Similarly, nsp3 and nsp6 delay IFN-$\gamma$ activation, further promoting immune evasion [86]. Zinc, however, has been shown to enhance IL production and support antiviral immunity, making it a potential modulator in host defense against SARS-CoV-2 [87].

## 4.4 EFFECT OF THE DISEASE ON THE IMMUNOLOGICAL SYSTEM

The immune response is broadly divided into two complementary arms: the innate immune system which provides rapid and non-specific defense through physical barriers, phagocytic cells, and inflammatory mediators; and the adaptive immune system which involves a slower yet highly specific response involving T and B lymphocytes capable of establishing immunological memory [88].

In the context of viral infections such as COVID-19, epithelial pattern recognition receptors (PRRs) detect pathogen-associated molecular patterns (PAMPs) derived from SARS-CoV-2, including double- and single-stranded RNA as well as structural viral proteins [89]. This recognition initiates a coordinated cascade of cytokine and chemokine release. The five major PRR families, Toll-like receptors (TLRs), RIG-I-like receptors (RLRs), NOD-like receptors (NLRs), C-type lectin receptors, and AIM2-like receptors, act in concert to trigger early immune activation, driving inflammatory signaling through mediators such as IL-1β, IL-6, tumor necrosis factor-alpha (TNF-α), IFN-β, and IFN-γ [90].

While these responses are essential for restricting viral replication, sustained or excessive cytokine production commonly referred to as a "cytokine storm" can result in immune exhaustion, sepsis, and multi-organ dysfunction. In COVID-19, this hyperinflammatory state is strongly associated with poor clinical prognosis and increased mortality risk. Counter-regulatory mechanisms involving anti-inflammatory cytokines attempt to restore immune balance by promoting the recruitment and activity of regulatory T cells, thereby limiting tissue damage and preventing chronic inflammation [91].

Innate immune cells, including monocytes, macrophages, neutrophils, and dendritic cells, function as antigen-presenting cells (APCs), thereby linking innate sensing to adaptive immune activation. Within the adaptive compartment, $CD4^+$ and $CD8^+$ T lymphocytes coordinate cellular immune responses, while B lymphocytes mediate humoral immunity through antibody production. SARS-CoV-2-specific $CD4^+$ T cells particularly those recognizing S, M, and N antigens are more frequently detected than $CD8^+$ T cells and their magnitude correlates with viral load and disease resolution. Although less prevalent, $CD8^+$ T cells play a critical role in the elimination of infected cells and are consistently associated with improved clinical outcomes [92].

To evade immune surveillance, SARS-CoV-2 encodes multiple proteins that interfere with PRR signaling pathways and suppress IFN production. The virus further protects its RNA by sequestering it within DMVs, methylating its transcripts, and inhibiting RIG-I/MDA5 activation as well as stress granule formation. These evasion strategies delay early antiviral responses, allowing viral replication to proceed largely unchecked during the initial stages of infection. The resulting immune dysregulation promotes the recruitment of pro-inflammatory cells while suppressing IFN signaling, thereby contributing to extensive pulmonary inflammation. Notably, emerging evidence indicates that this complex virus–host interplay including PRR signaling, cytokine balance, and antiviral effector mechanisms is modulated by essential metal ions [90].

The infectious process initiated by SARS-CoV-2 disrupts a broad spectrum of cellular and metabolic pathways, including those involved in lipid, protein, and carbohydrate turnover, all of which are tightly integrated with immune function. Within these interconnected metabolic networks, biometals – a term encompassing essential metallic and metalloid elements such as zinc, iron, copper, selenium, manganese, calcium, and magnesium – play a fundamental role. These elements act not only as structural or functional cofactors for viral components, as previously discussed in this chapter, but also as key regulators of host immune responses and systemic homeostasis. Despite increasing recognition of their importance, the regulatory interplay between biometals and immune function remains only partially understood in the field of life sciences. This is particularly relevant given that biometals are indispensable for the activity of metalloproteins involved in nucleic acid metabolism, enzymatic catalysis, and signal transduction. Consequently, alterations in their bioavailability whether due to excess, deficiency, or dysregulated distribution during infection can profoundly impact immune competence and ultimately influence clinical outcomes [93].

As outlined in the previous section, respiratory viruses initiate a well-defined sequence of epithelial invasion and immune activation. Building on this framework, the following sections examine how biometals modulate antiviral defenses through their roles in both innate and adaptive immunity. This integrated perspective enables the connection of metal-mediated molecular regulation with the clinical manifestations observed in respiratory viral infections [94], as summarized in Figure 4.1.

Zinc is a central regulator of antiviral immunity, exerting both immunomodulatory and direct antiviral effects. Within the innate immune response, zinc preserves epithelial barrier integrity, modulates cytokine gene transcription, and promotes type I IFN signaling through the activation of IFN regulatory factors IRF3 and IRF7. Simultaneously, under homeostatic conditions, zinc constrains excessive inflammation by inhibiting nuclear factor-kappa B (NF-κB) nuclear translocation. Dynamic intracellular Zn fluxes further enhance the antimicrobial activity of macrophages and neutrophils and optimize pathogen recognition by pattern recognition receptors and cytosolic viral sensors [95].

Beyond its immunoregulatory functions, zinc exhibits direct antiviral activity by inhibiting key viral enzymes involved in RNA virus replication, including RdRp and viral proteases. These effects are markedly potentiated in the presence of zinc ionophores, which increase intracellular Zn availability. Zinc deficiency compromises epithelial barrier function, impairs natural killer cell cytotoxicity, and promotes dysregulated cytokine release, thereby facilitating viral dissemination and systemic inflammation [96]. Within adaptive immunity, zinc is indispensable for effective T-cell receptor signaling, clonal expansion of $CD4^+$ and $CD8^+$ T lymphocytes, balanced IL-2 and IFN-γ production, and B-cell maturation with antibody class switching. Clinically, reduced zinc bioavailability has been consistently associated with lymphopenia, impaired antibody responses, increased disease severity, and adverse outcomes in COVID-19 across multiple independent cohorts [97].

Copper plays a relevant regulatory role in both innate and adaptive immune responses. It supports the differentiation and activation of innate immune cells, particularly through redox regulation mediated by Cu/Zn-superoxide dismutase and alpha-protein kinase 1 (ALPK1) signaling pathways [98]. During inflammation,

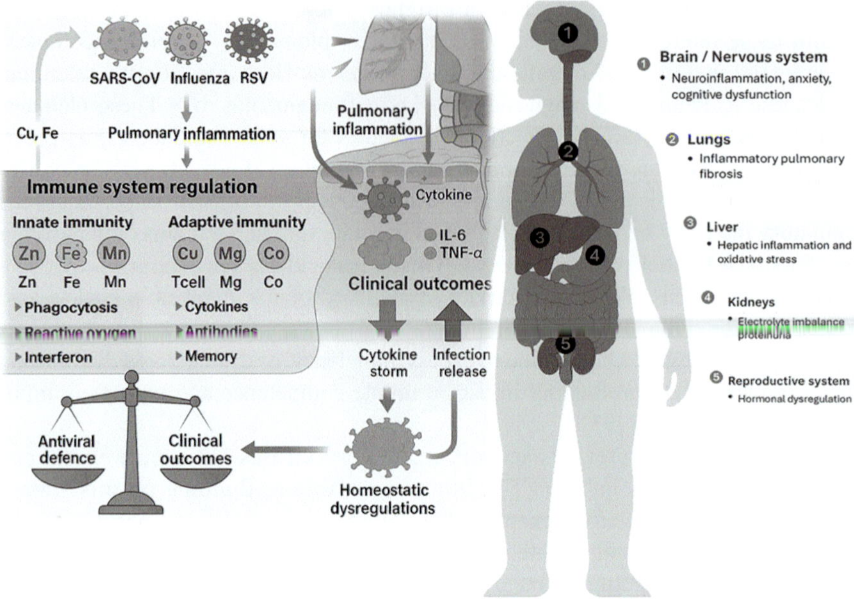


**FIGURE 4.1 Metal-mediated modulation of immune responses and systemic outcomes in respiratory viral infections**. Viral pathogens such as severe acute respiratory syndrome coronavirus 2 (SARS-CoV-2), influenza virus, and respiratory syncytial virus (RSV) induce pulmonary inflammation and trigger both innate and adaptive immune responses. These processes are tightly modulated by essential trace elements, including zinc (Zn), iron (Fe), manganese (Mn), copper (Cu), magnesium (Mg), cobalt (Co), calcium (Ca), and selenium (Se). These metals regulate key immunological functions such as phagocytosis, redox signaling, type I interferon (IFN) production, T cell activation, cytokine secretion, antibody generation, and immunological memory. Disruptions in metal homeostasis impair antiviral defense and promote dysregulated inflammation, including excessive cytokine release (e.g., interleukin-6 [IL-6], tumor necrosis factor-alpha [TNF-α]), cytokine storm, and systemic inflammatory responses. Resulting clinical manifestations span multiple organs, including the brain (neuroinflammation), lungs (alveolar damage), liver (oxidative stress), kidneys (electrolyte imbalance), and reproductive system, highlighting the critical interplay between trace metal status and immune regulation during viral infection. (Illustration created by the author.)

copper is mobilized by ceruloplasmin, which also modulates iron homeostasis through Fe oxidation and has been implicated in the release of pro-inflammatory mediators during SARS-CoV-2 infection. In adaptive immunity, copper contributes to T-cell proliferation, IL-2 production, natural killer cell activity, and antibody synthesis, thereby supporting effective immune coordination [99]. Clinically, multiple studies have reported elevated serum copper concentrations in COVID-19 patients, with levels increasing proportionally with disease severity [100]. In contrast, urinary copper excretion appears reduced, suggesting retention or renal dysregulation [13]. This pattern of high serum level and low urinary excretion of copper indicates a systemic imbalance in copper metabolism during infection, reflecting inflammatory redistribution or impaired elimination [101].

Iron plays a dual and tightly regulated role in immunity: it is essential for immune cell metabolism and redox homeostasis yet must be carefully sequestered to prevent pathogen proliferation. In innate immunity, iron supports oxidative burst mechanisms through NADPH oxidase and inducible nitric oxide synthases in macrophages and neutrophils, whereas excess labile iron promotes oxidative damage via Fenton chemistry. To counteract this, the host activates the hepcidin–ferroportin axis, inducing intracellular iron sequestration within ferritin and reducing plasma iron availability (hypoferremia), a hallmark of the acute phase response [102]. In adaptive immunity, iron supports T cell and B cell proliferation, DNA synthesis, and chromatin remodeling through iron-dependent dioxygenases. Iron deficiency impairs Th1 and $CD8^+$ responses, reduces antibody production, and compromises long-term humoral memory [103]. Clinically, hyperferritinemia is a defining feature of severe COVID-19, with ferritin concentrations exceeding 4500 ng/mL strongly associated with respiratory failure and mortality [104]. Ferritin is now recognized not merely as a passive marker but as a pro-inflammatory damage-associated molecular pattern (DAMP). Its degradation increases labile intracellular iron, promoting ferroptosis, a regulated form of cell death linked to lung, liver, and kidney injury. Accordingly, targeting iron metabolism has emerged as a potential therapeutic strategy to mitigate immune-mediated tissue damage.

Manganese (Mn) has recently emerged as an important immunoregulatory micronutrient in viral infections. In innate immunity, Mn serves as a cofactor for mitochondrial superoxide dismutase (MnSOD2) facilitating the detoxification of reactive oxygen species during early inflammatory responses. It also activates cytosolic pattern recognition receptors, particularly RIG-I and the cGAS-STING pathway, thereby initiating type I IFN responses against viral RNA and DNA [105]. In addition, Mn enhances endolysosomal acidification through mTORC1 activation, promoting virion degradation and antiviral mechanism relevant to SARS-CoV-2 infection. Within adaptive immunity, Mn supports mitochondrial metabolism in activated T cells, preserves lymphocyte viability under oxidative stress, and modulates antigen presentation by dendritic cells, indirectly strengthening both cellular and humoral responses. In COVID-19, reduced Mn concentrations have been reported in severe cases (median 8.9 μg/L) compared with moderate disease (11.1 μg/L), with inverse correlations observed between Mn levels and levels of inflammatory cytokines such as IL-6 and TNF-$\alpha$ [106].

Selenium is a redox-active trace element that supports antiviral defense primarily through its incorporation into selenoproteins, including glutathione peroxidases, thioredoxin reductases, and selenoprotein P (SELENOP). These enzymes maintain redox homeostasis during viral infections by reducing peroxides and limiting reactive oxygen species accumulation, thereby preserving membrane integrity and supporting phagocytic function in epithelial cells, macrophages, and neutrophils [107]. Adequate selenium status promotes balanced type I IFN responses and suppresses excessive NF-$\kappa$B activation, reducing the risk of progression toward hyperinflammation. In adaptive immunity, selenium supports T cell receptor signaling and clonal expansion, enhancing $CD4^+$ and $CD8^+$ T cell function as well as IL-2 and IFN-$\gamma$ production. It is also required for B cell maturation, class-switch recombination, and antibody secretion essential for durable humoral immunity [108]. Clinically, selenium deficiency is frequently observed in patients with severe or fatal COVID-19. In German and Iranian cohorts, non-survivors exhibited serum selenium concentrations as low as 30–40 μg/L, accompanied by reduced SELENOP levels, whereas survivors maintained concentrations above 50 μg/L [109]. These reductions correlated inversely with C-reactive protein (CRP), reflecting systemic inflammation. Additional cohorts from Russia, Belgium, and South Korea confirmed widespread selenium deficiency at hospital admission among fatal cases [110]. Beyond SARS-CoV-2, selenium depletion has been shown to enhance viral mutagenesis and increase disease severity in influenza and coxsackievirus infections, underscoring its critical role in antiviral defense, inflammation control, and clinical recovery.

In viral infections, cobalt acquires immunological relevance primarily through its incorporation into cobalamin (vitamin B12) where it serves as a central atom essential for DNA synthesis, chromatin regulation, and mitochondrial function. Although cobalt itself does not play a direct physiological role in human immunity, the cobalt moiety within vitamin B12 is indispensable for effective antiviral defense mechanisms [111].

Within the innate immune system, cobalamin sustains redox homeostasis by supporting the methionine cycle and regenerating intracellular glutathione, thereby limiting peroxide accumulation and mitigating oxidative stress in epithelial cells, macrophages, and neutrophils. These antioxidant effects preserve membrane integrity and phagocytic capacity, preventing premature immune cell exhaustion. Experimental models demonstrate that vitamin B12 deficiency impairs macrophage and neutrophil function, disrupts mitochondrial homeostasis, and lowers the activation threshold of NF-κB, promoting dysregulated cytokine release [112]. In vitro studies further indicate that exogenous cobalt modulates macrophage activation in a dose-dependent manner, with low concentrations promoting anti-inflammatory transcriptional profiles and higher levels stimulating TNF-α and IL-1β secretion [113]. In the adaptive immune compartment, adequate B12 availability is essential for T cell proliferation, IL-2 signaling, and the development of effective $CD4^+$ and $CD8^+$ cytotoxic responses. Its deficiency leads to reduced clonal expansion, attenuated IFN-γ production, and impaired B cell maturation and antibody generation, collectively weakening cellular and humoral immunity [114].

During acute viral inflammation, cobalt-related biomarkers have emerged as indirect indicators of systemic oxidative stress. Among these, ischemia-modified albumin (IMA) is particularly informative, as it inversely reflects albumin's cobalt-binding capacity. In a cohort of 100 hospitalized COVID-19 patients, IMA levels (mean ≈ 0.73 ± 0.17 ABSU) were significantly elevated compared with healthy controls (0.42 ± 0.11 ABSU), while cobalt-binding activity was markedly reduced, indicating heightened oxidative stress [115]. Reduced cobalt-binding capacity correlated strongly with CRP (r ≈ 0.78) and ferritin levels, highlighting a close link between cobalt–albumin dysfunction and systemic inflammation. Additional studies reported elevated serum cobalt concentrations in COVID-19 patients (mean 3.2–3.5 μg/L) relative to healthy individuals (1.6–1.8 μg/L), with positive correlations observed with IL-6 and neutrophil counts and negative correlations with lymphocyte numbers. These alterations likely reflect acute phase redistribution, oxidative albumin modification, and altered renal handling rather than nutritional cobalt overload.

Importantly, biomarkers such as IMA and the albumin–cobalt binding index may serve as surrogate indicators of disease severity, tracking respiratory compromise, inflammatory burden, and progression toward critical illness. Their integration into multi-marker clinical panels may improve early risk stratification in patients with severe viral pneumonia [116].

Calcium (Ca) functions as a universal second messenger regulating numerous cellular processes essential to immune defense. In innate immunity, calcium influx is required for pattern recognition receptor activation and downstream signaling pathways governing cytokine release and inflammasome assembly. These calcium-dependent mechanisms contribute to endothelial dysfunction, increased vascular permeability, and pulmonary edema – hallmark features of severe SARS-CoV-2 infection. Calcium signaling is also indispensable for T-cell receptor engagement and cytokine production, positioning $Ca^{2+}$ as a critical regulator of both innate and adaptive immune responses. Dysregulated calcium homeostasis has been implicated in cytokine storms and hyperinflammatory states observed in critical COVID-19 cases [117]. Clinically, abnormal serum calcium levels have been associated with coagulopathy, disease severity, and poorer prognosis in viral pneumonia [118].

Magnesium (Mg) acts as a cofactor in more than 600 enzymatic reactions and plays a fundamental role in immunoregulation. It stabilizes ATP, supports DNA replication, and is required for the synthesis and activation of vitamin D – a hormone crucial for both innate and adaptive immune responses. Magnesium enhances T cell responsiveness, particularly by modulating IL-2 and IFN-$\gamma$ signaling and contributes to the maintenance of endothelial barrier integrity. In viral infections, hypomagnesemia has been associated with oxidative stress, elevated inflammatory markers such as CRP, and impaired T cell function. In SARS-CoV-2 infection, reduced serum Mg levels correlate with increased disease severity, prolonged hospitalization, and greater need for ventilatory support [119]. Conversely, adequate Mg level appears to support immune regulation and is associated with reduced mortality, reinforcing its potential protective role in the inflammatory context of viral disease.

Taken together, the evidence presented in this section underscores the multifaceted roles of biometals in antiviral defense and immune regulation in COVID-19. Although each element exhibits distinct molecular and clinical associations, they converge as critical modulators of disease progression, severity, and recovery. To facilitate comparative interpretation, Table 4.2 summarizes the principal antiviral and immunological functions, clinical evidence, and proposed biomarkers for the key biometals discussed in this chapter.

## 4.5 CHALLENGES AND PERSPECTIVES

Despite the growing interest in the role of metals/metalloids during viral infections, consistent challenges hinder the consolidation of biometals as robust clinical indicators or therapeutic targets in diseases such as COVID-19. A central limitation lies in the heterogeneity of existing studies which differ considerably in sample size, patient selection, analytical techniques, and clinical outcomes. This variability contributes to conflicting results in literature, particularly for elements such as Ca, Mg, Zn, Fe, and Se and highlights the need for standardized methodologies and multicenter validation.

**TABLE 4.2**
**Comparative Overview of Biometals in the Immune Response and Clinical Progression of COVID-19**

| Metal | Direct Antiviral Function | Immunomodulatory Role | Main Mechanism | Clinical Evidence | Suggested Biomarkers | Ref. |
|---|---|---|---|---|---|---|
| **Zn** | Inhibits RdRp and viral proteases (with ionophores) | Stimulates type I IFN, stabilizes TCR, enhances NK and $CD8^+$ activity | NF-κB inhibition, IRF3/7 activation, epithelial barrier support | Deficiency linked to severe cases, lymphopenia, increased ventilation time | Serum Zn, Cu/Zn ratio | [95] |
| **Cu** | Virucidal properties (ex vivo) | Cofactor for SOD, regulates PRRs and inflammatory response | ALPK1 activation, ceruloplasmin regulation, oxidative burst support | Increased in serum, decreased in urine in severe cases | Serum and urinary Cu, ceruloplasmin | [98–101] |
| **Fe** | Required for viral replication but sequestered by immunity | Supports oxidative burst, lymphocyte epigenetic regulation | Hepcidin–ferroportin axis, ferritin iron storage | Hyperferritinemia linked to severity and mortality | Ferritin, hepcidin, serum iron | [102–104] |
| **Mn** | Potential viral restriction via STING pathway | Cofactor for MnSOD, supports type I IFN response | cGAS-STING activation, endosomal acidification, mitochondrial support | Lower in severe COVID-19; emerging role in long COVID | Total Mn, plasma Mn | [120] |
| **Se** | Inhibits viral mutagenesis and replication | Redox support via GPx, TrxR, favors type I IFN | Selenoproteins (GPx, SELENOP), NF-κB and IL-2 modulation | Severe deficiency in non-survivors; inverse correlation with CRP | Serum Se, SELENOP | [121] |
| **Co** | Experimental antiviral activity (non-physiological) | Methylation and GSH support, essential for IFN-γ and lymphocytes | Cobalamin cofactor: DNA regulation, TCR and differentiation | Elevated IMA and Co levels; correlated with inflammation | IMA (albumin-Co binding), serum Co) | [122] |
| **Ca** | Non-specific | Cofactor for TCR, inflammasome, cytokine signaling | Ca channel regulation and intercellular signaling | Hypocalcemia associated with disease severity | Total/ionized serum Ca | [123] |
| **Mg** | Non-specific | Enzymatic cofactor, regulates IL-2 and IFN-γ, supports endothelial barrier | Vitamin D activation, TCR modulation, inflammation regulation | Hypomagnesemia linked to mechanical ventilation and higher mortality | Total serum Mg | [124] |

*Abbreviations*: CRP, C-reactive protein; IFN, interferon; IL, interleukin; IMA, ischemia-modified albumin; PRR, pattern recognition receptor; TCR, T cell receptor.

An important and rarely addressed issue is the lack of analytical rigor and harmonization in studies of this magnitude. Many investigations fail to report critical aspects, such as matrix matching, calibration protocols, figures of merit, and sample preparation methods, particularly relevant for analytes prone to redox instability or protein binding. Without validated and reproducible workflows, especially for biologically labile species, the interpretation of data remains vulnerable to artifacts and bias. In addition, matrix-specific interferences and the absence of certified reference materials severely compromise the comparability of data across studies, limiting their translational value.

Curiously, an additional obstacle is the absence of a curated inorganic element database, analogous to those used in metabolomics or proteomics, which would enable researchers to contextualize measured concentrations across different sample types (e.g., serum, urine, plasma), different instrumentation and analytical conditions, patient statuses (e.g., disease severity, comorbidities), and demographic variables (e.g., age, sex, region). The development of such a platform would support the creation of predictive models linking metal status to clinical trajectories, enabling hypothesis-driven investigations rather than post hoc associations. It would also facilitate meta-analyses and the identification of trace element patterns associated with specific disease phenotypes or immune responses.

Within this fragmented landscape, some emerging trends offer promising directions. Elements such as copper and cobalt have shown relevant associations with COVID-19 severity across different independent cohorts' copper being elevated in serum [101] and suppressed in urine [13], while cobalt albumin binding activity decreases in parallel with inflammation and oxidative stress [115, 116]. These markers may serve as indirect indicators of systemic damage or redox imbalance, even in the absence of classical deficiency.

Looking ahead, one of the most urgent priorities is the investigation of long COVID, a condition characterized by persistent symptoms and multisystem dysfunction beyond the acute phase. Although the pathophysiology of long COVID remains unclear, recent hypotheses implicate immune dysregulation, mitochondrial dysfunction, and residual inflammation. However, longitudinal studies assessing biometal status over extended recovery periods are virtually nonexistent. The design of such studies presents logistical and methodological challenges, including participant retention, time-resolved biospecimen collection, and analytical consistency. Nonetheless, understanding how metal dynamics involved in long COVID could provide insight into chronic immune activation, post-viral fatigue, and neuroinflammation, as well as correlations with other post-COVID-19 pathologies, offers new biomarkers and therapeutic targets for a condition that continues to affect millions of people worldwide. One of the few studies addressing post-acute COVID-19 sequelae reported significantly elevated levels of copper, cobalt, and manganese, as well as an increased Cu/Zn ratio in patients with long COVID. Metal concentrations were found to correlate with markers of inflammation (CRP), disease duration, age, and body weight. Importantly, manganese emerged as a promising biomarker, showing 93.3% sensitivity and specificity for long COVID classification [125]. These findings underscore the value of metal profiling in post-viral syndromes and highlight the need for longitudinal, multicentric studies with rigorous analytical frameworks.

In summary, the field of biometal research in viral diseases is rich in biological plausibility, yet it faces difficulties due to methodological inconsistency and lack of integration. Overcoming these barriers will require interdisciplinary efforts, uniting clinicians, analytical chemists, and systems biologists to build validated and interoperable frameworks capable of supporting precision diagnostics and personalized interventions based on metal-dependent immune regulation.

## 4.6 CONCLUSIONS

In summary, understanding the interplay between SARS-CoV-2 and chemical elements is essential for fully elucidating the mechanisms underlying viral infection, immune response, and disease progression. While significant advances have been made in characterizing the virus at the molecular and immunological levels, the role of essential and non-essential metals remains an underrepresented, but crucial, component of COVID-19 research. By examining how these elements influence viral stability, entry, replication, and host defenses, as well as how metal dysregulation may contribute to severe disease and long-term complications, this chapter highlights an important yet often overlooked dimension of COVID-19 pathogenesis. Ultimately, integrating knowledge from virology, immunology, and bioinorganic chemistry not only deepens our scientific understanding but also opens new avenues for the development of diagnostics, targeted therapies, and preventive strategies that address the multifaceted nature of SARS-CoV-2 infection.

## REFERENCES

1. Zhu N et al. A novel coronavirus from patients with pneumonia in China, 2019. New Engl J Med. 2020;382:727–733.
2. V'kovski P, Kratzel A, Steiner S, Stalder H, Thiel V. Coronavirus biology and replication: Implications for SARS-CoV-2. Nature Rev Microbiol. 2021;19:155–170.
3. Read SA, Obeid S, Ahlenstiel C, Ahlenstiel G. The role of zinc in antiviral immunity. The role of zinc in antiviral immunity. Adv Nutr. 2019;10:696–710.
4. Prasad AS. Zinc in human health: Effect of zinc on immune cells. Mol Med. 2008;14: 353–357.
5. Chaturvedi UC, Shrivastava R. Interaction of viral proteins with metal ions: Role in maintaining the structure and functions of viruses. FEMS Immunol Med Microbiol. 2005;43(2):105–114.
6. Andreini C, Arnesano F, Rosato A. The zinc proteome of SARS-CoV-2. Metallomics. 2022;14(7):mfac047.
7. Osipiuk J, Azizi SA, Dvorkin S, Endres M, Jedrzejczak R, Jones KA, Kang S, Kathayat RS, Kim Y, Lisnyak VG, Maki SL, Nicolaescu V, Taylor CA, Tesar C, Zhang YA, Zhou Z, Randall G, Michalska K, Snyder SA, Dickinson BC, Joachimiak A. Structure of papain-like protease from SARS-CoV-2 and its complexes with non-covalent inhibitors. Nat Commun. 2021;12(1):743.
8. Tolbatov I, Storchi L, Marrone A. Structural reshaping of the zinc-finger domain of the SARS-CoV-2 nsp13 protein using bismuth(III) ions: A multilevel computational study. Inorg Chem. 2022;61(39):15664–15677.
9. Inniss NL, Rzhetskaya M, Ling-Hu T, Lorenzo-Redondo R, Bachta KE, Satchell KJF, Hultquist JF. Activity and inhibition of the SARS-CoV-2 omicron nsp13 R392C variant using RNA duplex unwinding assays. SLAS Discov. 2024;29(3):100145.

10. Maio N, Raza MK, Li Y, Zhang DL, Bollinger JM Jr, Krebs C, Rouault TA. An iron-sulfur cluster in the zinc-binding domain of the SARS-CoV-2 helicase modulates its RNA-binding and -unwinding activities. Proc Natl Acad Sci U S A. 2023;120(33):e2303860120.
11. Pelucelli A, Peana M, Orzeł B, Piasta K, Gumienna-Kontecka E, Medici S, Zoroddu MA. $Zn^{2+}$ and $Cu^{2+}$ interaction with the recognition interface of ACE2 for SARS-CoV-2 spike protein. Int J Mol Sci. 2023;24(11):9202.
12. Skalny AV, Timashev PS, Aschner M, Aaseth J, Chernova LN, Belyaev VE, Grabeklis AR, Notova SV, Lobinski R, Tsatsakis A, Svistunov AA, Fomin VV, Tinkov AA, Glybochko PV. Serum zinc, copper, and other biometals are associated with COVID-19 severity markers. Metabolites. 2021;11(4):244.
13. Leal KNDS, Santos da Silva AB, Fonseca EKB, Moreira OBO, de Lemos LM, Leal de Oliveira MA, Stewart AJ, Arruda MAZ. Metallomic analysis of urine from individuals with and without covid-19 infection reveals extensive alterations in metal homeostasis. J Trace Elem Med Biol. 2024;86:127557.
14. Wang Q, Su Z, Zhang J, Yan H, Zhang J. Unraveling the copper-death connection: Decoding COVID-19's immune landscape through advanced bioinformatics and machine learning approaches. Hum Vaccin Immunother. 2024;20(1):2310359.
15. Govind V, Bharadwaj S, Sai Ganesh MR, Vishnu J, Shankar KV, Shankar B, Rajesh R. Antiviral properties of copper and its alloys to inactivate covid-19 virus: A review. Biometals. 2021;34(6):1217–1235.
16. van Doremalen N, Bushmaker T, Morris DH, Holbrook MG, Gamble A, Williamson BN, Tamin A, Harcourt JL, Thornburg NJ, Gerber SI, Lloyd-Smith JO, de Wit E, Munster VJ. Aerosol and surface stability of SARS-CoV-2 as compared with SARS-CoV-1. N Engl J Med. 2020;382(16):1564–1567.
17. Viswanathan T, Misra A, Chan SH, Qi S, Dai N, Arya S, Martinez-Sobrido L, Gupta YK. A metal ion orients SARS-CoV-2 mRNA to ensure accurate 2'-o methylation of its first nucleotide. Nat Commun. 2021;12(1):3287.
18. Fan L, Zhu X, Zheng Y, Zhang W, Seidner DL, Ness R, Murff HJ, Yu C, Huang X, Shrubsole MJ, Hou L, Dai Q. Magnesium treatment on methylation changes of transmembrane serine protease 2 (TMPRSS2). Nutrition. 2021;89:111340.
19. Guerrero-Romero F, Micke O, Simental-Mendía LE, Rodríguez-Morán M, Vormann J, Iotti S, Banjanin N, Rosanoff A, Baniasadi S, Pourdowlat G, Nechifor M. Importance of magnesium status in COVID-19. Biology. 2023;12(5):735.
20. van Kempen TATG, Deixler E. SARS-CoV-2: Influence of phosphate and magnesium, moderated by vitamin D, on energy (ATP) metabolism and on severity of COVID-19. Am J Physiol Endocrinol Metab. 2021;320(1):E2–E6.
21. Eskander M, Razzaque MS. Can maintaining optimal magnesium balance reduce the disease severity of COVID-19 patients? Front Endocrinol. 2022;13:843152.
22. Bouhaddou M, Memon D, Meyer B, White KM, Rezelj VV, Correa Marrero M, Polacco BJ, Melnyk JE, Ulferts S, Kaake RM, Batra J, Richards AL, Stevenson E, Gordon DE, Rojc A, Obernier K, Fabius JM, Soucheray M, Miorin L, Moreno E, Koh C, Tran QD, Hardy A, Robinot R, Vallet T, Nilsson-Payant BE, Hernandez-Armenta C, Dunham A, Weigang S, Knerr J, Modak M, Quintero D, Zhou Y, Dugourd A, Valdeolivas A, Patil T, Li Q, Hüttenhain R, Cakir M, Muralidharan M, Kim M, Jang G, Tutuncuoglu B, Hiatt J, Guo JZ, Xu J, Bouhaddou S, Mathy CJP, Gaulton A, Manners EJ, Félix E, Shi Y, Goff M, Lim JK, McBride T, O'Neal MC, Cai Y, Chang JCJ, Broadhurst DJ, Klippsten S, De Wit E, Leach AR, Kortemme T, Shoichet B, Ott M, Saez-Rodriguez J, tenOever BR, Mullins RD, Fischer ER, Kochs G, Grosse R, García-Sastre A, Vignuzzi M, Johnson JR, Shokat KM, Swaney DL, Beltrao P, Krogan NJ. The global phosphorylation landscape of SARS-CoV-2 infection. Cell. 2020;182(3):685–712.e19.

23. Neufurth M, Wang X, Tolba E, Lieberwirth I, Wang S, Schröder HC, Müller WEG. The inorganic polymer, polyphosphate, blocks binding of SARS-CoV-2 spike protein to ACE2 receptor at physiological concentrations. Biochem Pharmacol. 2020;182:114215.
24. Chaubey GK, Dilawari R, Modanwal R, Talukdar S, Dhiman A, Raje CI, Raje M. Excess iron aggravates the severity of COVID-19 infection. Free Radic Biol Med. 2023;208:186–193.
25. Del Nonno F, Nardacci R, Colombo D, Visco-Comandini U, Cicalini S, Antinori A, Marchioni L, D'Offizi G, Piacentini M, Falasca L. Hepatic failure in COVID-19: Is iron overload the dangerous trigger? Cells. 2021;10(5):1103.
26. Carolin A, Frazer D, Yan K, Bishop CR, Tang B, Nguyen W, Helman SL, Horvat J, Larcher T, Rawle DJ, Suhrbier A. The effects of iron deficient and high iron diets on SARS-CoV-2 lung infection and disease. Front Microbiol. 2024;15:1441495.
27. Madu AJ, Ughasoro MD. Anaemia of chronic disease: An in-depth review. Med Princ Pract. 2017;26(1):1–9.
28. Wrighting DM, Andrews NC. Interleukin-6 induces hepcidin expression through STAT3. Blood. 2006;108(9):3204–9.
29. Ganz T, Nemeth E. Iron sequestration and anemia of inflammation. Semin Hematol. 2009;46(4):387–93.
30. Gupta Y, Maciorowski D, Medernach B, Becker DP, Durvasula R, Libertin CR, Kempaiah P. Iron dysregulation in COVID-19 and reciprocal evolution of SARS-CoV-2: Natura nihil frustra facit. J Cell Biochem. 2022;123(3):601–619.
31. Anghel L, Manole C, Nechita A, Tatu AL, Ştefănescu BI, Nechita L, Buşilă C, Zainea P, Baroiu L, Muşat CL. Calcium, phosphorus and magnesium abnormalities associated with COVID-19 infection, and beyond. Biomedicines. 2023;11(9):2362.
32. Al-Fartusie FS, Kader SI, Mohammed SJ, Farhan MN, Mahmood FM, Algaber AA. A comparative study of serum Zn, Cu, Mg, Mn, Cr, and Fe levels and their association with the vulnerability of Iraqi COVID-19 patients. J Trace Elem Med Biol. 2023;79:127242.
33. Sahihi M, Faraudo J. Computer simulation of the interaction between SARS-CoV-2 spike protein and the surface of coinage metals. Langmuir. 2022;38(48):14673–85.
34. Zhou P, Yang X-L, Wang X-G, Hu B, Zhang L, Zhang W, Si H-R, Zhu Y, Li B, Huang C-L, et al. A pneumonia outbreak associated with a new coronavirus of probable bat origin. Nature. 2020;579:270–273. doi: 10.1038/s41586-020-2012-7.
35. Donoghue M et al. A novel angiotensin-converting enzyme-related carboxypeptidase (ACE2) cinonverts angiotensin I to angiotensin 1–9. Circ Res. 2000;87:E1–E9.
36. Vickers C, Hales P, Kaushik V, Dick L, Gavin J, Tang J, Godbout K, Parsons T, Baronas E, Hsieh F, Acton S, Patane M, Nichols A, Tummino P. Hydrolysis of biological peptides by human angiotensin-converting enzyme-related carboxypeptidase. J Biol Chem. 2002;277(17):14838–43. doi: 10.1074/jbc.M200581200.
37 in. Towler P, Staker B, Prasad SG, Menon S, Tang J, Parsons T, Ryan D, Fisher M, Williams D, Dales NA et al. ACE2 X-ray structures reveal a large hinge-bending motion important for inhibitor binding and catalysis. J Biol Chem. 2004;279:17996–18007.
38. Jothimani D, Kailasam E, Danielraj S, Nallathambi B, Ramachandran H, Sekar P, Manoharan S, Ramani V, Narasimhan G, Kaliamoorthy I et al. COVID-19: Poor outcomes in patients with zinc deficiency. Int J Infect Dis. 2020;100:343–349.
39. Jackson CB, Farzan M, Chen B et al. Mechanisms of SARS-CoV-2 entry into cells. Nat Rev Mol Cell Biol. 2022;23:3–20. https://doi.org/10.1038/s41580-021-00418-x
40. Walls AC, Tortorici MA, Frenz B. Glycan shield and epitope masking of a coronavirus spike protein observed by cryo-electron microscopy. Nat Struct Mol Biol. 2016;23:899–905.
41. Watanabe Y, Allen JD, Wrapp D, McLellan JS, Crispin M. Site-specific glycan analysis of the SARS-CoV-2 spike. Science. 2020;369:330–333.

42. Tortorici MA, Veesler D. Structural insights into coronavirus entry. Adv Virus Res. 2019;105:93–116.
43. Bosch BJ, van der Zee R, de Haan CA. The coronavirus spike protein is a class I virus fusion protein: Structural and functional characterization of the fusion core complex. J Virol. 2003;77:8801–8811.
44. Bangaru S et al. Structural analysis of full-length SARS-CoV-2 spike protein from an advanced vaccine candidate. Science. 2020;370:1089–1094.
45. Gui M, Song W, Zhou H. Cryo-electron microscopy structures of the SARS-CoV spike glycoprotein reveal a prerequisite conformational state for receptor binding. Cell Res. 2017; 27:119–129.
46. Li F, Li W, Farzan M, Harrison SC. Structure of SARS coronavirus spike receptor-binding domain complexed with receptor. Science. 2005;309:1864–1868.
47. Belouzard S, Chu VC, Whittaker GR. Activation of the SARS coronavirus spike protein via sequential proteolytic cleavage at two distinct sites. Proc Natl Acad Sci U S A. 2009;106:5871–5876.
48. Szabo R, Bugge TH. Membrane-anchored serine proteases in vertebrate cell and developmental biology. Annu Rev Cell Dev Biol. 2011;27:213–235.
49. Shaik, M. M. et al. Structural basis of coreceptor recognition by HIV-1 envelope spike. Nature 565, 318–323 (2019).
50. Glende J, Schwegmann-Wessels C, Al-Falah M, Pfefferle S, Qu X, Deng H, Drosten C, Naim HY, Herrler G. Importance of cholesterol-rich membrane microdomains in the interaction of the S protein of SARS-coronavirus with the cellular receptor angiotensin-converting enzyme 2. Virology. 2008;381:215–221. doi: 10.1016/J.VIROL.2008.08.026.
51. Straus MR, Tang T, Lai AL, Flegel A, Bidon M, Freed JH, Daniel S, Whittaker GR. $Ca^{2+}$ ions promote fusion of Middle East respiratory syndrome coronavirus with host cells and increase infectivity. J Virol. 2020;94(13):e00426–20. doi: 10.1128/jvi.00426-20.
52. Wang D, Jiang A, Feng J, Li G, Guo D, Sajid M, Wu K, Zhang Q, Ponty Y, Will S, et al. The SARS-CoV-2 subgenome landscape and its novel regulatory features. Mol Cell. 2021;81(10):2135–47.
53. Kim D, Lee JY, Yang JS, Kim JW, Kim VN, Chang H. The architecture of SARS-CoV-2 transcriptome. Cell. 2020;181(4):914–21.
54. Tvarogova J, Madhugiri R, Bylapudi G, Ferguson LJ, Karl N, Ziebuhr J. Identification and characterization of a human coronavirus 229E nonstructural protein 8-associated RNA 3′-terminal adenylyltransferase activity. J Virol. 2019;93(12):e00291-19.
55. Roingeard P, Eymieux S, Burlaud-Gaillard J et al. The double-membrane vesicle (DMV): A virus-induced organelle dedicated to the replication of SARS-CoV-2 and other positive-sense single-stranded RNA viruses. Cell Mol Life Sci. 2022;79:425. https://doi.org/10.1007/s00018-022-04469-x
56. Cortese M, Lee JY, Cerikan B, Neufeldt CJ, Oorschot VMJ, Köhrer S, Hennies J, Schieber NL, Ronchi P, Mizzon G, Romero-Brey I, Santarella-Mellwig R, Schorb M, Boermel M, Mocaer K, Beckwith MS, Templin RM, Gross V, Pape C, Tischer C, Frankish J, Horvat NK, Laketa V, Stanifer M, Boulant S, Ruggieri A, Chatel-Chaix L, Schwab Y, Bartenschlager R. Integrative imaging reveals SARS-CoV-2-induced reshaping of subcellular morphologies. Cell Host Microbe. 2020;28:853–866.e5.
57. Ricciardi S et al. The role of NSP6 in the biogenesis of the SARS-CoV-2 replication organelle. Nature. 2022;606:761–768.
58. Dias SSG, Soares VC, Ferreira AC, Sacramento CQ, Fintelman-Rodrigues N, Temerozo JR, Teixeira L, Nunes da Silva MA, Barreto E, Mattos M, de Freitas CS, Azevedo-Quintanilha IG, Manso PPA, Miranda MD, Siqueira MM, Hottz ED, Pão CRR, Bou-Habib DC, Barreto-Vieira DF, Bozza FA, Souza TML, Bozza PT. Lipid droplets fuel SARS-CoV-2 replication and production of inflammatory mediators. PLoS Pathog. 2020;16:e1009127.

59. Schubert K et al. SARS-CoV-2 Nsp1 binds the ribosomal mRNA channel to inhibit translation. Nat Struct Mol Biol. 2020;27:959–966.
60. Subissi L, Posthuma CC, Collet A, Zevenhoven-Dobbe JC, Gorbalenya AE, Decroly E, Snijder EJ, Canard B, Imbert I. One severe acute respiratory syndrome coronavirus protein complex integrates processive RNA polymerase and exonuclease activities. Proc Natl Acad Sci U S A. 2014;111(37):E3900–E3s909.
61. Snijder EJ, Decroly E, Ziebuhr J. The nonstructural proteins directing coronavirus RNA synthesis and processing. Adv Virus Res. 2016;96:59–126.
62. Gao Y et al. Structure of the RNA-dependent RNA polymerase from COVID-19 virus. Science. 2020;368:779–782.
63. Chen J, Malone B, Llewellyn E, Grasso M, Shelton PM, Olinares PDB, Maruthi K, Eng ET, Vatandaslar H, Chait BT et al. Structural basis for helicase-polymerase coupling in the SARS-CoV-2 replication-transcription complex. Cell. 2020;182:1560–1573.
64. Lin S et al. Crystal structure of SARS-CoV-2 nsp10 bound to nsp14–ExoN domain reveals an exoribonuclease with both structural and functional integrity. Nucleic Acids Res. 2021;49:5382–5392.
65. Sargsyan K, Lin CC, Chen T, Grauffel C, Chen YP, Yang WZ, et al. Multi-targeting of functional cysteines in multiple conserved SARS-CoV-2 domains by clinically safe Zn-ejectors Chem Sci. 2020;11(36):9904–9909. doi:10.1039/d0sc02646h.
66. Velthuis AJ, van den Worm SH, Sims AC, Baric RS, Snijder EJ, van Hemert MJ. Zn(2+) inhibits coronavirus and arterivirus RNA polymerase activity in vitro and zinc ionophores block the replication of these viruses in cell culture. PLoS Pathog. 2010 Nov 4;6(11):e1001176. doi: 10.1371/journal.ppat.1001176.
67. Cubuk J et al. The SARS-CoV-2 nucleocapsid protein is dynamic, disordered, and phase separates with RNA. Nat Commun. 2021;12:1936.
68. Yan W, Zheng Y, Zeng X et al. Structural biology of SARS-CoV-2: Open the door for novel therapies. Sig Transduct Target Ther. 2022;7:26. https://doi.org/10.1038/s41392-022-00884-5
69. Leroy H, Han M, Woottum M, Bracq L, Bouchet J, Xie M, Benichou S. Virus-mediated cell-cell fusion. Int J Mol Sci. 2020;21:9644. https://doi.org/10.3390/ijms21249644.
70. Astuti I, Ysrafil. Severe acute respiratory syndrome coronavirus 2 (SARS-CoV-2): An overview of viral structure and host response. Diabetes Metab Syndr. 2020;14, 407–412.
71. Hartenian E, Nandakumar D, Lari A, Ly M, Tucker JM, Glaunsinger BA. The molecular virology of coronaviruses. J Biol Chem. 2020;295(37):12910–34.
72. Neuman B et al. A structural analysis of M protein in coronavirus assembly and morphology. J Struct Biol. 2011;174:11–22.
73. Arya R, Kumari S, Pandey B, Mistry H, Bihani SC, Das A, Prashar V, Gupta GD, Panicker L, Kumar M. Structural insights into SARS-CoV-2 proteins. J Mol Biol. 2021;433(2):166725.
74. Nieto-Torres JL et al. Severe acute respiratory syndrome coronavirus e protein transports calcium ions and activates the NLRP3 inflammasome. Virology. 2015;485:330–39.
75. Schoeman D, Fielding B. Coronavirus envelope protein: Current knowledge. Virol J. 2019;16:69.
76. Chai J et al. Structural basis for SARS-CoV-2 envelope protein recognition of human cell junction protein PALS1. Nat Commun. 2021;12:3433.
77. Cheng N, Liu M, Li W, Sun B, Liu D, Wang G, Shi J, Li L. Protein post-translational modification in SARS-CoV-2 and host interaction. Front Immunol. 2023 Jan 13;13:1068449. doi: 10.3389/fimmu.2022.1068449.
78. Steiner S, Kratzel A, Barut GT et al. SARS-CoV-2 biology and host interactions. Nat Rev Microbiol. 2024;22:206–225. https://doi.org/10.1038/s41579-023-01003-z.

79. He YX et al. Mapping of antigenic sites on the nucleocapsid protein of the severe acute respiratory syndrome coronavirus. J Clin Microbiol. 2004;42:5309–5314.
80. Lu X, Pan JA, Tao J, Guo D. SARS-CoV nucleocapsid protein antagonizes IFN-β response by targeting initial step of IFN-β induction pathway, and its C-terminal region is critical for the antagonism. Virus Genes. 2011;42:37–45.
81. Chen RE et al. Resistance of SARS-CoV-2 variants to neutralization by monoclonal and serum-derived polyclonal antibodies. Nat Med. 2021;27:717–726.
82. Liu J et al. The membrane protein of severe acute respiratory syndrome coronavirus acts as a dominant immunogen revealed by a clustering region of novel functionally and structurally defined cytotoxic T-lymphocyte epitopes. J Infect Dis. 2010;202:1171–80.
83. Thoms M et al. Structural basis for translational shutdown and immune evasion by the Nsp1 protein of SARS-CoV-2. Science. 2020;369:1249–55.
84. Gao B et al. Inhibition of anti-viral stress granule formation by coronavirus endoribo nuclease nsp15 ensures efficient virus replication. PLoS Pathog. 2021;17:e1008690.
85. Arshad N et al. SARS-CoV-2 accessory proteins ORF7a and ORF3a use distinct mechanisms to down-regulate MHC-I surface expression. Proc Natl Acad Sci U S A. 2023;120:e2208525120.
86. Lei X et al. Activation and evasion of type I interferon responses by SARS-CoV-2. Nat Commun. 2020;11:3812.
87. Khachfe HH, Chahrour M, Sammouri J, Salhab H, Makki BE, Fares M. An epidemiological study on COVID-19: A rapidly spreading disease. Cureus. 2020;12(3) doi: 10.7759/cureus.7313
88. Wang R, Lan C, Benlagha K, Camara NOS, Miller H, Kubo M, Heegaard S, Lee P, Yang L, Forsman H, Li X, Zhai Z, Liu C. The interaction of innate immune and adaptive immune system. MedComm (2020). 2024 Sep 15;5(10):e714. doi: 10.1002/mco2.714. PMID: 39286776; PMCID: PMC11401974.
89. Hadjadj J, Yatim N, Barnabei L, Corneau A, Boussier J, Smith N, Péré H, Charbit B, Bondet V, Chenevier-Gobeaux C, Breillat P, Carlier N, Gauzit R, Morbieu C, Pène F, Marin N, Roche N, Szwebel T-A, Merkling SH, Treluyer J-M. Impaired type I interferon activity and inflammatory responses in severe COVID-19 patients. Science. 2020;369(6504):718–724. https://doi.org/10.1126/science.abc6027.
90. Minkoff JM, tenOever B. Innate immune evasion strategies of SARS-CoV-2. Nat Rev Microbiol. 2023;21(3):1–17. https://doi.org/10.1038/s41579-022-00839-1.
91. Sherwood ER, Toliver-Kinsky T. Mechanisms of the inflammatory response. Best Pract Res Clin Anaesthesiol. 2004;18(3):385–405. https://doi.org/10.1016/j.bpa.2003.12.002.
92. Lineburg KE, Grant EJ, Swaminathan S, Chatzileontiadou DSM, Szeto C, Sloane H, Panikkar A, Raju J, Crooks P, Rehan S, Nguyen AT, Lekieffre L, Neller MA, Tong ZWM, Jayasinghe D, Chew KY, Lobos CA, Halim H, Burrows JM, Riboldi-Tunnicliffe A, Chen W, D'Orsogna L, Khanna R, Short KR, Smith C, Gras S. $CD8^+$ T cells specific for an immunodominant SARS-CoV-2 nucleocapsid epitope cross-react with selective seasonal coronaviruses. Immunity. 2021 May 11;54(5):1055–1065.e5. doi: 10.1016/j.immuni.2021.04.006. Epub 2021 Apr 13. PMID: 33945786; PMCID: PMC8043652.
93. Khan N, Chen X, Geiger JD. Role of endolysosomes in severe acute respiratory syndrome coronavirus-2 infection and coronavirus disease 2019 pathogenesis: Implications for potential treatments. Front Pharmacol. 2020;11. https://doi.org/10.3389/fphar.2020.595888.
94. Ackermann M, Verleden SE, Kuehnel M et al. Pulmonary vascular endothelialitis, thrombosis, and angiogenesis in COVID-19. N Engl J Med. 2020;383:120–128.
95. Maywald M, Rink L. Zinc signals and immunity. Int J Mol Sci. 2022;23(4) https://doi.org/10.3390/ijms18102222.

96. Wessels I, Haase H, Engelhardt G, Rink L, Uciechowski P. Zinc deficiency induces production of the proinflammatory cytokines IL-1β and TNFα in promyeloid cells via epigenetic and redox-dependent mechanisms. Front Immunol. 2018;9:1957. doi: 10.3389/fimmu.2018.01947.
97. Reinhold D, Brocke S. The differential roles of zinc in immune responses and their potential implications in antiviral immunity against SARS-CoV-2. Clin Nutr. 2021 Feb;40(2):652. doi: 10.1016/j.clnu.2020.12.005. Epub 2020 Dec 10.
98. Lu J, Liu X, Li X, Li H, Shi L, Xia X, He BL, Meyer TF, Li X, Sun H, Yang X. Copper regulates the host innate immune response against bacterial infection via activation of ALPK1 kinase. Proc Natl Acad Sci U S A. 2024 Jan 23;121(4):e2311630121. doi: 10.1073/pnas.2311630121. Epub 2024 Jan 17. PMID: 38232278; PMCID: PMC10823219.
99. Dharmalingam K, Birdi A, Tomo S *et al.* Trace elements as immunoregulators in SARS-CoV-2 and other viral infections. Ind J Clin Biochem. 2021;36:416–426. https://doi.org/10.1007/s12291-021-00961-6
100. Zeng HL, Zhang B, Wang X, Yang Q, Cheng L. Urinary trace elements in association with disease severity and outcome in patients with COVID-19. Environ Res. 2021;194:110670. doi: 10.1016/j.envres.2020.110670. Epub 2020 Dec 30. PMID: 33387537; PMCID: PMC7772999.
101. Ivanova ID, Pal A, Simonelli I, Atanasova B, Ventriglia M, Rongioletti M, Squitti R. Evaluation of zinc, copper, and Cu:Zn ratio in serum, and their implications in the course of COVID-19. J Trace Elem Med Biol. 2022;71:126944. doi: 10.1016/j.jtemb.2022.126944. Epub 2022 Feb 8. PMID: 35168023; PMCID: PMC8820953.
102. Nairz M, Schroll A, Sonnweber T, Weiss G. The struggle for iron: A metal at the host-pathogen interface. Cell Microbiol. 2010 Dec;12(12):1691–702. doi: 10.1111/j.1462-5822.2010.01529.x. Epub 2010 Oct 21. PMID: 20964797.
103. Martinez S, Hausinger RP. Catalytic mechanisms of fe(II)- and 2-oxoglutarate-dependent oxygenases. J Biol Chem. 2015 Aug 21;290(34):20702–20711. doi: 10.1074/jbc.R115.648691. Epub 2015 Jul 7. PMID: 26152721; PMCID: PMC4543632.
104. Ramonfaur D, Aguirre-García GM, Diaz-Garza CA, Torre-Amione G, Sanchez-Nava VM, Lara-Medrano R, Ramírez-Elizondo MT, Esparza-Sandoval AC, Ortega-Hernández FJ, Martínez-Reséndez MF. Early increase of serum ferritin among COVID-19 patients is associated with need of invasive mechanical ventilation and with in-hospital death. Infect Dis (Lond). 2022 Nov;54(11):810–818. doi: 10.1080/23744235.2022.2101691. Epub 2022 Jul 19. PMID: 35854671.
105. Sun S, Xu Y, Qiu M, Jiang S, Cao Q, Luo J, Zhang T, Chen N, Zheng W, Meurens F et al. Manganese mediates its antiviral functions in a cGAS-STING pathway independent manner. Viruses. 2023;15:646. https://doi.org/10.3390/v15030646.
106. Zeng HL, Yang Q, Yuan P, Wang X, Cheng L. Associations of essential and toxic metals/metalloids in whole blood with both disease severity and mortality in patients with COVID-19. FASEB J. 2021 Mar;35(3):e21392. doi: 10.1096/fj.202002346RR. PMID: 33577131; PMCID: PMC7995111.
107. Šķesters A, Lece A, Kustovs D, Zolovs M. Selenium status and oxidative stress in SARS-CoV-2 patients. Medicina. 2023;59:527. https://doi.org/10.3390/medicina59030527.
108. Wang W, Jiang QL, Xu Q, Zeng Y, Jiang R, Jiang J. Selenium regulates t cell differentiation in experimental autoimmune thyroiditis in mice. Int Immunopharmacol. 2023 Nov;124(Pt B):110993. doi: 10.1016/j.intimp.2023.110993. Epub 2023 Sep 28. PMID: 37776772.
109. Moghaddam A, Heller RA, Sun Q, Seelig J, Cherkezov A, Seibert L, Hackler J, Seemann P, Diegmann J, Pilz M, Bachmann M, Minich WB, Schomburg L. Selenium deficiency is associated with mortality risk from COVID-19. Nutrients. 2020 Jul 16;12(7):2098. doi: 10.3390/nu12072098. PMID: 32708526; PMCID: PMC7400921.
110. Im JH, Je YS, Baek J, Chung MH, Kwon HY, Lee JS. Nutritional status of patients with COVID-19. Int J Infect Dis. 2020;100:390–393. doi: 10.1016/j.ijid.2020.08.018. Epub 2020 Aug 11. PMID: 32795605; PMCID: PMC7418699.

111. Okamoto S, Eltis LD. The biological occurrence and trafficking of cobalt. Metallomics. 2011;3,(10):963–970. https://doi.org/10.1039/c1mt00056j.
112. Devcic J, Dussol M, Collin-Faure V, Pérard J, Fenel D, Schoehn G, Carrière M, Rabilloud T, Dalzon B. Immediate and sustained effects of cobalt and zinc-containing pigments on macrophages. Front Immunol. 2022 Jul 19;13:865239. doi: 10.3389/fimmu.2022.865239. PMID: 35928812; PMCID: PMC9343594.
113. Díez-Tercero L, Delgado LM, Bosch-Rué E et al. Evaluation of the immunomodulatory effects of cobalt, copper and magnesium ions in a pro inflammatory environment. Sci Rep. 2021;11:11707. https://doi.org/10.1038/s41598-021-91070-0.
114. Amerikanou C, Kleftaki SA, Karavoltsos S, Tagkouli D, Sakellari A, Valsamidou E, Gioxari A, Kalogeropoulos N, Kaliora AC. Vanadium, cobalt, zinc, and rubidium are associated with markers of inflammation and oxidative stress in a Greek population with obesity. Front Endocrinol (Lausanne). 2023 Nov 22;14:1265310. doi: 10.3389/fendo.2023.1265310. PMID: 38075040; PMCID: P.MC10703041.m
115. Chatterjee M, Sinha S, Mukhopadhyay M, Sinha S, Chakraborty I. Assessment of cobalt binding activity of albumin in COVID-19 patients and its correlation with CRP. Indian J Public Health Res Dev. 2024;15(2):213–6.
116. Wozniak H, Le Terrier C, Primmaz S, Suh N, Lenglet S, Thomas A, Vuilleumier N, Pagano S, de Watteville A, Stringhini S, Guessous I, Quintard H, Heidegger CP, Pugin J. Association of trace element levels with outcomes in critically ill COVID-19 patients. Nutrients. 2023 Jul 26;15(15):3308. doi: 10.3390/nu15153308. PMID: 37571249; PMCID: PMC10421129.
117. Samakai E, Go C, Soboloff J. Defining the roles of $Ca^{2+}$ signals during T cell activation. In: Soboloff J, Kappes DJ, editors. Signaling Mechanisms Regulating T Cell Diversity and Function. Boca Raton (FL): CRC Press/Taylor & Francis; 2018. Chapter 10. doi: 10.1201/9781315371689-10.
118. Liu J, Han P, Wu J, Gong J, Tian D. Prevalence and predictive value of hypocalcemia in severe COVID-19 patients. J Infect Public Health. 2020 Sep;13(9):1224–1228. doi: 10.1016/j.jiph.2020.05.029. Epub 2020 Jun 22. PMID: 32622796; PMCID: PMC7306733.
119. Sadikan MZ, Lambuk L, Hairi HA, Mohamud R. Molecular impact of magnesium-mediated immune regulation in diseases. Scientifica (Cairo). 2025; Sep 8;2025:4211238. doi: 10.1155/sci5/4211238. PMID: 40959529; PMCID: PMC12436024.
120. Li C, Zhou HM. The role of manganese superoxide dismutase in inflammation defense. Enzyme Res. 2011;2011:387176. doi: 10.4061/2011/387176. Epub 2011 Oct 3. PMID: 21977313; PMCID: PMC3185262.
121. Zoidis E, Seremelis I, Kontopoulos N, Danezis GP. Selenium-dependent antioxidant enzymes: Actions and properties of selenoproteins. Antioxidants (Basel). 2018 May 14;7(5):66. doi: 10.3390/antiox7050066. PMID: 29758013; PMCID: PMC5981252.
122. Genchi G, Lauria G, Catalano A, Carocci A, Sinicropi MS. Prevalence of cobalt in the environment and its role in biological processes. Biology (Basel). 2023 Oct 16;12(10):1335. doi: 10.3390/biology12101335. PMID: 37887045; PMCID: PMC10604320.
123. Racioppi L, Means AR. Calcium/calmodulin-dependent kinase IV in immune and inflammatory responses: Novel routes for an ancient traveller. Trends Immunol. 2008 Dec;29(12):600–7. doi: 10.1016/j.it.2008.08.005. Epub 2008 Oct 17. PMID: 18930438.
124. Matek Sarić M, Sorić T, Juko Kasap Ž, Lisica Šikić N, Mavar M, Andruškienė J, Sarić A. Magnesium: Health effects, deficiency burden, and future public health directions. Nutrients. 2025 Nov 20;17(22):3626. doi: 10.3390/nu17223626. PMID: 41305676; PMCID: PMC12655508.
125. Masoodi WTA, Radhi SW, Al-Hakeim HK, Abdalsada HK. Trace element imbalance as a possible factor in long-COVID pathophysiology: Links to disease duration and inflammation. Am J Med Sci. 2025;371:259–266. https://doi.org/10.1016/j.amjms.2025.10.013.

# 5 The Role of ICP-MS for Determining Chemical Species in Biofluids

## *Selenoproteins and Selenometabolites*

*S. Sánchez-Espirilla, B. Callejón-Leblic, A. Arias-Borrego, R.F. Vélez-Pérez, and T. García-Barrera*

## ABBREVIATIONS

The abbreviations used throughout this chapter are listed below:

| | |
|---|---|
| **AD** | Alzheimer's disease |
| **AE** | anion exchange |
| **AEC** | anion-exchange chromatography |
| **AF** | affinity chromatography |
| **AFS** | atomic fluorescence spectroscopy |
| **CE** | capillary electrophoresis |
| **COPD** | chronic obstructive pulmonary disease |
| **CSF** | cerebrospinal fluid |
| **DRC** | dynamic reaction cell |
| **eGPx** | extracellular glutathione peroxidase |
| **ETAAS** | electrothermal atomic absorption spectroscopy |
| **FIA** | flow injection analysis |
| **FPLC** | fast protein liquid chromatography |
| **HGAAS** | hydride generation atomic absorption spectrometry |
| **HG** | hydride generation |
| **HPLC** | high-performance liquid chromatography |
| **ICP-MS** | inductively coupled plasma mass spectrometry |
| **IDA** | isotopic dilution analysis |
| **IEC** | ion-exchange chromatography |
| **IPRP** | ion-pair reversed-phase |
| **LA** | laser ablation |
| **LC** | lung cancer |
| **MCI** | mild cognitive impairment |


DOI: 10.1201/9781003632092-6

| | |
|---|---|
| **MAF** | micro-affinity chromatography |
| **MS/MS** | tandem mass spectrometry |
| **NAA** | neutron activation analysis |
| **nanoLC-MS/MS** | nano-liquid chromatography–tandem mass spectrometry |
| **PD** | Parkinson's disease |
| **PVDF** | polyvinylidene fluoride |
| **RDA** | recommended dietary allowance |
| **RP** | reversed phase |
| **SAX** | strong anion exchange |
| **SDS-PAGE** | sodium dodecyl sulfate polyacrylamide gel electrophoresis |
| **SEC** | size-exclusion chromatography |
| **Se** | selenium |
| **SeAlb** | selenoalbumin |
| **SeCys** | selenocysteine |
| **SeCys2** | selenocystine |
| **SeEt** | selenoethionine |
| **SeMet** | selenomethionine |
| **SeSug** | selenosugars |
| **SEN** | selenoneine |
| **SECA** | selenocystamine |
| **SeUr** | selenourea |
| **SelenoP** | selenoprotein P |
| **SPE** | solid-phase extraction |
| **TMSe** | trimethylselenonium |
| **TrxR** | thioredoxin reductase |
| **UL** | tolerable upper intake level |
| **UV/Vis** | ultraviolet–visible spectroscopy |

## 5.1 INTRODUCTION

Selenium (Se) belongs to the same chemical group as sulfur, which it can replace in several molecules, such as in the 21st amino acid selenocysteine (SeCys), the most important biologically active form of Se [1]. Se is a micronutrient with a very narrow range of recommended dietary allowance (RDA) of 55–70 μg/day [2]. The tolerable upper limit (UL) is 400 μg Se/day which is eight-fold higher than the requirement of the body. Se is present in different forms in the environment (e.g., Se (0), Se (-II), Se (IV), Se (VI), $SeO_2$, $SeO_3$, dimethyl selenium, and dimethyldiselenium) [3]; in food (e.g., Se (IV), Se (VI), selenomethionine (SeMet), SeCys, methylselenocysteine (MetSeCys)); and in the body (selenometabolites and selenoproteins) [4]. Each chemical form of Se has different bioavailability, bioactivity, biological role, and a range of essential/toxic concentrations [5].

In humans, 25 selenoprotein genes with diverse biological functions have been identified; 24 are also present in rodents [6]. Among these, only a few selenoproteins (or selenoprotein genes) have been identified in biofluids such as serum and plasma [7], cerebrospinal fluid (CSF) [8, 9], or human milk [10, 11]. The selenoprotein that accounts for the highest amount of Se is selenoprotein P (SELENOP; more than 50%

of total Se in human plasma), which serves as the main Se transport protein [12–14], followed by extracellular glutathione peroxidase (eGPx; ~15% in human plasma), a key antioxidant enzyme [15]. Small amounts of other selenoproteins can also be detected in biofluids. Among them, thioredoxin reductase (TrxR) is a redox-active enzyme involved in multiple cellular processes, including growth regulation, apoptosis, and cytoprotection [16]. On the other hand, selenoalbumin (SeAlb; ~15–20% in human plasma) is considered a Se-containing protein since Se is introduced as a form of selenomethionyl residues instead of as selenocysteinyl residues during polypeptide chain biosynthesis. However, SeAlb has a key role in the transport and storage of Se in the body [17].

Similarly, in other mammals, 24–26 selenoproteins have been described, reflecting a highly conserved selenoproteome across species [18]. In birds also, the number is comparable, with 23–25 selenoproteins identified to date [19].

Regarding the methodology, heteroatom-tagged proteomics has been proven to be an excellent analytical method for screening essential heteroelements (elements other than C, H, N, O, or F) in biological samples and for the absolute quantification of a known panel of targeted proteins (e.g., selenoproteins). This approach is based on the use of a sensitive and selective atomic detector, such as inductively coupled plasma mass spectrometry (ICP-MS) that uses the element as a "tag" to detect the molecules after their chromatographic or electrophoretic separation [20]. The absolute quantification of the heteroatom-containing molecules can be performed by isotope dilution analysis (IDA) which overcomes instrumental drift problems and matrix effects, and simplifies the laboratory work since preconcentration or dilution factors do not need to be taken into account [21]. Figure 5.1 presents a schematic workflow of IDA applied to heteroatom proteomics for the accurate quantification of selenoproteins in biofluids.

In IDA, quantification involves the addition of known amounts of an isotopically enriched element to the sample. After mixing with the sample, the isotopic

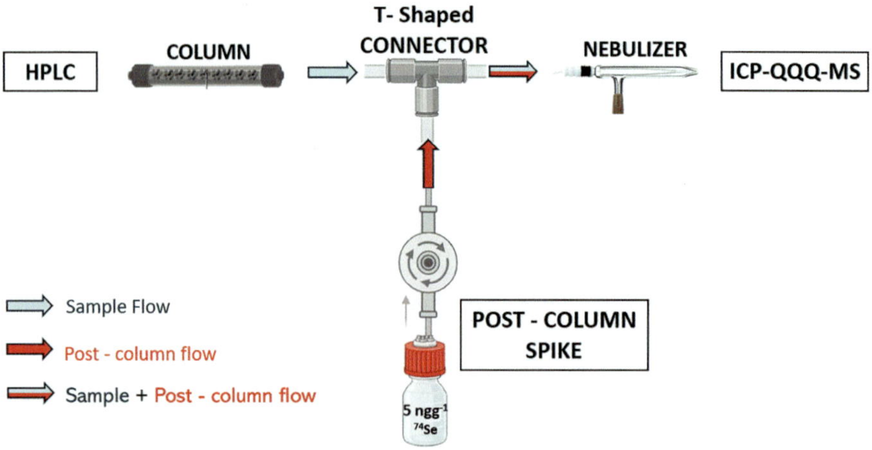


**FIGURE 5.1** Schematic diagram of the T-shaped connector used for isotopic dilution analysis by inductively coupled plasma triple quadrupole mass spectrometry (ICP-QQQ-MS).

enrichment of the standard is diluted and the isotopic ratios are measured by ICP-MS, and then these are used to calculate the concentration. Thus, this specific quantification method of internal standardization uses signal ratios instead of signal intensity used in traditional methods [21]. The concentration of selenoproteins can be calculated from Se concentration determined in the chromatographic peak since the number of atoms in the selenoproteins is known. For example, the human SELENOP contains ten Se atoms (10 SeCys, $C_3H_7NO_2Se$) [22]. An alternative method could be selenotranscriptomics which is very sensitive to determining transcripts, but there is no direct connection between the transcript and the protein. Western blot analysis allows only the relative quantification of selenoprotein and requires specific antibodies for each protein, but it provides information about the molecular weight. Finally, ultraviolet–visible spectroscopy (UV/vis) allows determining the enzymatic activity and the quantification of total protein, but not all selenoproteins are enzymes [23].

The workflow for heteroatom-tagged proteomics consists of several steps which usually include: (i) extraction of selenoproteins and selenometabolites from food; (ii) separating the proteins by using one or more orthogonal chromatographic steps (e.g. size-exclusion chromatography [SEC] and affinity chromatography [AF]) or by capillary electrophoresis (CE); (iii) detection of selenoproteins and selenometabolites by ICP-MS; and (iv) absolute quantification of selenoproteins and selenometabolites by IDA. Several of these methodologies allow the determination of the concentration of total selenometabolites, but they can also be determined independently in biofluids by coupling chromatography or CE to ICP-MS.

Herein, we review the analytical approaches for selenometabolites and selenoproteins in biofluids based on ICP-MS detection, as well as their advantages and pitfalls.

## 5.2 SPECIATION OF SELENOPROTEINS IN BIOFLUIDS

Specific characterization of selenoproteins in biofluids is a central topic in modern bioanalytical chemistry. Because these proteins exhibit structural complexity and occur at relatively low concentrations, speciation approaches must offer both high selectivity and strong analytical sensitivity.

The main strategies developed for identifying and quantifying selenoproteins in diverse biological matrices rely on high-performance liquid chromatography (HPLC), including SEC [24], ion-exchange chromatography (IEC) [25], and AF [25–28], as well as micro-affinity chromatography (MAF) [26] or combining SEC with AF [29–33].

On the other hand, electrophoretic methods, although less sensitive than chromatographic techniques, provide a straightforward and widely used approach for separating proteins by their size. These methods can be combined with radioisotope labeling or staining to detect selenoproteins and have also been proposed for their determination in biofluids from different organisms [34–38]. In this context, the most common electrophoretic approaches for determining selenoproteins in biofluids is sodium dodecyl sulfate polyacrylamide gel electrophoresis (SDS-PAGE), often in combination with complementary techniques such as radioactivity measurements [37], electrothermal atomic absorption spectroscopy (ETAAS) [38], or laser ablation (LA) coupled to ICP-MS [35].

### 5.2.1 Chromatographic Methods

In general, techniques based on HPLC coupled to ICP-MS detection constitute the main analytical approach for the separation, characterization, and quantification of selenoproteins and other Se species in biological samples [25–30, 39–44]. Table 5.1 describes numerous chromatographic configurations coupled to ICP-MS that have been applied for absolute quantification of selenoproteins in biofluids, including serum/plasma, human milk, and CSF.

**TABLE 5.1**
**Chromatography Methods, Extraction Procedure, Species, and Biological Fluids Analyzed for the Determination of Selenoproteins**

| Sample | Organism | Analytical Technique | Se Species | References |
|---|---|---|---|---|
| CSF | Human | IEC-ICP-MS | SELENOP, SeAlb, and SeMet | [25] |
| Milk | Human | 2SEC-AF-ICP-MS | SELENOP | [10] |
| Paired serum and CSF | Human | SEC and SAX-ICP-DRC-MS | SELENOP, eGPx, SeAlb, and TrxR | [45] |
| Plasma | Human | AF-ICP-MS | SELENOP | [28] |
| Plasma | Human | MAF-ICP-MS | SELENOP | [26] |
| Plasma | Human | SEC-AF-ICP-MS | SELENOP, eGPx, SeAlb, selenometabolites | [29] |
| Plasma | Mice | SEC-AF-ICP-MS | SELENOP, eGPx, SeAlb, selenometabolites | [46] |
| Blood | Human | IEC-ICP-MS | SELENOP, TrxR, eGPx, and SeAlb | [47] |
| Serum | Human | 2AF-ICP-MS | SELENOP | [40] |
| Serum | Human | SEC-ICP-MS | SELENOP and SeAlb | [24] |
| Serum | Human | AF-ICP-MS | SELENOP | [27] |
| Serum | Human | SEC-AF-ICP-MS | SELENOP, eGPx, and SeAlb | [30] |
| Serum | Human | SEC-AF-ICP-MS | SELENOP, eGPx, and SeAlb | [33] |
| Serum | Human | SEC-AF-AEC-ICP-MS | SELENO, eGPx, and SeAlb | [31] |
| Serum | Human | 2AF-HG-AFS | SELENOP and eGPx | [48] |
| Serum | Human | IEC-ICP-MS | SELENOP, TrxR, eGPx, and SeAlb | [49] |
| Serum | Mice | SEC-AF-ICP-MS | SELENOP, eGPx, and SeAlb | [50] |
| Serum | Mice | SEC-AF-ICP-MS | SELENOP, eGPx, and SeAlb | [51] |
| Serum | Mice | SEC-AF-ICP-MS | SELENOP, eGPx, and SeAlb | [52] |
| Serum | Mice | SEC-AF-ICP-MS | SELENOP, eGPx, and SeAlb | [53] |
| Serum | Human | IDA-AE-FPLC-ICP-MS | SELENOP, eGPx, and SeAlb | [54] |

*Abbreviations*: AEC, anion-exchange chromatography; AF, affinity chromatography; AFS, atomic fluorescence spectroscopy; CSF, cerebrospinal fluid; DRC, dynamic reaction cell; eGPx, extracellular glutathione peroxidase; FPLC, fast protein liquid chromatography; HG, hydride generation; ICP, inductively coupled plasma; IEC, ion-exchange chromatography; MAF, micro-affinity chromatography; MS, mass spectrometry; SAX, strong anion exchange; SeAlb, selenoalbumin; SEC, size-exclusion chromatography; SELENOP, selenoprotein P; SeMet, selenomethionine; TrxR, thioredoxin reductase.

#### 5.2.1.1 Affinity Chromatography and Micro-affinity Chromatography

Affinity-based chromatographic methods are among the earliest approaches for selective selenoprotein separation. Specifically, SELENOP has been analyzed in diluted serum samples using fast AF coupled to ICP-MS [27]. Similarly, SELENOP in human plasma has been determined by immobilized metal ion AF coupled to ICP-MS following solid-phase extraction (SPE), offering rapid, high-throughput analysis with a reusable, non-biological sorbent that maintains consistent chromatographic performance across multiple runs [28]. Moreover, anion-exchange (AE)-SPE has been carried out before double-column AF-HPLC hyphenated with ICP-MS, and described as one of the most feasible approaches for speciation analysis of eGPx, SelenoP, and SeAlb in human serum [40]. In addition, MAF coupled with ICP-MS has been successfully employed to quantify SELENOP in very small plasma volumes, demonstrating superior selectivity and sensitivity compared to SEC [26]. In addition, double AF columns coupled to hydride generation and atomic fluorescence spectroscopy (HG-AFS) have been applied to determine serum Se-containing proteins, including SELENOP and eGPx, in individuals exposed to high concentrations of mercury (Hg) [54].

#### 5.2.1.2 Ion-Exchange Chromatography

IEC has also been applied to biological samples for the separation and analysis of selenoproteins. Specifically, Se species characterization, including SELENOP and SeAlb, has been determined using IEC coupled to ICP-MS to investigate whether speciation analysis could detect organic and inorganic binding forms of Se in the CSF of patients with Parkinson's disease (PD) [25]. In addition, SELENOP, TrxR, eGPx, and SeAlb have been determined by IEC-ICP-MS in human serum to analyze the correlation between SeAlb and other Se species [48]. Other authors have combined AF with IEC to quantify selenoproteins in human serum by AF coupled to post-column isotope dilution analysis by ICP-MS [53].

#### 5.2.1.3 Size-Exclusion Chromatography and Combined Techniques

SEC is commonly used to fractionate proteins based on their molecular size. For the determination of selenoproteins, some authors have described analyzing SEC fractions containing Se proteins in serum samples from patients with Alzheimer's disease (AD) and mild cognitive impairment (MCI) which were quantified by diluting an aliquot of each fraction 1:10 with 1% $HNO_3$ and measuring Se by ICP-MS [24]. While SEC alone may lack the selectivity required for complex samples, its combination with AF (SEC–AF) or with additional chromatographic modes in multidimensional setups allows for improved resolution of selenoproteins and separation from low-molecular-weight selenometabolites. For example, SEC separation combined with strong anion-exchange (SAX) chromatography has been applied to paired CSF and serum samples for the determination of SELENOP, eGPx, SeAlb, and TrxR in both fluids [45].

Figure 5.2 illustrates an optimized column-switching strategy that integrates SEC and AF chromatography with ICP-MS and IDA which has been widely applied for the characterization of major selenoproteins (SELENOP, eGPx, and SeAlb) and total low-molecular-weight selenometabolites in mouse plasma, supporting both

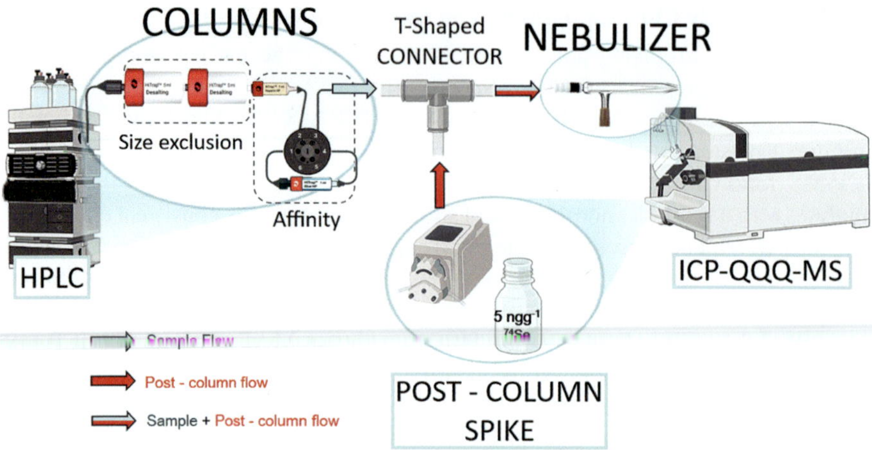


**FIGURE 5.2** IDA-SEC–AF-HPLC-ICP-QQQ-MS method applied for the analysis of selenoproteins and selenometabolites. IDA, isotopic dilution analysis; HPLC, high-performance liquid chromatography; ICP-QQQ-MS, inductively coupled plasma mass spectrometry.

clinical assessments of Se status and environmental or toxicological studies in mammals [50].

This multidimensional IDA-SEC–AF-HPLC-ICP-MS workflow has also been used to investigate how antibiotic-induced microbiota disruption and subsequent Se supplementation modulate the selenoproteome [49], as well as to evaluate changes after exposure to chemical mixtures [52]. In addition, it has proven essential for elucidating the detoxification mechanisms of Hg and Cd through Se-dependent pathways by monitoring selenoprotein responses upon metal exposure [46, 51]. Beyond animal models, this analytical platform has been successfully applied to human serum from mothers and cord blood, revealing selective placental transfer of Se species[33]. Its capacity to distinguish between selenoproteins and small selenometabolites has also provided new insights into Se metabolism and its potential implications in lung cancer (LC) and chronic obstructive pulmonary disease (COPD) [29, 30]. In addition, a column-switching system combining SEC, AF, and anion-exchange chromatography (AEC) allows for the simultaneous separation of selenoproteins (SELENOP, eGPx, and SeAlb) and selenite (Se(IV)) and selenate (Se(VI)) species in serum with analysis times under 35 minutes [31]. Similar methods have been applied to human milk to quantify SELENOP for the first time, demonstrating the versatility of this methodology for diverse matrices [10].

### 5.2.2 Electrophoretic Methods

Electrophoretic techniques represent a powerful and complementary strategy to chromatographic approaches for the separation and analysis of selenoproteins, exploiting intrinsic differences in their physicochemical properties, such as net charge, molecular weight, and structural conformation. Among these techniques, SDS-PAGE is widely employed due to its exceptional resolving capacity, which allows for the high-resolution fractionation of complex protein mixtures and facilitates subsequent downstream analyses. The analytical performance of electrophoresis is substantially enhanced when integrated with advanced elemental detection methods such as LA-ICP-MS. This coupling enables not only the precise quantification of Se within separated protein bands but also the direct investigation of Se speciation in biological fluids, which is critical given the diverse biochemical roles of selenoproteins in antioxidant defense, redox regulation, and metabolic homeostasis [35, 55].

Despite these methodological advantages, the application of electrophoretic techniques for the targeted analysis of selenoproteins in biofluids has been relatively limited, with only a few studies systematically addressing their potential. For instance, SELENOP has been successfully quantified in human serum by combining SDS-PAGE separation with LA-ICP-MS detection, providing an accurate measure of Se incorporation and protein abundance [55]. Beyond quantification, structural and functional characterization of selenoproteins such as SELENOP and eGPx has been achieved through a multi-step approach involving SDS-PAGE separation, electroblotting onto polyvinylidene fluoride (PVDF) membranes, elemental analysis via LA-ICP-MS, and definitive protein identification through nano-liquid chromatography–tandem mass spectrometry (nanoLC-MS/MS) [35].

### 5.2.3 Applications in Human and Animal Samples

#### 5.2.3.1 Applications in Human Samples

Se speciation in human samples is highly informative for assessing nutritional status [56], investigating disease-related alterations [57], and evaluating exposure scenarios [47]. In serum, SELENOP is the main Se-containing fraction, often accounting for half or more of circulating Se [58]. In healthy mother–infant pairs, SELENOP concentrations were 42.5 $ng \cdot g^{-1}$ in maternal serum and 28.1 $ng \cdot g^{-1}$ in cord serum, while GPx was 11.2 and 10.5 $ng \cdot g^{-1}$, respectively, measured using IDA-SEC–AF-ICP-MS [33] which highlighting maternal-fetal differences and placental transfer.

Clinical applications have used absolute quantification via IDA-SEC–AF-ICP-MS [29, 30], SEC-ICP-MS [24], or IEC-ICP-MS [25] to compare patients and controls. In LC patients, eGPx was 21.2 $ng \cdot g^{-1}$ as opposed to 17.0 $ng \cdot g^{-1}$ in healthy subjects, and SeAlb was 49.6 $ng \cdot g^{-1}$ as opposed to 38.3 $ng \cdot g^{-1}$ [29]. Higher eGPx levels were also observed in COPD patients [30]. SEC-ICP-MS identified SELENOP-containing fractions in AD and MCI patients, confirmed by tryptic digestion and LC-MS/MS [24]. In CSF of PD patients, SELENOP and SeAlb were 1923 and 592 ng $Se \cdot L^{-1}$, respectively, with no significant differences from controls [25]. On the other hand, electrophoretic separation using isoelectric focusing coupled to LA-ICP-MS enables qualitative and semi-quantitative mapping of SELENOP, eGPx, and SeAlb [43].

In human milk, SELENOP represents a substantial Se fraction in colostrum (~18–20 ng $Se \cdot g^{-1}$) [10], crucial for neonatal Se supply. Multidimensional chromatography with column-switching SEC-AF-AEC-HPLC-ICP-MS and IDA allows simultaneous analysis of selenoproteins and low-molecular-weight Se species (Se(IV) and Se(VI)) with detection limits of 0.2–1.3 ng Se, confirming SELENOP at 4–5 μg/mL [31]. Figure 5.3 illustrates the methodology based on 2D-SEC-SEC × SEC-AF-ICP-MS column switching approach to quantify selenoproteins in human milk.

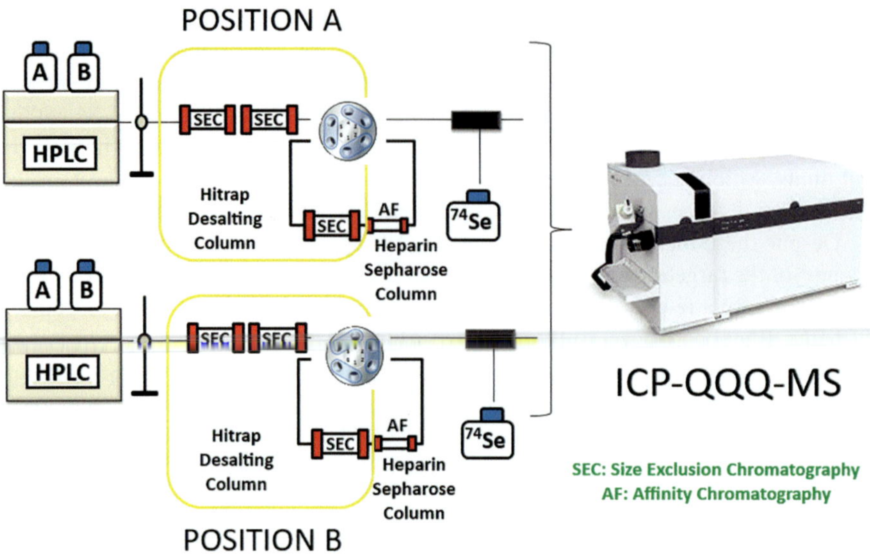


**FIGURE 5.3** Schematic diagram of the 2D-SEC-SEC × SEC-AF-ICP-MS column switching arrangement for the determination of selenoproteins in human breast milk. SEC, Size-exclusion chromatography; AF, affinity chromatography; ICP-MS, inductively coupled plasma mass spectrometry.

While serum and plasma are well studied, human milk is underexplored but critical for maternal–infant Se transfer. Milk matrix poses analytical challenges, and the chemical form of Se affects bioavailability, metabolism, and infant selenoprotein incorporation. Concentrations are influenced by birth weight, gestational age [59], and maternal Se intake or nutritional status [60]. Early studies on Se in human milk mainly focused on total content or limited speciation. Capillary zone electrophoresis coupled to ICP-MS (CZE-ICP-MS), sometimes combined with SEC and electrothermal vaporization (ETV) ICP-MS, allowed the detection of low-molecular-weight Se species but did not provide full characterization of selenoproteins [61]. The use of SEC-ICP-MS later enabled the evaluation of Se distribution among protein-associated fractions in human milk and infant formulas [62]. Additional studies used HG-AFS for total Se determination [63], while more advanced HPLC column-switching approaches coupled to ICP-MS and quadrupole time of flight (QTOF) allowed the direct identification of SELENOP in human colostrum [10].

Comparative studies of human milk and commercial infant formulas from different countries, using ICP-MS and LC-ICP-MS showed variability in Se content and distribution [64]. These results are in agreement with previous reports indicating the influence of the protein matrix, interactions with other trace elements, and maternal diet on Se levels in human milk [65–69].

Overall, the studies summarized in Table 5.2 show the progression from total Se determination to more selective speciation-based methods. Recent developments in ICP-MS and HPLC column-switching techniques allow the direct determination of

**TABLE 5.2**
**Analytical Techniques for the Determination of Selenoproteins in Human Milk and Infant Formulas**

| Sample | Organism | Analytical Techniques | Se Species | Reference |
|---|---|---|---|---|
| Milk | Human | SEC + ETV-ICP-MS + CZE-ICP-MS | eGPx, SECA, SeCys2, and SeMet | [56] |
| Milk | Human | SEC-ICP-MS | Se(IV), Se(VI), SeMet, and SeCys | [57] |
| Colostrum milk | Human | HPLC column-switching + ICP-MS/QTOF | SelenoP, Selenometabolites, eGPx, and SECA | [10] |
| Commercial infant formulas and milks | Commercial | ICP-MS and LC-ICP-MS | Se(IV), Se(VI), and SeMet | [58] |

*Abbreviations:* AEC, anion-exchange chromatography; CZE, capillary zone electrophoresis; eGPx, extracellular glutathione peroxidase; ETV, electrothermal vaporization; HG, hydride generation; HPLC, high-performance liquid chromatography; ICP, inductively coupled plasma; LC, liquid chromatography; MS, mass spectrometry; NAA, neutron activation analysis; QTOF, Quadrupole time-of-flight; SEC, size-exclusion chromatography; SECA, selenocystamine; SeCys, selenocysteine; SeCys2, selenocystine; SELENOP, selenoprotein P; SeMet, selenomethionine.

SELENOP in human milk [10], supporting the application of these methods in nutritional and clinical studies [53, 62, 70].

#### 5.2.3.2 Applications in Animal Models

Selenoproteins have been identified in rat serum using SEC-ICP-MS, a method complemented by enzyme activity measurements and Western blotting [59]. SEC-ICP-MS has also been applied to selenoproteins in mouse urine, serum, red blood cells, and organs [60]. Selenoproteins and total selenometabolites have also been determined in mouse serum by IDA-SEC-AF-ICP-MS [31, 50]. Using this analytical method, the effect of Hg exposure on mouse plasma selenoproteome was described along with the crucial role of Se in Hg detoxification [51]. Likewise, Se speciation in mouse plasma under Cd exposure using SEC-AF-HPLC-ICP-MS demonstrated the involvement of SELENOP in Cd detoxification/excretion and highlighted the importance of Se intake after Cd exposure [46].

Advances in Se speciation methodologies have enabled the quantification of selenoproteins under diverse physiological and experimental conditions. For example, in mouse plasma, the combination of heteroatom-tagged proteomics with SEC-AF-HPLC and ICP-MS allowed the absolute quantification of the plasma selenoproteome. This approach revealed that antibiotic-induced microbiota depletion followed by Se supplementation increased eGPx and SeAlb concentrations compared with controls, as determined by SEC-AF-SUID-ICP-QQQ-MS [49]. Same authors applied this methodology to mouse serum under antibiotic-induced microbiota depletion and Se supplementation when mouse were additionally exposed to a chemical cocktail.

They observed a slight increase in plasma eGPx, whereas SELENOP and SeAlb showed reduced concentrations after the exposure [52].

On the other hand, a recent study on avian serum reported a new method for the determination of SELENOP, eGPx, SeAlb, various selenometabolites, and a SELENOW-containing fraction using the SEC-AF-ICP-MS chromatographic approach complemented with HPLC-ESI-QTOF-MS [61].

## 5.3 SPECIATION OF SELENOMETABOLITES IN BIOFLUIDS

Speciation of selenometabolites, including the separation and identification of inorganic species, selenoamino acids, and low molecular weight selenocompounds, is essential to understand Se biotransformation and its nutritional relevance [62, 70]. Modern approaches rely mainly on chromatographic techniques coupled to ICP-MS or molecular MS, enabling sensitive detection of diverse Se metabolites in biological fluids such as blood [47, 71], serum [47, 48, 71–73], plasma [63, 64], urine [44, 62, 65–67, 70], or milk [10]. Among the chromatographic methods, IEC and RP chromatography are the most commonly used methods for this purpose. However, some authors have also employed electrophoretic methods to achieve the same objective.

### 5.3.1 Chromatographic Methods

LC coupled with ICP-MS has emerged as a powerful and widely applied technique for the speciation and quantitative determination of selenometabolites in biofluids for clinical, epidemiological, and toxicological studies [29, 44, 47, 48, 62–67, 70–72]. This approach allows the simultaneous separation and detection of inorganic Se species, such as Se(IV) and Se(VI), and organic metabolites, including SeMet, methylselenocysteine (MeSeCys), SeCys, SeCys2, selenoethionine (SeEt), selenosugars (SeSug), selenoneine (SEN), selenocystamine (SECA), selenourea (SeUr), trimethylselenonium (TMSe), and other minor Se compounds (Table 5.3). Table 5.3 describes the most used analytical techniques for the determination of Se species and selenometabolites in different biofluids.

In blood and serum, IEC-ICP-MS has been widely used to separate Se species, including Se(IV), Se(VI), SeMet, and SeCys, enabling the assessment of Se status in population studies [47] and the evaluation of correlations between specific selenometabolites and selenoprotein levels [48]. IEC-ICP-MS has also been applied to study these Se compounds, including MeSeCys, in serum and urine from humans after Se supplementation [72].

Anion-exchange approaches, particularly AEC-ICP-MS, have been used to determine SeMet, MeSeCys, SeCys2, and inorganic forms in serum from LC patients [29]. Similarly, strong AEC (SAEC-ICP-MS) has been applied to serum from cystic fibrosis patients for the analysis of SeMet, SeEt, Se(IV), SeCys2, and Se(VI) [73].

RP and ion-pairing RP (IP-RP) chromatography have been employed to achieve high-resolution separation of low-molecular-weight selenometabolites, such as MeSeCys, SeMet, SeSug, and L-$\gamma$-glutamyl-Se-methylseleno-L-cysteine in plasma from cancer patients [63]. These approaches have also facilitated the study of Se metabolism following pharmacological Se(IV) treatment, enabling the analysis of

**TABLE 5.3**
**Chromatographic Methods for the Determination of Selenometabolites in Biofluids**

| Sample | Organism | Analytical Technique | Se Species | Reference |
|---|---|---|---|---|
| Blood | Human | IEC-ICP-MS | Se(IV), Se(VI), SeMet, SeCys | [47] |
| Milk | Human | SEC-AF-ICP-MS | SECA | [10] |
| Plasma | Human | IP-RP-HPLC-ICP-MS | SeMet, MeSeCys, SeSug, and L-γ-glutamyl-Se-methylseleno-L-cysteine | [67] |
| Plasma | Human | RP-LC-ICP-MS | MeSeCys, SeMet, SeSug, and Se (IV) | [68] |
| Serum | Human | AEC-ICP-MS | SeMet, MeSeCys, $SeCys_2$, Se(IV), and Se (VI) | [29] |
| Serum | Human | IEC-ICP-MS | Se (IV), Se (VI), SeCys, and SeMet | [49] |
| Serum | Human | SAEC-ICP-MS | SeMet, SeEt, Se(IV), $SeCys_2$, and Se(VI) | [66] |
| Serum and urine | Human | IEC-ICP-MS | Se (IV), Se (VI), SeMet, and MeSeCys | [65] |
| Urine | Human | RP-ICP-MS and IEC-ICP-MS | SeMet, MeSeCys, SeEt, Se (VI), Se (IV), SeSug, and TMSe | [63] |
| Urine | Human | RP-ICP-MS and IEC-ICP-MS | Se(IV), DL-SeMet, and L-SeMet | [62] |
| Urine | Human | RP-ICP-MS and IEC-ICP-MS | TMSe, SeMet, and SeSug | [72] |
| Urine | Human | RP-LC-ICP-MS | MeSeCys, Se (IV), Se (VI), D-SeMet, DL-SeMet | [69] |
| Urine | Human | AE-ICP-MS and RP-ICP-MS | Se(VI), Se(IV) and SeSug | [44] |
| Urine | Human | HPLC-ICP-MS | Se(VI), Se(IV), MeSeCys, TMSe, SeUr, SeEt, and SeMet | [70] |

*Abbreviations:* AE, anion exchange; AEC, anion-exchange chromatography; AF, affinity chromatography; HPLC, high-performance liquid chromatography; ICP, inductively coupled plasma; IEC, ion-exchange chromatography; IP, ion-pairing; LC, liquid chromatography; MeSeCys, methylselenocysteine; MS, mass spectrometry; RP, reversed phase; SAEC, strong anion-exchange chromatography; Se, selenium; SEC, size-exclusion chromatography; SECA, selenocystamine; SeCys, selenocysteine; SeCys2, selenocystine; SeEt, selenoethionine; SeMet, seleniomethionine; SeMeSeCys, selenomethylselenocysteine; SeSug, selenosugars; SeUr, selenourea; TMSe, trimethylselenonium.

MeSeCys, SeMet, SeGal, and Se(IV) in plasma [64]. In urine, RP-ICP-MS has provided insights into Se excretion and biotransformation, revealing MeSeCys, Se(IV), Se(VI), L-SeMet, and D-SeMet [65]. In addition, combinations of RP and IEC chromatography with ICP-MS have been used to quantify various small Se species in urine, including Se(IV), Se(VI), SeMet, TMSe, MeSeCys, SeEt, SeUr, and various

SeSug [62, 66]. SEC coupled with AF and ICP-MS has been employed for the determination of SECA in human milk [10].

Overall, chromatographic ICP-MS techniques, including IEC, AEC, RP, IP-RP, SEC, and AF, provide powerful and versatile tools for the simultaneous separation and quantification of both inorganic and organic selenometabolites, highlighting their critical role in Se research in human biofluids.

### 5.3.2 Electrophoretic Methods

Electrophoretic methods, particularly CE, offer a valuable complementary approach for certain selenometabolites, providing high separation efficiency, minimal sample consumption, and rapid screening of polar or charged Se species. When coupled with ICP-MS, CE enables element-specific detection of Se at trace levels [67, 69]. Michalke and Schramel demonstrated the feasibility of CE-ICP-MS by separating six Se species, including inorganic Se(IV) and Se(VI), SeMet, SeCys, glutathione-bound Se, and SECA, in a single 10-minute run with detection limits around 10 μg Se $L^{-1}$ [69].

Subsequent developments have improved sensitivity and applicability to biological samples. Lu et al. used in-capillary enrichment to speciate small Se-containing molecules in urine and cell lysates, achieving detection limits of 0.29–0.52 μg $L^{-1}$ and resolving multiple organic species, including SeMet, SeCys, SECA, methylseleno compounds, and SeSug [67]. Bendahl et al. introduced a direct-injection nebulizer for CE-ICP-MS, allowing for continuous sample uptake, sub-μg $L^{-1}$ detection limits, and baseline separation of Se(IV), Se(VI), SeCys2, and SeMet within minutes [74]. CE-ICP-MS combines the high separation efficiency of CE with ICP-MS element-specific sensitivity, enabling comprehensive speciation of both inorganic and organic Se species using minimal sample volumes, ideal for limited-volume matrices such as urine, plasma, or cell lysates [69, 75]. ICP-MS eliminates the need for molecular-specific detectors or derivatization, and methodological enhancements, such as optimized nebulizers or stacking/enrichment procedures, further increase sensitivity.

Challenges remain, including the mismatch between low flow rates of CE (<1 μL $min^{-1}$) and standard ICP-MS nebulizers, which can reduce sensitivity even with sheath-flow interfaces [75, 76]. Complex biofluids may cause capillary clogging or polyatomic interferences (e.g., 80Se from argon species) requiring collision/reaction-cell ICP-MS, high-resolution instruments, isotopic dilution, or careful sample cleanup [76]. The stability of labile Se species during handling is also critical, and despite minimal manipulation in CE protocols, pretreatments like centrifugation, ultrafiltration, and pH adjustments are necessary. In addition, small injection volumes in CE may limit sensitivity compared with HPLC-ICP-MS for very low-abundance Se species [69, 76]. Despite these limitations, CE-ICP-MS has proven to be a powerful and complementary tool for Se speciation in biological fluids. It is particularly suitable for analyzing polar, charged, or low-molecular-weight selenometabolites that are often poorly retained or co-eluted in conventional chromatographic separations. When applied thoughtfully, electrophoretic methods occupy a defined niche within comprehensive Se speciation strategies.

### 5.3.3 Comparative Evaluation of ICP-MS-Based Separation Strategies for Selenoprotein Analysis

As discussed in previous sections, several ICP-MS-based separation strategies have been used for selenoprotein speciation in biofluids, mainly based on SEC-AF-ICP-MS, IEC-ICP-MS, and CE-ICP-MS.

SEC-AF-ICP-MS is the most comprehensive approach for complex matrices because it fractionates proteins by size while preserving their native structure, and the affinity step selectively isolates major selenoproteins like SELENOP and eGPx. Coupled with online isotope dilution, it allows accurate absolute quantification without protein standards and is robust against matrix effects. This technique has been applied in nutritional, toxicological, and clinical studies [10, 29–31, 50]. Its main drawbacks are complex instrumentation and lower throughput.

On the other hand, IEC-ICP-MS separates proteins based on charge and generally offers higher resolution and faster analysis than SEC, and can measure both selenoproteins and low-molecular-weight Se species. However, it is sensitive to matrix composition and requires careful optimization to prevent co-elution of proteins with similar charge. Despite this, IEC-ICP-MS is valuable for targeted and comparative studies in serum and CSF [25, 47, 48].

Finally, CE-ICP-MS offers high separation efficiency, short analysis times, and minimal sample consumption. Its use for intact selenoproteins is limited due to interface and sensitivity constraints, so it is mainly applied to low-molecular-weight selenometabolites [69]. Nonetheless, it can serve as a useful complementary technique. In summary, SEC-AF-ICP-MS is preferred for comprehensive selenoprotein profiling, IEC-ICP-MS balances resolution and throughput, and CE-ICP-MS plays a specialized confirmatory role. Using these techniques together enhances the reliability of Se speciation in biofluids.

### 5.3.4 Conclusion

The analysis of selenoproteins and selenometabolites in biofluids by ICP-MS after a separation dimension presents several important advantages, especially when using IDA. Thus, absolute quantification is possible using Se as a "tag" in the atomic detector and protein standards are not required. This method has high sensitivity, accuracy, and precision.

Future developments in Se speciation in biofluids are expected to focus on several areas. Integration with high-throughput metabolomics, including selenometabolomics and selenoproteomics, will allow more detailed analysis of Se species using advanced MS techniques. Improvements in miniaturization and analytical efficiency, such as microfluidics, chip-based separations, and better interference control, will enhance sensitivity while reducing sample requirements. Finally, the discovery of novel Se metabolites, including additional SeSug and methylated derivatives, continues to expand our understanding of Se metabolism. Overall, combining chromatographic or electrophoretic separations with ICP-MS has greatly advanced knowledge of selenometabolites and selenoproteins, offering potential applications to find biomarkers for nutritional assessment, disease risk, and the development of supplementation strategies or detection of nutritional deficiency, despite remaining challenges in sensitivity, complex matrices, and throughput.

## ACKNOWLEDGMENTS

Projects PID2024-155422NB-C21 and PID2021-123073NB-C21 from the Ministry of Science, Innovation and Universities (MCIN/AEI/10.13039/501100011033/FEDER, UE). Project I+D+i PCM_00023 "Monitorización ambiental mediante metodologías ómicas y analíticas utilizando aves y sus hábitats como bioindicadores de calidad ambiental (MONAMBIO)", subsidized by the Administration of the Junta de Andalucía, Ministry of University, Research and Innovation, within the framework of the Complementary Plan for Marine Sciences and the Recovery, Transformation and Resilience Plan, in its call for the 2023 financial year (Measure: C17. I1). Aid co-financed 100% by the European Union through the European Union's Recovery and Resilience Facility. "Recovery, Transformation and Resilience Plan – Financed by the European Union – NextGenerationEU". Project TED2021-130361B-I00 Ministerio de Ciencia e Innovación (MCIN). Knowledge Generation. MCIN/AEI/10.13039/501100011033 and European Union "NextGenerationEU"/PRTR». The authors are grateful to the FEDER Grant UNHU13-1E-161. RFVP gratefully acknowledges the support of the Ministry of Science and Innovation (MCIN) for a FPI predoctoral fellowship (PRE2021-102165). S. Sánchez thanks the Ministry of Science, Innovation and Universities for a predoctoral contract (FI24/00310).

## REFERENCES

1. Ogra, Y.; Anan, Y. Selenometabolomics Explored by Speciation. *Biol Pharm Bull* 2012, *35* (11), 1863–1869. https://doi.org/10.1248/bpb.b212016.
2. Institute of Medicine (US) Panel on Dietary Antioxidants and Related Compounds. *Dietary Reference Intakes for Vitamin C, Vitamin E, Selenium, and Carotenoids: A Report of the Panel on Dietary Antioxidants and Related Compounds, Subcommittees on Upper Reference Levels of Nutrients and of Interpretation and Use of Dietary Reference Intakes, and the Standing Committee on the Scientific Evaluation of Dietary Reference Intakes, Food and Nutrition Board, Institute of Medicine*; National Academy Press, 2000.
3. Szpunar, J.; Łobiński, R. Speciation in the Environmental Field – Trends in Analytical Chemistry. *Fresenius J Anal Chem* 1999, *363* (5), 550–557. https://doi.org/10.1007/s002160051245.
4. Lobinski, R.; Edmonds, J. S.; Suzuki, K. T.; Uden, P. C. Species-Selective Determination of Selenium Compounds in Biological Materials (Technical Report). *Pure Appl Chem* 2000, *72* (3), 447–461. https://doi.org/10.1351/pac200072030447.
5. Gómez-Jacinto, V.; Navarro-Roldán, F.; Garbayo-Nores, I.; Vílchez-Lobato, C.; Borrego, A. A.; García-Barrera, T. In Vitro Selenium Bioaccessibility Combined with in Vivo Bioavailability and Bioactivity in Se-Enriched Microalga (*Chlorella sorokiniana*) to Be Used as Functional Food. *J Funct Foods* 2020, *66*, 103817. https://doi.org/10.1016/j.jff.2020.103817.
6. Reeves, M. A.; Hoffmann, P. R. The Human Selenoproteome: Recent Insights into Functions and Regulation. *Cellular and Molecular Life Sciences* 2009, *66* (15), 2457–2478. https://doi.org/10.1007/s00018-009-0032-4.

7. Brown, K. M.; Arthur, J. R. Selenium, Selenoproteins and Human Health: A Review. *Public Health Nutr* 2001, *4* (2b), 593–599. https://doi.org/10.1079/PHN2001143.
8. Schweizer, U.; Schomburg, L. Selenium, Selenoproteins and Brain Function. In *Selenium: Its Molecular Biology and Role in Human Health*; Hatfield, D. L., Berry, M. J., Gladyshev, V. N., Eds.; Springer US: Boston, MA, 2006; pp 233–248. https://doi.org/10.1007/0-387-33827-6_21.
9. Urbano, T.; Filippini, T.; Cardoso, B. R.; Wise, L. A.; Zamboni, G.; Chiari, A.; Vinceti, G.; Tondelli, M.; Marti, A.; Malavolti, M.; Vinceti, M.; Michalke, B. Biomarkers of Total Selenium and Selenium Species in Paired Serum and Cerebrospinal Fluid Samples. *J Trace Element Med Biol* 2025, *92*, 127765. https://doi.org/10.1016/j.jtemb.2025.127765.
10. Arias-Borrego, A.; Callejón-Leblic, B.; Rodríguez-Moro, G.; Velasco, I.; Gómez-Ariza, J. L.; García-Barrera, T. A Novel HPLC Column Switching Method Coupled to ICP MS/QTOF for the First Determination of Selenoprotein P (SELENOP) in Human Breast Milk. *Food Chem* 2020, *321*, 126692. https://doi.org/10.1016/j.foodchem.2020.126692.
11. Milner, J. A.; Sherman, L.; Picciano, M. F. Distribution of Selenium in Human Milk. *Am J Clin Nutr* 1987, *45* (3), 617–624. https://doi.org/10.1093/ajcn/45.3.617.
12. Busto, M. E. del C.; Oster, C.; Cuello-Nuñez, S.; Deitrich, C. L.; Raab, A.; Konopka, A.; Lehmann, W. D.; Goenaga-Infante, H.; Fisicaro, P. Accurate Quantification of Selenoproteins in Human Plasma/Serum by Isotope Dilution ICP-MS: Focus on Selenoprotein P. *J Anal At Spectrom* 2016, *31* (9), 1904–1912. https://doi.org/10.1039/C6JA00122J.
13. Mostert, V. Selenoprotein P: Properties, Functions, and Regulation. *Arch Biochem Biophys* 2000, *376* (2), 433–438. https://doi.org/10.1006/abbi.2000.1735.
14. Minich, W. B. Selenium Metabolism and Biosynthesis of Selenoproteins in the Human Body. *Biochemistry (Moscow)* 2022, *87* (1), S168–S177. https://doi.org/10.1134/S0006297922140139.
15. Comhair, S. A. A.; Erzurum, S. C. The Regulation and Role of Extracellular Glutathione Peroxidase. *Antioxid Redox Signal* 2004, *7* (1–2), 72–79. https://doi.org/10.1089/ars.2005.7.72.
16. Arnér, E. S. J. Focus on Mammalian Thioredoxin Reductases: Important Selenoproteins with Versatile Functions. *Biochimica et Biophysica Acta (BBA) - General Subjects* 2009, *1790* (6), 495–526. https://doi.org/10.1016/j.bbagen.2009.01.014.
17. Haratake, M.; Hongoh, M.; Miyauchi, M.; Hirakawa, R.; Ono, M.; Nakayama, M. Albumin-Mediated Selenium Transfer by a Selenotrisulfide Relay Mechanism. *Inorg Chem* 2008, *47* (14), 6273–6280. https://doi.org/10.1021/ic800310j.
18. Kryukov, G. V; Castellano, S.; Novoselov, S. V; Lobanov, A. V; Zehtab, O.; Guigó, R.; Gladyshev, V. N. Characterization of Mammalian Selenoproteomes. *Science (1979)* 2003, *300* (5624), 1439–1443. https://doi.org/10.1126/science.1083516.
19. Li, S.; Gao, F.; Huang, J.; Wu, Y.; Wu, S.; Lei, X. G. Regulation and Function of Avian Selenogenome. *Biochim Biophys Acta – General Subjects* 2018, *1862*, 2473–2479. https://doi.org/10.1016/j.bbagen.2018.03.029.
20. Sanz-Medel, A. Heteroatom(Isotope)-Tagged Proteomics via ICP-MS: Screening and Quantification of Proteins and Their Post-Translational Modifications. *Anal Bioanal Chem* 2008, *391* (3), 885–894. https://doi.org/10.1007/s00216-008-2083-z.
21. Rodríguez-González, P.; Marchante-Gayón, J. M.; García Alonso, J. I.; Sanz-Medel, A. Isotope Dilution Analysis for Elemental Speciation: A Tutorial Review. *Spectrochim Acta Part B At Spectrosc* 2005, *60* (2), 151–207. https://doi.org/10.1016/j.sab.2005.01.005.
22. Hill, K. E.; Lloyd, R. S.; Yang, J. G.; Read, R.; Burk, R. F. The CDNA for Rat Selenoprotein P Contains 10 TGA Codons in the Open Reading Frame. *J Biol Chem* 1991, *266* (16), 10050–10053. https://doi.org/10.1016/S0021-9258(18)99185-4.

23. Ramírez-Acosta, S.; Arias-Borrego, A.; Navarro-Roldán, F.; Selma-Royo, M.; Calatayud, M.; Collado, M. C.; Huertas-Abril, P. V; Abril, N.; Barrera, T. G. Omic Methodologies for Assessing Metal(-Loid)s-Host-Microbiota Interplay: A Review. *Anal Chim Acta* 2021, *1176*, 338620. https://doi.org/10.1016/j.aca.2021.338620.
24. Cardoso, Bárbara R; Hare, Dominic J; Bush, Ashley I; Li, Qiao-Xin; Fowler, Christopher J; Masters, Colin L; Martins, Ralph N; Ganio, Katherine; Lothian, Amber; Mukherjee, Soumya; Kapp, Eugene A; Roberts, Blaine R. Selenium Levels in Serum, Red Blood Cells, and Cerebrospinal Fluid of Alzheimer's Disease Patients: A Report from the Australian Imaging, Biomarker & Lifestyle Flagship Study of Ageing (AIBL). *J Alzheimer Dis* 2017, *57* (1), 183–193. https://doi.org/10.3233/JAD-160622.
25. Maass, F.; Michalke, B.; Willkommen, D.; Schulte, C.; Tönges, L.; Boerger, M.; Zerr, I.; Bähr, M.; Lingor, P. Selenium Speciation Analysis in the Cerebrospinal Fluid of Patients with Parkinson's Disease. *J Trace Element Med Biol* 2020, *57*, 126412. https://doi.org/10.1016/j.jtemb.2019.126412.
26. Shigeta, K.; Sato, K.; Furuta, N. Determination of Selenoprotein P in Submicrolitre Samples of Human Plasma Using Micro-Affinity Chromatography Coupled with Low Flow ICP-MS. *J Anal At Spectrom* 2007, *22* (8), 911–916. https://doi.org/10.1039/B701206C.
27. Heitland, P.; Köster, H. D. Biomonitoring of Selenoprotein P in Human Serum by Fast Affinity Chromatography Coupled to ICP-MS. *Int J Hyg Environ Health* 2018, *221* (3), 564–568. https://doi.org/10.1016/j.ijheh.2018.02.006.
28. Bendahl, L.; Sidenius, U.; Gammelgaard, B. Determination of Selenoprotein P in Human Plasma by Solid Phase Extraction and Inductively Coupled Plasma Mass Spectrometry. *Anal Chim Acta* 2000, *411* (1), 103–108. https://doi.org/10.1016/S0003-2670(00)00731-5.
29. Callejón-Leblic, B.; Rodríguez-Moro, G.; Arias-Borrego, A.; Pereira-Vega, A.; Gómez-Ariza, J. L.; García-Barrera, T. Absolute Quantification of Selenoproteins and Selenometabolites in Lung Cancer Human Serum by Column Switching Coupled to Triple Quadrupole Inductively Coupled Plasma Mass Spectrometry. *J Chromatogr A* 2020, *1619*, 460919. https://doi.org/10.1016/j.chroma.2020.460919.
30. Callejón-Leblic, B.; Sánchez Espirilla, S.; Gotera-Rivera, C.; Santana, R.; Díaz-Olivares, I.; María Marín Trigo, J.; Casanova Macario, C.; Cosio, B. G.; Fuster, A.; Solanes García, I.; de-Torres, J. P.; Feu Collado, N.; Cabrera Lopez, C.; Amado Diago, C.; Romero Plaza, A.; Padrón Fraysse, L. A.; Márquez Martín, E.; Marín Royo, M.; Balcells Vilarnau, E.; Llunell Casanovas, A.; Martínez González, C.; Bautista Galdíz Iturri, J.; Lacárcel Bautista, C.; Gómez-Ariza, J. L.; Pereira-Vega, A.; Seijo, L.; López-Campos, J. L.; Peces-Barba, G.; García-Barrera, T. Heteroatom-Tagged Proteomics of Lung Cancer and Chronic Obstructive Pulmonary Disease Human Serum Reveal Alterations in Selenoproteins. *Microchem J* 2024, *199*, 110033. https://doi.org/10.1016/j.microc.2024.110033.
31. García-Sevillano, M. A.; García-Barrera, T.; Gómez-Ariza, J. L. Development of a New Column Switching Method for Simultaneous Speciation of Selenometabolites and Selenoproteins in Human Serum. *J Chromatogr A* 2013, *1318*, 171–179. https://doi.org/10.1016/j.chroma.2013.10.012.
32. Ramírez-Acosta, S.; Selma-Royo, M.; Collado, M. C.; Navarro-Roldán, F.; Abril, N.; García-Barrera, T. Selenium Supplementation Influences Mice Testicular Selenoproteins Driven by Gut Microbiota. *Sci Rep* 2022, *12* (1), 4218. https://doi.org/10.1038/s41598-022-08121-3.
33. Santos, C.; García-Fuentes, E.; Callejón-Leblic, B.; García-Barrera, T.; Gómez-Ariza, J. L.; Rayman, M. P.; Velasco, I. Selenium, Selenoproteins and Selenometabolites in Mothers and Babies at the Time of Birth. *Br J Nutr* 2017, *117* (9), 1304–1311. https://doi.org/10.1017/S0007114517001155.

34. Hollenbach, B.; Morgenthaler, N. G.; Struck, J.; Alonso, C.; Bergmann, A.; Köhrle, J.; Schomburg, L. New Assay for the Measurement of Selenoprotein P as a Sepsis Biomarker from Serum. *J Trace Element Med Biol* 2008, *22* (1), 24–32. https://doi.org/10.1016/j.jtemb.2007.11.003.
35. Ballihaut, G.; Kilpatrick, L. E.; Kilpatrick, E. L.; Davis, W. C. Detection and Characterization of Selenoproteins by Tandem Mass Spectrometry Assisted by Laser Ablation Inductively Coupled Plasma Mass Spectrometry: Application to Human Plasma Selenoproteins. *J Anal At Spectrom* 2011, *26* (2), 383–394. https://doi.org/10.1039/C0JA00115E.
36. Read, R.; Bellew, T.; Yang, J. G.; Hill, K. E.; Palmer, I. S.; Burk, R. F. Selenium and Amino Acid Composition of Selenoprotein P, the Major Selenoprotein in Rat Serum. *J Biol Chem* 1990, *265* (29), 17899–17905. https://doi.org/10.1016/S0021-9258(18)38248-6.
37. Motchnik, P. A.; Tappel, A. L. Rat Plasma Selenoprotein P Properties and Purification. *Biochim Biophys Acta (BBA) – General Subjects* 1989, *993* (1), 27–35. https://doi.org/10.1016/0304-4165(89)90138-4.
38. Plecko, T.; Nordmann, S.; Rükgauer, M.; Kruse-Jarres, J. D. Determination and Distribution of Human Plasma Selenoproteins. *Fresenius J Anal Chem* 1999, *363* (5), 517–519. https://doi.org/10.1007/s002160051236.
39. Kokarnig, S.; Kuehnelt, D.; Stiboller, M.; Hartleb, U.; Francesconi, K. A. Quantitative Determination of Small Selenium Species in Human Serum by HPLC/ICPMS Following a Protein-Removal, Pre-Concentration Procedure. *Anal Bioanal Chem* 2011, *400* (8), 2323–2327. https://doi.org/10.1007/s00216-011-4992-5.
40. Jitaru, P.; Prete, M.; Cozzi, G.; Turetta, C.; Cairns, W.; Seraglia, R.; Traldi, P.; Cescon, P.; Barbante, C. Speciation Analysis of Selenoproteins in Human Serum by Solid-Phase Extraction and Affinity HPLC Hyphenated to ICP-Quadrupole MS. *J Anal At Spectrom* 2008, *23* (3), 402–406. https://doi.org/10.1039/b712693j.
41. Lamarche, J.; Ronga, L.; Szpunar, J.; Lobinski, R. Characterization and Quantification of Selenoprotein P: Challenges to Mass Spectrometry. *Int J Mol Sci* 2021, *22* (12). https://doi.org/10.3390/ijms22126283.
42. Oveisgharan, S.; Yu, L.; Wang, T.; Schneider, J. A.; Bennett, D. A.; Buchman, A. S. Neurodegenerative and Cerebrovascular Brain Pathologies Are Differentially Associated with Declining Grip Strength and Gait in Older Adults. *J Gerontol: Series A* 2023, *78* (3), 504–513. https://doi.org/10.1093/gerona/glac128.
43. Bianga, J.; Touat-Hamici, Z.; Bierla, K.; Mounicou, S.; Szpunar, J.; Chavatte, L.; Lobinski, R. Speciation Analysis for Trace Levels of Selenoproteins in Cultured Human Cells. *J Proteomics* 2014, *108*, 316–324. https://doi.org/10.1016/j.jprot.2014.05.025.
44. Gammelgaard, B.; Stürup, S.; Christensen, M. V. Human Urinary Excretion and Metabolism of 82Se-Enriched Selenite and Selenate Determined by LC-ICP-MS. *Metallomics* 2012, *4* (2), 149–155. https://doi.org/10.1039/C2MT00163B.
45. Solovyev, N.; Berthele, A.; Michalke, B. Selenium Speciation in Paired Serum and Cerebrospinal Fluid Samples. *Anal Bioanal Chem* 2013, *405* (6), 1875–1884. https://doi.org/10.1007/s00216-012-6294-y.
46. García-Sevillano, M. A.; García-Barrera, T.; Navarro, F.; Gómez-Ariza, J. L. Cadmium Toxicity in Mus Musculus Mice Based on a Metallomic Study. Antagonistic Interaction between Se and Cd in the Bloodstream. *Metallomics* 2014, *6* (3), 672–681. https://doi.org/10.1039/C3MT00350G.
47. Vinceti, M.; Grill, P.; Malagoli, C.; Filippini, T.; Storani, S.; Malavolti, M.; Michalke, B. Selenium Speciation in Human Serum and Its Implications for Epidemiologic Research: A Cross-Sectional Study. *J Trace Element Med Biol* 2015, *31*, 1–10. https://doi.org/10.1016/j.jtemb.2015.02.001.

48. Filippini, T.; Urbano, T.; Grill, P.; Malagoli, C.; Ferrari, A.; Marchesi, C.; Natalini, N.; Michalke, B.; Vinceti, M. Human Serum Albumin-Bound Selenium (Se-HSA) in Serum and Its Correlation with Other Selenium Species. *J Trace Element Med Biol* 2023, *79*, 127266. https://doi.org/10.1016/j.jtemb.2023.127266.
49. Callejón-Leblic, B.; Selma-Royo, M.; Collado, M. C.; Abril, N.; García-Barrera, T. Impact of Antibiotic-Induced Depletion of Gut Microbiota and Selenium Supplementation on Plasma Selenoproteome and Metal Homeostasis in a Mouse Model. *J Agric Food Chem* 2021, *69* (27), 7652–7662. https://doi.org/10.1021/acs.jafc.1c02622.
50. García-Sevillano, M. A.; García-Barrera, T.; Gómez-Ariza, J. L. Simultaneous Speciation of Selenoproteins and Selenometabolites in Plasma and Serum by Dual Size Exclusion-Affinity Chromatography with Online Isotope Dilution Inductively Coupled Plasma Mass Spectrometry. *Anal Bioanal Chem* 2014, *406* (11), 2719–2725. https://doi.org/10.1007/s00216-014-7629-7.
51. García-Sevillano, M. A.; Rodríguez-Moro, G.; García-Barrera, T.; Navarro, F.; Gómez-Ariza, J. L. Biological Interactions between Mercury and Selenium in Distribution and Detoxification Processes in Mice under Controlled Exposure. Effects on Selenoprotein. *Chem Biol Interact* 2015, *229*, 82–90. https://doi.org/10.1016/j.cbi.2015.02.001.
52. Arias-Borrego, A.; Selma-Royo, M.; Collado, M. C.; Abril, N.; García-Barrera, T. Impact of "Chemical Cocktails" Exposure in Shaping Mice Gut Microbiota and the Role of Selenium Supplementation Combining Metallomics, Metabolomics, and Metataxonomics. *J Hazard Mater* 2022, *438*, 129444. https://doi.org/10.1016/j.jhazmat.2022.129444.
53. Hinojosa Reyes, L.; Marchante-Gayón, J. M.; García Alonso, J. I.; Sanz-Medel, A. Quantitative Speciation of Selenium in Human Serum by Affinity Chromatography Coupled to Post-Column Isotope Dilution Analysis ICP-MS. *J Anal At Spectrom* 2003, *18* (10), 1210–1216. https://doi.org/10.1039/B305455A.
54. Chen, C.; Yu, H.; Zhao, J.; Li, B.; Qu, L.; Liu, S.; Zhang, P.; Chai, Z. The Roles of Serum Selenium and Selenoproteins on Mercury Toxicity in Environmental and Occupational Exposure. *Environ Health Perspect*. February 2006, pp 297–301. https://doi.org/10.1289/ehp.7861.
55. Ballihaut, G.; Kilpatrick, L. E.; Kilpatrick, E. L.; Davis, W. C. Multiple Forms of Selenoprotein P in a Candidate Human Plasma Standard Reference Material. *Metallomics* 2012, *4* (6), 533–538. https://doi.org/10.1039/C2MT20059G.
56. Michalke, B.; Schramel, P. Selenium Speciation in Human Milk with Special Respect to Quality Control. *Biol Trace Element Res* 1997, *59* (1), 45–56. https://doi.org/10.1007/BF02783229.
57. Sastre, J.; Colomer, B.; Pérez Solís, D.; Fernandez, M.; Remy, R. R.; Sanz-Medel, A. Determination and Speciation of Selenium in Human Milk and Commercial Formulae: Importance in Infant Nutrition. *Rev Esp Pediatr* 2009, *65*, 216–224.
58. de Paiva, E. L.; Ruttens, A.; Waegeneers, N.; Laing, G. Du; Morgano, M. A.; Cheyns, K.; Arisseto-Bragotto, A. P. Selenium in Selected Samples of Infant Formulas and Milk Commercialized in Belgium and Brazil: Total Content, Speciation and Estimated Intake. *Food Res Int* 2023, *164*, 112289. https://doi.org/10.1016/j.foodres.2022.112289.
59. Anan, Y.; Hatakeyama, Y.; Tokumoto, M.; Ogra, Y. Chromatographic Behavior of Selenoproteins in Rat Serum Detected by Inductively Coupled Plasma Mass Spectrometry. *Analytical Sciences* 2013, *29* (8), 787–792. https://doi.org/10.2116/analsci.29.787.
60. Suzuki, K. T.; Itoh, M.; Ohmichi, M. Detection of Selenium-Containing Biological Constituents by High-Performance Liquid Chromatography-Plasma Source Mass Spectrometry. *J Chromatogr B Biomed Sci Appl* 1995, *666* (1), 13–19. https://doi.org/10.1016/0378-4347(94)00561-I.

61. Callejón-Leblic, B.; Hachemi, M. A.; Cardoso, D.; García-Barrera, T. Development of a New Method for the Absolute Quantification of Selenoproteins in Chicken Serum by Heteroatom-Tagged Proteomics. *J Agric Food Chem* 2026, 74(8):7300.
62. Hildebrand, J.; Greiner, A.; Drexler, H.; Göen, T. Determination of Eleven Small Selenium Species in Human Urine by Chromatographic-Coupled ICP-MS Methods. *J Trace Elements Med Biol* 2020, *61*, 126519. https://doi.org/10.1016/j.jtemb.2020.126519.
63. del Castillo Busto, M. E.; Ward-Deitrich, C.; Evans, S. O.; Rayman, M. P.; Jameson, M. B.; Goenaga-Infante, H. Selenium Speciation Studies in Cancer Patients to Evaluate the Responses of Biomarkers of Selenium Status to Different Selenium Compounds. *Anal Bioanal Chem* 2024, *416* (11), 2835–2848. https://doi.org/10.1007/s00216-024-05141-y.
64. Flouda, K.; Dersch, J. M.; Gabel-Jensen, C., Stürup, S.; Misra, S.; Björnstedt, M.; Gammelgaard, B. Quantification of Low Molecular Weight Selenium Metabolites in Human Plasma after Treatment with Selenite in Pharmacological Doses by LC-ICP-MS. *Anal Bioanal Chem* 2016, *408* (9), 2293–2301. https://doi.org/10.1007/s00216-016-9325-2.
65. Moreno, F.; García-Barrera, T.; Gómez-Ariza, J. L. Simultaneous Analysis of Mercury and Selenium Species Including Chiral Forms of Selenomethionine in Human Urine and Serum by HPLC Column-Switching Coupled to ICP-MS. *Analyst* 2010, *135* (10), 2700–2705. https://doi.org/10.1039/C0AN00090F.
66. Eichler, Š.; Kaňa, A.; Kalousová, M.; Vosmanská, M.; Korotvička, M.; Zima, T.; Mestek, O. Speciation Analysis of Selenium in Human Urine by Liquid Chromatography and Inductively Coupled Plasma Mass Spectrometry for Monitoring of Selenium in Body Fluids. *Chem Spec Bioavail* 2015, *27* (3), 127–138. https://doi.org/10.1080/09542299.2015.1107502.
67. Lu, Y.; Men, X.; Wu, C.; Wei, X.; Chen, M.; Wang, J. Speciation of Selenium-Containing Small Molecules in Urine and Cell Lysate by CE-ICPMS with in-Capillary Enrichment. *Talanta* 2025, *281*, 126929. https://doi.org/10.1016/j.talanta.2024.126929.
68. Lu, Y.; Rumpler, A.; Francesconi, K. A.; Pergantis, S. A. Quantitative Selenium Speciation in Human Urine by Using Liquid Chromatography-Electrospray Tandem Mass Spectrometry. *Anal Chim Acta* 2012, *731*, 49–59.
69. Michalke, B.; Schramel, P. Selenium Speciation by Interfacing Capillary Electrophoresis with Inductively Coupled Plasma-Mass Spectrometry. *Electrophoresis* 1998, *19* (2), 270–275. https://doi.org/10.1002/elps.1150190221.
70. Kuehnelt, D.; Kienzl, N.; Traar, P.; Le, N. H.; Francesconi, K. A.; Ochi, T. Selenium Metabolites in Human Urine after Ingestion of Selenite, L-Selenomethionine, or DL-Selenomethionine: A Quantitative Case Study by HPLC/ICPMS. *Anal Bioanal Chem* 2005, *383* (2), 235–246. https://doi.org/10.1007/s00216-005-0007-8.
71. Mehta, R.; Qamar, H.; Monangi, N.; Zhang, G.; Combs, G.; Al Mahmud, A.; Islam, M.; Ahmed, T.; Muglia, L.; Belling, E.; Roth, D.; Landero, J. Selenium Speciation in Paired Whole Blood and Serum Samples From Pregnant Bangladeshi Women. *Curr Dev Nutr* 2022, *6*, 312. https://doi.org/10.1093/cdn/nzac053.053.
72. Kokarnig, S.; Tsirigotaki, A.; Wiesenhofer, T.; Lackner, V.; Francesconi, K. A.; Pergantis, S. A.; Kuehnelt, D. Concurrent Quantitative HPLC–Mass Spectrometry Profiling of Small Selenium Species in Human Serum and Urine after Ingestion of Selenium Supplements. *J Trace Elements Med Biol* 2015, *29*, 83–90. https://doi.org/10.1016/j.jtemb.2014.06.012.
73. Michalke, B. Selenium Speciation in Human Serum of Cystic Fibrosis Patients Compared to Serum from Healthy Persons. *J Chromatogr A* 2004, *1058* (1), 203–208. https://doi.org/10.1016/j.chroma.2004.08.063.

74. Bendahl, L.; Gammelgaard, B.; Jøns, O.; Farver, O.; Hansen, S. H. Interfacing Capillary Electrophoresis with Inductively Coupled Plasma Mass Spectrometry by Direct Injection Nebulization for Selenium Speciation. *J Anal At Spectrom* 2001, *16* (1), 38–42. https://doi.org/10.1039/B007137O.
75. Grenville Holland; Dmitry Bandura. Speciation of Selenium Compounds by Capillary Electrophoresis-ICP-MS. Evaluation of ICP-SF-MS/ICP-DRC-MS Detection of Different Quantification Methods. In *Plasma Source Mass Spectrometry*; Holland, G., Bandura, D., Holland, J. G., Bandura, D. R., Eds.; The Royal Society of Chemistry, 2005; pp 28–42. https://doi.org/10.1039/9781847552419-00028.
76. Prange, A.; Pröfrock, D. Application of CE–ICP–MS and CE–ESI–MS in Metalloproteomics: Challenges, Developments, and Limitations. *Anal Bioanal Chem* 2005, *383* (3), 372–389. https://doi.org/10.1007/s00216-005-3420-0.

# 6 Last Developments in Speciation and Bioavailability Studies

## *Important Regulatory Implications for Health Protection*

*Juan José López-Mayán, Ana Justo-Vega, Elena Peña-Vázquez, María Carmen Barciela-Alonso, Antonio Moreda-Piñeiro, and Pilar Bermejo-Barrera*

### 6.1 STATE OF THE ART

In recent years, speciation and bioavailability studies have received special attention due to the significant differences in toxicity, mobility, bioavailability, and reactivity among the various chemical forms of the same element. Speciation refers to the identification and quantification of the distinct chemical forms of an element, whereas bioavailability refers to the fraction of a species that can be absorbed by an organism. Together, these provide crucial information on environmental behaviour, human exposure, and health risks, beyond what total elemental concentrations can reveal.

Speciation analysis usually employs combined instrumental techniques and approaches, and has become a pillar of modern analytical chemistry, with broad applications in toxicology, environmental science, and food safety. Reflecting the growing regulatory relevance of speciation data, in 2025 the European Food Safety Authority (EFSA) issued a call for the continuous collection of occurrence data on chemical contaminants in food and feed, including arsenic (inorganic arsenic [iAs] and total arsenic [tAs]), mercury (methylmercury [MeHg] and total mercury [tHg]), cadmium (Cd), lead (Pb), inorganic tin (iSn), per- and polyfluoroalkyl substances (PFAS) and even nanomaterials (NMs), to support improved risk assessment and regulatory decision-making [1].

As illustrated in Figure 6.1, the number of scientific publications related to elemental speciation and bioavailability has grown over the past decades. According

DOI: 10.1201/9781003632092-7

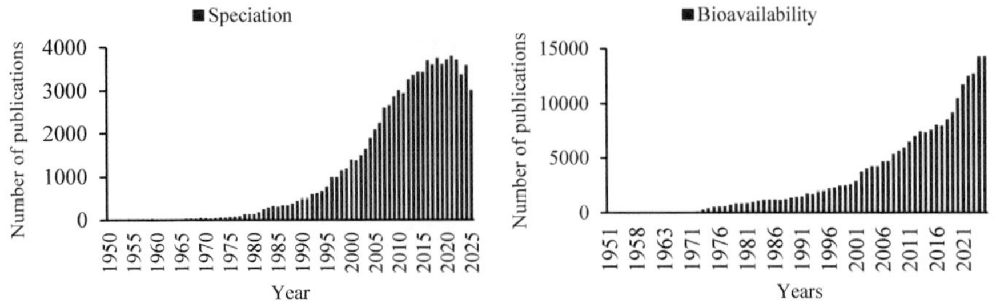


**FIGURE 6.1** Evolution of scientific publications related to elemental speciation and bioavailability over the past decades [3].

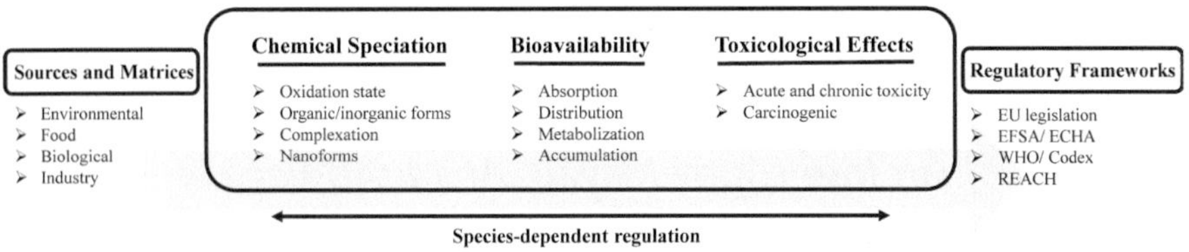


**FIGURE 6.2** Conceptual overview of the role of chemical speciation and bioavailability in regulatory risk assessment.

to the European Virtual Institute for Speciation Analysis (EVISA) this growth also underscores the expanding recognition of speciation in routine practices, research and regulatory frameworks [2, 3]. Despite this progress, most international regulations continue to rely on total element concentrations, as limited toxicological information has hindered the adoption of speciation-based regulation.

Trace elements such as As, Cr, and Hg are classic examples of how chemical form dictates their health risks and environmental behaviour. Other contaminants including Pb, organotin (oSn) compounds such as tributyltin (TBT), and ENMs, have emerged as substances of concern due to their widespread presence, potential for bioaccumulation, and still incompletely characterized toxicological profiles. Consequently, identifying the specific chemical species present in food, water, and environmental or clinical samples is essential for accurate risk assessment. Figure 6.2 provides a conceptual overview of how chemical speciation and bioavailability link contamination sources with toxicological outcomes and regulatory frameworks.

An overview of the main contaminants addressed in this chapter, including their predominant chemical species, major food matrices affected, principal toxic effects, and key regulatory limits in the EU, are summarised in Table 6.1.

## 6.2 ARSENIC

Arsenic (As) is a trace element ubiquitously present in air, water, soil, animals, and plants, and it is introduced into the environment via both natural and anthropogenic sources [4]. The toxicity and bioavailability of this element are strongly influenced by

**TABLE 6.1**
**Overview of Major Elements and Contaminants Discussed in This Chapter, Including Main Sources, Predominant Chemical Species, Toxicological Effects, and Key Regulatory Limits**

| Element/Compound | Main Environmental Sources | Major Food Matrices Affected | Predominant Species Forms | Main Toxic Effects | Key Regulatory Limits |
|---|---|---|---|---|---|
| Arsenic (As) | Natural weathering, volcanic activity, mining, fertilizers, pesticides | Rice, cereals, seafood, seaweed | iAs (III), iAs(V), DMA, MMA, AsB, arsenosugars | Carcinogenic (skin, lung, bladder), cardiovascular, genotoxic | EU: Water: 10 µg $L^{-1}$; rice/seafood 0.10–0.30 mg $kg^{-1}$ |
| Chromium (Cr) | Rocks, soils, volcanic emissions, dust, industrial emissions (metal plating, pigments, leather tanning, steel manufacturing), fertilizers, contaminated water | Cereals, vegetables, fruits, meat, fish, eggs, rice | Cr(III), Cr(VI) | Cr(VI): carcinogenic (lung, nasal cavity, paranasal, sinuses), dermal irritation, ulceration, sensitization Cr(III): essential micronutrient, low toxicity | EU Drinking water: 50 µg $L^{-1}$ (transitional to 25 µg $L^{-1}$ by 2036), food contact materials: SML Cr(III) 3.6 mg $kg^{-1}$; leather goods: 3 mg $kg^{-1}$ (2 mg $kg^{-1}$ for baby products); fertilizers: Cr(VI) max 2 mg $kg^{-1}$ dry matter; toys: Cr(VI) 0.02–0.053 mg $kg^{-1}$ depending on type |
| Mercury (Hg) | Natural: volcanic emissions, crustal degassing, rock weathering Anthropogenic: coal combustion, mining, waste incineration, dental amalgams, PVC production, fertilizers | Fish and seafood (predatory fish like tuna, swordfish, shark), and rice (especially in Hg-contaminated soils) | Elemental Hg ($Hg^0$), inorganic Hg ($Hg_2^{2+}$, $Hg^{2+}$), organic Hg (methylmercury MeHg, dimethylmercury $Me_2Hg$, phenylmercury PhHg), MeHg-Cys complexes | Neurological disorders, cardiovascular, renal, hepatic and immune system toxicity, developmental neurotoxicity (especially in infants), hormonal imbalances, skin lesions, vision impairment, reproductive toxicity, genotoxicity | EU: Fish and seafood: 0.5 mg $kg^{-1}$ wet weight (general), 1.0 mg $kg^{-1}$ (predatory fish), 0.3 mg $kg^{-1}$ (cephalopods, gastropods), 0.1 mg $kg^{-1}$ (food supplements, salt). Drinking water: 1 µg $L^{-1}$ |

*(Continued)*

**TABLE 6.1** *(Continued)*
**Overview of Major Elements and Contaminants Discussed in This Chapter, Including Main Sources, Predominant Chemical Species, Toxicological Effects, and Key Regulatory Limits**

| Element/Compound | Main Environmental Sources | Major Food Matrices Affected | Predominant Species Forms | Main Toxic Effects | Key Regulatory Limits |
|---|---|---|---|---|---|
| Lead (Pb) | Mining and smelting, coal combustion, batteries, pigments, gasoline residues, fertilizers, industrial emissions | Milk, infant formula, fruits, vegetables, cereals, seafood | $Pb^{2+}$, TML, TEL, TTEL | Neurotoxicity, developmental toxicity, immune and cardiovascular effects | Drinking water: 10 μg $L^{-1}$ (5 μg $L^{-1}$ future target); Food: 0.02–1.5 mg $kg^{-1}$ (matrix-dependent); EEE: ≤0.1% (w $w^{-1}$) |
| Organotin compounds (OTCs) | Antifouling paints, plastics manufacturing, industrial applications, contaminated sediments | Seafood, drinking water (PVC pipes) | MBT, TMT, TET, TBT, TPhT | Endocrine disruption, neurotoxicity, immunotoxicity, reproductive toxicity | TBT banned in antifouling paints; Restricted under REACH; migration limits apply depending on application |
| PFAS | Industrial manufacturing, firefighting foams, food-contact materials, textiles, coatings | Drinking water, food, food packaging | PFOA, PFOS, short- and long-chain PFAS | Hepatic, endocrine, immune, metabolic and reproductive toxicity; developmental effects | Drinking water: PFAS total 0.50 μg $L^{-1}$; Sum of PFAS 0.10 μg $L^{-1}$; PFOS restricted under POPs regulation |
| Engineered nanomaterials (ENMs) | Manufactured nanoproducts, industrial processes, consumer products, waste streams | Food additives, food-contact materials, drinking water, supplements | Nanoforms (metal/metal oxide NPs, carbon-based NMs, polymeric NPs) | Oxidative stress, inflammation, genotoxicity; effects depend on size, surface, solubility and persistence | Not concentration-based; regulation relies on nanoform characterization, particle size, solubility and surface properties (EFSA guidance, sector-specific legislation) |

*Abbreviations:* AsB, arsenobetaine; DMA, dimethylarsinic acid; MBT, monobutyltin; MMA, monomethylarsonic acid; PFAS, polyfluoroalkyl substances; PFOA, perfluorooctanoic acid; PFOS, perfluorooctanesulfonic acid; PVC, polyvinyl chloride; sTML, trimethyllead chloride; TEL, triethyllead chloride; TTEL, tetraethyllead; TMT, trimethyltin; TET, triethyltin; TBT, tributyltin; TPhT, triphenyltin.

its concentration and chemical form. The inorganic As species (iAs), which include trivalent arsenite [As(III)] and pentavalent arsenate [As(V)], are approximately two orders of magnitude more toxic than the organic forms (oAs), such as monomethylarsonic acid (MMA) and dimethylarsinic acid (DMA) or tetramethylarsonium. Other organic species, such as arsenobetaine (AsB) and arsenosugars, as well as arsenolipids exhibit low acute toxicity or still unclear toxicological significance, respectively [5]. Moreover, trivalent As is about 60 times more toxic than the pentavalent form [6].

The International Agency for Research on Cancer (IARC) has classified iAs compounds as carcinogenic to humans (Group 1). Certain oAs compounds, such as DMA, are classified as possibly carcinogenic to humans (Group 2B), whereas arsenobetaine (AsB) and other oAs species commonly found in seafood are classified in Group 3 (not classifiable as to their carcinogenicity to humans) [7].

EFSA has conducted several scientific evaluations on iAs in food, concluding that chronic exposure can increase the risk of skin, bladder, and lung cancer [8], because it induces genetic and epigenetic damage, mainly through oxidative stress, inhibition of DNA repair, and disruption of cell-cycle and apoptotic pathways [8]. Accordingly, Directive (EU) 2020/2184, sets a maximum allowable concentration of 10 $\mu g\ L^{-1}$ of As in water for human consumption [9]. Nevertheless, in countries with limited resources, the outdated 1963 World Health Organization (WHO) guideline value of 50 $\mu g\ L^{-1}$ is still applied [10].

Maximum allowed levels (MLs) are detailed in Commission Regulation (EU) 2023/915 [11] and were recently amended by Regulation (EU) 2025/1891 [12] to include MLs of iAs in fish and other seafood like seaweed [13]. EU legislation establishes MLs for iAs of 0.15 mg $kg^{-1}$ in non-parboiled milled rice, 0.25 mg $kg^{-1}$ in parboiled and husked rice and rice flour, and up to 0.30 mg $kg^{-1}$ in rice-based breakfast and snack products. For infant foods, limits range from 0.010 to 0.020 mg $kg^{-1}$ depending on the product type. In fish and seafood, MLs range from 0.050 to 0.50 mg $kg^{-1}$ depending on species, while specific crustaceans may reach up to 1.5 mg $kg^{-1}$. For bivalve molluscs, limits vary between 0.10 and 0.50 mg $kg^{-1}$, whereas, in cephalopods, the ML is set at 0.050 mg $kg^{-1}$. For edible salt, a ML for tAs of 0.50 mg $kg^{-1}$ applies. Beyond European legislation, international reference values have also been established by the Codex Alimentarius Commission. Codex has set MLs of iAs of 0.20 mg $kg^{-1}$ in polished rice and 0.35 mg $kg^{-1}$ in husked rice, serving as harmonized benchmarks for global food safety and international trade [8]. Although regulations establish limits for iAs and tAs in different foods, no specific legislation currently addresses oAs compounds [5].

From an analytical perspective, high-performance liquid chromatography coupled with inductively coupled plasma mass spectrometry (HPLC-ICP-MS) remains the most widely used technique for As speciation, with proposals of rapid, semi-quantitative approaches for a wide range of foodstuffs [14]. The combination of electrospray ionization mass spectrometry (ESI-MS) with ICP-MS has been employed to detect and quantify both known and previously unidentified As species. Charged As species can be separated using ion exchange chromatography, while hydrophilic interaction liquid chromatography (HILIC) allows the simultaneous separation of neutral, cationic and anionic species, offering advantages for comprehensive As speciation in complex matrices [15]. More recently, non-chromatographic approaches have also been explored for rapid iAs speciation in the food industry. Polyethylene glycol-coated silver nanoparticles (NPs) combined with UV-Vis detection have been applied to rice, yielding results comparable to those obtained by chromatographic methods [16]. Most studies report iAs, MMA, and DMA as the predominant species in rice [17], although novel oAs compounds have also been identified [16]. The bioavailability of As species has been studied using an in vitro dialysis method simulating gastrointestinal digestion. Changes in species distribution were observed, including a decrease in As(III), which lowered the risk associated with rice consumption, and an increase in As(V) [18].

Arsenic speciation has been performed in seaweed species used for human consumption and animal feed [19, 20]. While tAs concentrations may be high, speciation studies often reveal marked interspecies variability in iAs content, as observed among different Laminaria species [20, 21]. Arsenosugars are mainly present in seaweed, particularly in brown algae, whereas arsenolipids are generally detected at low concentrations (<10%) in fish, crustaceans, molluscs, and seaweed [5, 22]. Bioavailability studies showed that 45–80% of iAs(III) and iAs(V) can be absorbed from the gastrointestinal tract and distributed to different tissues [8]. Organoarsenic compounds predominate, with AsB typically accounting for approximately 90% of tAs in seafood. However, in vitro gastrointestinal studies, showed that only 12–16% of tAs in edible seaweed is bioavailable, and dialyzable As was present mostly as arsenosugars [23].

Finally, As is routinely studied in human urine for biomonitoring purposes. Speciation analysis using anion-exchange chromatography coupled with ICP-MS has been widely applied to assess occupational and environmental exposure [24]. Blood, hair, and nails have also been proposed as complementary biomarkers, with blood often providing the most reliable indicator of recent exposure in occupational settings [25].

## 6.3 CHROMIUM

Chromium (Cr) is a naturally occurring metal found throughout the environment, including rocks, soils, volcanic emissions, and dust. Cr(III) is considered an essential trace element and is mainly supplied through the diet, with a bioavailability of approximately 1–2% [26]. Cr(III) is also present as an active component in various commercially available formulations, including dietary supplements and pharmaceutical products [26]. In contrast, Cr(VI) is primarily generated through industrial activities and typically enters drinking water because of anthropogenic contamination [27]. In general, Cr(VI) compounds are much more toxic than Cr(III) compounds, with toxicity levels estimated to be 100 to 300 times higher [28]. The IARC has classified Cr(VI) as carcinogenic to humans (Group 1), while Cr(III) is classified as non-carcinogenic (Group 3) [29]. The toxicological behaviour of Cr(VI) further depends on the solubility of the compounds, with poorly soluble chromates ($ZnCrO_4$, $BaCrO_4$, and $PbCrO_4$) generally regarded as more hazardous than soluble forms, since the latter are more readily reduced to Cr(III) in biological systems. Consequently, quantification of total Cr(VI), including both soluble and insoluble fractions, is essential [30]. Environmental persistence and mobility are largely attributed to Cr(VI) which exhibits higher solubility and cellular uptake than Cr(III), whose oral toxicity is limited by poor bioavailability.

Cr is regulated under the Directive (EU) 2020/2184 on the quality of water used for human consumption [9]. During a transitional period pending WHO toxicological reassessment, the parametric value remains at 50 µg $L^{-1}$ but will be reduced to 25 µg $L^{-1}$ by January 2036 [9]. Discharge limits for Cr in wastewater are set at the national level, with reported permissible Cr(VI) concentrations ranging from 0.05 to 2 mg $L^{-1}$ across Europe [31]. EU Member States must also report emissions exceeding 50 kg $year^{-1}$ to water and 100 kg $year^{-1}$ to air from individual facilities [32].

Cr is further regulated in materials and articles in contact with food. Regulation (EU) 2020/1245 established that the migration of total Cr must not exceed the limit of detection (0.1 mg $kg^{-1}$ in food or food simulants) unless manufacturers demonstrate that only Cr(III) is present, in which case a specific migration limit of 3.6 mg $kg^{-1}$ applies [33]. Regulation (EU) 2018/725 limits Cr(VI) migration from children's toys, with a maximum allowable level depending on toy type [34].

In response to the recognised hazards, the EU has continued to strengthen regulatory controls. In 2024, the European Chemicals Agency (ECHA) was mandated to expand the Registration, Evaluation, Authorisation and restriction of Chemicals (REACH) restriction proposal to include additional Cr(VI) compounds widely used in industrial applications. In 2025, ECHA proposed a new EU-wide restriction aimed at replacing the authorisation system, potentially preventing up to 17 tonnes of Cr(VI) emissions annually and avoiding nearly 200 cancer cases per year across the EU [35]. Cr levels in organic fertilizers are regulated under Regulation (EU) 2019/1009 [36], which sets a maximum Cr(VI) concentration of 2 mg $kg^{-1}$ dry matter.

Speciation studies in environmental matrices such as water and soil are essential for assessing Cr distribution and exposure pathways. Numerous studies have reported preferential occurrence of Cr(VI) in groundwater compared to surface waters [27], and rapid monitoring tools, including fluorescence sensor arrays, are under active development for drinking water applications [37]. Cr speciation is also relevant in atmospheric studies, where interconversion during sampling can lead to overestimation of Cr(VI), but improved extraction and analytical protocols have been developed to minimise such artefacts [30, 38]. For instance, Spinazzè et al. developed a 48-hour Cr extraction method for airborne particulates in Cr(III)-rich occupational environments using an isotope-enriched spike addition technique and HPLC-ICP-MS [30]. Speciation studies have been also applied to assess Cr release from tanned leather using chromatographic techniques coupled with ICP-MS [39].

Urinary monitoring remains the primary approach for assessing Cr exposure, particularly for Cr(VI). Recent developments include point-of-care testing devices offering non-invasive and cost-effective tools for occupational and clinical monitoring [40]. In blood, Cr(III) is predominantly protein-bound, whereas Cr(VI) may circulate as exchangeable species [41]. Selective and sensitive methodologies have been successfully developed for Cr speciation in blood, demonstrating the feasibility of accurate assessment in complex biological matrices [42].

## 6.4 MERCURY

Mercury (Hg) originates from both natural and anthropogenic sources (Table 6.1), and it is recognized by the WHO as one of the top ten chemicals of major public health concern due to its persistence, bioaccumulation potential, and high toxicity. Hg occurs in multiple chemical forms, including elemental mercury ($Hg^0$), inorganic species (iHg: mercurous Hg(I) and mercuric Hg(II)), and organic derivatives such as MeHg, dimethylmercury ($Me_2Hg$), and phenylmercury (PhHg), with MeHg being the most toxic and environmentally relevant form [43, 44]. Regulatory agencies classify these species differently. The United States Environmental Protection Agency (USEPA) categorises $Hg^0$ as Group D and iHg and MeHg as Group C, whereas the IARC classifies $Hg^0$ and iHg as Group 3 (not classifiable as to its carcinogenicity to humans) and MeHg as Group 2B (possibly carcinogenic to humans) [29].

Once released into the environment, Hg enters ecosystems primarily as elemental $Hg^0$ and Hg(II) that can be transformed by microorganisms into MeHg, which is subsequently taken up by multicellular organisms as MeHg-cysteinate complexes [45]. Human exposure to MeHg mainly occurs through the consumption of fish and seafood [44] and historical cases like the Minamata disaster highlight the severe neurodevelopmental and systemic effects of MeHg [43]. Bioavailability studies show that approximately 90–95% of dietary MeHg is absorbed in the gastrointestinal tract and binds to haemoglobin, enabling it to cross the central nervous system and the placenta. In contrast, iHg is generally less toxic, as it is rapidly excreted via feces, with minimal urinary and biliary elimination [46].

The Commission Regulation (EU) 2023/915 [11], defines current MLs for Hg in foodstuffs. These limits primarily apply to fish and fishery products, with 0.5 mg $kg^{-1}$ wet weight for most species and 1.0 mg $kg^{-1}$ for predatory fish such as tuna, swordfish, and shark, reflecting their bioaccumulation potential. Additional limits include 0.3 mg $kg^{-1}$ for cephalopods and marine gastropods and 0.1 mg $kg^{-1}$ for food supplements and salt [11]. For drinking water, Directive (EU) 2020/2184 [9] establishes a maximum admissible concentration of 1 µg $L^{-1}$ of tHg, which also applies to natural mineral waters under Directive 2003/40/EC [47]. Analytical performance criteria for Hg determination in foodstuffs are specified in Regulation (EU) 836/2011 [48]. At the international level, the Codex Alimentarius has established guideline values for total MeHg of 0.5 mg $kg^{-1}$ for non-predatory fish and 1 mg $kg^{-1}$ for predatory species [49].

Mercury speciation is commonly carried out using chromatographic methods (GC, HPLC, and capillary electrophoresis) coupled to detectors including MS, AAS, AFS, cold-vapour systems, or ICP-MS, providing high sensitivity and selectivity [43]. Extensive studies on Hg speciation and bioavailability in fish tissues have been carried out to improve risk assessment models and ensure seafood [46, 50]. Recent studies have shown that some species, such as marlins, may contain high proportions of iHg, suggesting species-specific metabolic pathways and in vivo demethylation processes that reduce MeHg toxicity through the formation of stable Hg-selenium complexes [50].

In clinical contexts, biomonitoring studies have identified Hg species in breast milk, urine, blood, and hair by HPLC-ICP-MS [51–54]. Recent high-throughput methods allow simultaneous measurement of Hg(II) and MeHg, showing no clear correlation between MeHg levels in breast milk and maternal seafood consumption [51].

Finally, Hg speciation is equally critical in industrial contexts, e.g. in processes such as polyvinyl chloride (PVC) production and petroleum refining. Thus, Wang et al. developed an advanced Hg speciation platform based on the Ontario Hydro Method (OHM), enabling real-time, simultaneous detection of elemental and divalent Hg to support monitoring and risk management in chemical manufacturing [55].

## 6.5 OTHER ELEMENTS

### 6.5.1 Lead

Lead (Pb) is a soft, malleable, and ductile metal with broad industrial applications. In the environment, Pb has no known biological function and is considered one of the most abundant and toxic trace metals in soils. Organic Pb compounds (oPb), such as trimethyllead chloride (TML), triethyllead chloride (TEL) and tetraethyllead (TTEL), are significantly more toxic than Pb(II). Human exposure occurs mainly through the food chain, causing adverse effects on the immune, digestive and nervous systems, particularly in young children. At the cellular level, it induces neuronal damage, apoptosis, disruption of neurotransmitter storage and release, and lipid peroxidation [56–58].

In response to these health risks, the adopted Directive (EU) 2024/869 updated the occupational exposure limits for Pb and its inorganic compounds in the workplace. The inhalable exposure limit was lowered from 0.15 mg $m^{-3}$ to 0.03 mg $m^{-3}$ for improving worker protection [59]. Drinking water quality is also regulated by Directive (EU) 2020/2184 [9], which established Pb as a controlled chemical parameter. In line with WHO recommendations, the current parametric value of 10 μg $L^{-1}$ may be maintained during a transitional period until 2036, after which a binding limit of 5 μg $L^{-1}$ will apply to all new installations and water-contact materials, strengthening long-term public health protection [9]. In addition, the use of Pb in electrical and electronic equipment is restricted under the Directive (EU) 2011/65, which limits Pb content to a maximum of 0.1% by weight in each homogeneous material, with exemptions permitted only where technical substitution is not feasible [60]. Maximum levels of Pb in food are regulated under Regulation (EU) 2023/915 to minimise exposure, particularly in vulnerable populations such as infants and children. The limits are set at 0.020 mg $kg^{-1}$ for raw milk, 0.020–0.050 mg $kg^{-1}$ for infant formula, 0.10–0.20 mg $kg^{-1}$ for fruits, 0.10 mg $kg^{-1}$ for root and tuber vegetables, 0.50 mg $kg^{-1}$ for crustaceans, and 1.50 mg $kg^{-1}$ for bivalve molluscs, with similar values established for other food categories [11].

Currently, analytical approaches for Pb speciation rely on separation techniques such as HPLC and GC, coupled with element-specific detection systems like ICP-MS. HPLC is often preferred to GC since it operates at room temperature, without derivatization, and has simplified sample preparation [58]. These methods are widely applied in environmental monitoring of waters [61] and soils [57, 62], as well as in food safety studies involving fish and seafood [56] and crops like rice [58], providing detailed information on the distribution of Pb(II), TML, TEL, and TTEL, to support risk assessment and mitigation strategies.

### 6.5.2 Tin

Tin (Sn) occurs in inorganic forms ($Sn^{2+}$, $Sn^{4+}$) and organic forms (oSn) including monobutyltin (MBT), trimethyltin (TMT), triethyltin (TET), tributyltin (TBT), and triphenyltin (TPhT). These compounds have been widely used as biocides, particularly in antifouling paints, as well as in agricultural and industrial applications [63].

TBT was banned under Regulation (EC) 782/2003 [64], complemented by the International Maritime Organization's (IMO) Anti-Fouling Systems (AFS) Convention, in force since 2008. The oSn compounds are also regulated under the REACH Regulation (EC) 1907/2006, which controls their manufacture, marketing, and use to protect human health and the environment [65].

Analytical methods like capillary electrophoresis, GC coupled with tandem mass spectrometry (MS/MS), and HPLC combined with ICP-MS are employed in oSn speciation analysis for sediments [66], seafood [56, 67], consumer products such as tobacco [68], workplace air [63], and drinking water distributed through PVC piping systems [69].

### 6.5.3 Perfluoroalkyl and Polyfluoroalkyl Substances

Perfluoroalkyl and polyfluoroalkyl substances (PFASs) are a large group of synthetic fluorinated compounds with numerous applications (Table 6.1) and characterised by strong carbon–fluoride bonds, which confer exceptional chemical stability, resistance to degradation and persistence. This group includes historically used compounds such as perfluorooctanoic acid (PFOA), and perfluorooctanesulfonic acid (PFOS), as well as numerous short-chain and polymeric PFAS. They have been associated with adverse effects on hepatic, endocrine, immune, metabolic, and reproductive systems, as well as developmental outcomes.

Several PFAS, including PFOA and PFOS, are regulated under the EU Persistent Organic Pollutants Regulation (EU) 2019/1021, implementing obligations under the Stockholm Convention. PFOS is subject to strict concentrations thresholds (typically <10 mg $kg^{-1}$ in substances or mixtures and <0.1% (w $w^{-1}$) in articles), effectively prohibiting its use except for limited essential applications, such as anti-mist agents in closed-loop hard-chrome plating, under stringent emission controls [70]. Furthermore, the EU Drinking Water Directive (EU) 2020/2184 [9] introduced parametric values for PFAS of: 0.50 μg $L^{-1}$ for "PFAS Total", covering the full chemical class (to become enforceable once monitoring guidelines are finalised), and 0.10 μg $L^{-1}$ for the "Sum of PFAS", covering a defined subset of perfluoroalkyl and perfluoroalkyl ether compounds of ≥3 carbons [9]. Several Member States have adopted stricter national measures, including France, which enacted legislation in 2025 restricting PFAS-containing consumer products and expanding drinking water monitoring [71].

Despite these regulations, significant analytical challenges remain, and a wide range of methods has been developed, with water samples being the preferred matrix treated by solid-phase extraction. Detection is commonly performed by HPLC or UHPLC coupled to tandem or high-resolution mass spectrometry [72]. These techniques have been applied to PFAS speciation in water [73], soil [74], air [75], food packaging [76], PFAS-treated textiles [77], and biological samples such as bird feathers [78], human hair [79], plants [80], and aquatic organisms [81].

## 6.6 NANOMATERIALS

NMs are defined in the EU according to the Commission Recommendation 2022/C229/01, as incidental or manufactured materials (engineered nanomaterials [ENMs]) composed of particles, present as individual particles, aggregates or agglomerates, of which at least 50% have one or more external dimension in the nanometric range (1–100 nm) [82]. Owing to their unique properties [83, 84], nanotechnology is increasingly integrated into everyday products, resulting in a rapid rise in human and environmental exposure. By the end of 2025, the Nanotechnology Products Database (NPD) listed more than 11,000 nanoproducts worldwide [85].

ENMs are classified as inorganic, organic, and carbon-based materials. Common examples include metal and metal oxide NPs (e.g., Ag, Au, $TiO_2$, ZnO), carbon-based NMs (e.g., fullerenes, carbon nanotubes, and graphene derivatives), semiconductor quantum dots, and polymeric or silica-based nanostructures [84].

The toxicity of NMs depends on multiple parameters, including composition, particle size, shape, surface charge, functionalization, solubility, stability, and bioavailability. Inorganic NPs such as silver, gold, titanium dioxide, and iron, may induce oxidative stress, inflammation, DNA damage, and organ-specific toxicity, particularly at high doses or smaller particles sizes. Carbon-based NMs including graphene, graphene oxide, fullerenes, and carbon quantum dots, may generate reactive oxygen species (ROS) and cause DNA and mitochondrial damage. In contrast, organic NPs generally exhibit lower toxicities, although exposure to high doses may sometimes result in adverse effects. Surface modification strategies, such as PEGylation or biocompatible coatings, can reduce these effects [84].

Although regulatory authorities increasingly recognize the need to address nanoscale-specific properties, harmonized regulatory approaches remain limited, due to the rapid evolution of nanotechnology, the absence of universally validated metrological standards, and the difficulty of defining exposure threshold applicable across highly diverse NM classes. Within the EU, NMs are explicitly addressed in key regulations that include REACH (EU 2018/1881) [86], the Medical Devices Regulation (EU 2017/745) [87], the Biocidal Products Regulation (EC 528/2012) [88], the Novel Foods Regulation (EU 2015/2283) [89], and the Cosmetic Products Regulation (EC 1223/2009) [90]. In the food and feed sector, EFSA has issued specific guidance documents on the risk assessment of NMs and on the technical requirements to establish the presence of small particles, including NPs, in regulated products [91, 92]. They provide detailed criteria for the characterization, exposure assessment, and toxicological evaluation of NMs in food and feed, reflecting the application of the EC NM definition in a sector-specific context. EFSA has further evaluated several NMs used as food additives and colourants, including silver (E174) [93], gold (E175) [94], iron oxides (E172) [95], silicon dioxide (E551) [96], and titanium dioxide (E171) [97]. Among these, E171 was concluded to be no longer safe and has been banned in the EU, while the remaining substances remain authorized under strict conditions related to particle size, solubility, and surface properties. At the international level, the Organisation for Economic Co-operation and Development (OECD) has developed test guidelines specifically adapted for NMs, addressing physicochemical properties, environmental fate, and toxicological effects, while ISO is developing complementary standards to support harmonized testing and regulatory compliance [98].

From an analytical perspective, sample preparation is a major challenge, as NMs may aggregate, interact with complex matrices, or undergo transformations that affect measurement accuracy and reproducibility. Commonly applied techniques for characterization and speciation include electron microscopy (TEM, SEM) for direct imaging, dynamic light scattering (DLS), nanoparticle tracking analysis (NTA) for hydrodynamic sizing, and single-particle inductively coupled plasma mass spectrometry (sp-ICP-MS) for particle number concentration and size determination. The coupling separation techniques such as field-flow fractionation (FFF) or hydrodynamic chromatography (HDC) to ICP-MS or light-scattering detectors enables the differentiation of NPs from ions and agglomerates, improving quantification and speciation in environmental and biological samples [99].

Due to their unique behaviour and potential toxicity, numerous studies have investigated speciation, environmental fate, bioavailability, and toxicology of NMs across diverse matrices. In soils, metal oxide NPs have been shown to influence the chemical speciation of co-occurring elements and their bioavailability in soils [100]. In aquatic systems, studies with fish demonstrate that nanoscale $TiO_2$, ZnO, and $CeO_2$ can be taken up from the water column and accumulate in tissues such as gills, confirming bioaccumulation and direct bioavailability of particulate forms to aquatic organisms [101]. In mammalian models, inhalation exposure experiments in rats with metal-based NPs such as $TiO_2$ and $CeO_2$ have shown that NP-derived elements can translocate from the lungs to secondary organs, including liver, spleen, kidneys, blood, and excreta. These findings demonstrate that NM bioavailability depends on particle persistence, partial dissolution, and biological interactions, highlighting the need to consider both particulate and transformed species when assessing internal exposure [102]. Methodologies combining sp-ICP-MS and single-cell ICP-MS (sc-ICP-MS) have been recently developed to study the bioavailability and cellular transport of Ag and $TiO_2$ NPs in seaweed and mussels [103]. On the other hand, the use of ZnO and Ag NPs have been explored for aquaculture wastewater treatment, where their application altered pollutant profiles and microbial populations, transforming environmental chemistries and potentially affecting the bioavailability of contaminants [104].

## 6.7 FINAL REMARKS

Chemical speciation and bioavailability have emerged as central concepts in modern risk assessment, providing insights that cannot be obtained from total concentration measurements alone. As demonstrated throughout this chapter, the toxicity, mobility, and persistence of many contaminants are strongly dependent on their chemical form, making speciation analysis indispensable for accurately assessing human and environmental exposure. Elements such as As, Cr, and Hg exemplify how differences in oxidation state or molecular structure translate into profoundly different toxicological outcomes, while compounds such as PFAS and ENMs highlight the growing complexity of emerging contaminants.

Regulatory frameworks, particularly within the EU, are increasingly acknowledging the importance of speciation and nanoform characterization. However, most legislation still relies primarily on total concentrations, reflecting persistent gaps in toxicological knowledge, analytical standardization, and harmonized testing protocols.

Ultimately, the systematic consideration of chemical speciation and bioavailability will enhance the capacity to manage both traditional and emerging contaminants, supporting safer environmental practices, improved food safety, and the responsible development of new technologies.

## REFERENCES


1. EFSA. 2025. Annual Call for Continuous Collection of Chemical Occurrence Data in Food and Feed. Accessed December 26, 2025. https://www.efsa.europa.eu/en/call/annual-call-continuous-collection-chemical-contaminants-occurrence-data-food-and-feed.
2. EVISA. 2025. EVISA. Accessed October 8, 2025. https://speciation.net/Public/Linklists/EVISA.html.
3. Scopus Database. 2025. Scopus Database. Accessed October 8, 2025 https://www-scopus-com.ezbusc.usc.gal/pages/home?display=basic#basic.
4. Chen, Q. Y., and M. Costa. 2025. Arsenic: A Global Environmental Challenge. *Annual Review of Pharmacology and Toxicology* 61: 47–63. https://doi.org/10.1146/annurev-pharmtox-030220-013418.
5. Knutsen, H. K., A. Åkesson, V. Bampidis, M. Bignami, L. Bodin, J. K. Chipman, G. Degen, et al. 2024. Risk Assessment of Complex Organoarsenic Species in Food. *EFSA Journal* 22: e9112 (1–78). https://doi.org/10.2903/j.efsa.2024.9112.
6. Li, Y., C. Wang, J. Zhao, X. Liu, M. Ma, A. Gao, and K. Sun. 2025. A Systematic Review of Separation/Preconcentration and Detection Techniques for Speciation Analysis of Arsenic and Selenium in Water. *Analytical Methods* 17: 5807–5830. https://doi.org/10.1039/D5AY00626K.
7. International Agency for Research on Cancer. 2004. Some Drinking-Water Disinfectants and Contaminants, Including Arsenic. *IARC Monographs on the Evaluation of Carcinogenic Risks to Humans*, Vol. 84. Lyon, France: IARC Press.
8. Schrenk, D., M. Bignami, L. Bodin, J. K. Chipman, J. del Mazo, B. Grasl-Kraupp, C. Hogstrand, et al. 2024. Update of the Risk Assessment of Inorganic Arsenic in Food. *EFSA Journal* 22 (1): e8488. https://doi.org/10.2903/j.efsa.2024.8488.
9. European Parliament and Council. 2020. Directive (EU) 2020/2184 of the European Parliament and of the Council of 16 December 2020 on the quality of water intended for human consumption (recast). *Official Journal of the European Union* L 435, 23 December 2020.
10. Frisbie, S. H., and E. J. Mitchell. 2022. Arsenic in Drinking Water: An Analysis of Global Drinking Water Regulations and Recommendations for Updates to Protect Public Health. *PLoS ONE* 17 (4): e0263505 https://doi.org/10.1371/journal.pone.0263505.
11. European Commission. 2023. Commission Regulation (EU) 2023/915 of 25 April 2023 on maximum levels for certain contaminants in food and repealing Regulation (EC) No 1881/2006. *Official Journal of the European Union* L 119, 5 May 2023.

12. European Commission. 2025. Commission Regulation (EU) 2025/1891 of 17 September 2025 amending Regulation (EU) 2023/915 as regards maximum levels of inorganic arsenic in fish and other seafood. *Official Journal of the European Union* L 237, 18 September 2025.
13. EFSA. 2009. Scientific Opinion on Arsenic in Food. 2009. *EFSA Journal* 7 (10). https://doi.org/10.2903/j.efsa.2009.1351.
14. Jones, R. R., J. M. Chaparro, K. Maloley, J. C. Evans, S. Boyko, M. Gruszczynski, and J. E. Prenni. 2025. Rapid, Semi-Quantitative Arsenic Speciation across Diverse Food Matrices via Reverse Phase HPLC-ICP-MS. *Journal of Food Composition and Analysis* 147: 108000 (1–7). https://doi.org/10.1016/j.jfca.2025.108000.
15. Visciano, P. 2025. Arsenic in Water and Food: Toxicity and Human Exposure. *Foods* 14 (13): 2229 (1–15). https://doi.org/10.3390/foods14132229.
16. Lima, F. E. H., F. L. Fonsêca da Silva, A. A. Akhdhar, P. B. Almeida-Fechine, A. R. de Araujo-Nogueira, and W. Oliveira-Matos. 2025. Non-Chromatographic Inorganic Arsenic Speciation in Rice by Spectrophotometry Using Silver Nanoparticles as Platform Sensor. *Food Chemistry* 476: 143321 (1–8). https://doi.org/10.1016/j.foodchem.2025.143321.
17. Navaretnam, R., A. Z. Aris, M. F. A. Shukor, M. N. F. Norrrahim, V. F. Knight, T. T. Chin, N. M. Isa, and L. J. Looi. 2025. Arsenic Speciation Using HPLC–ICP–MS in White and Brown Rice and Health Risk Assessment. *Environmental Geochemistry and Health* 47: 411 (1–24). https://doi.org/10.1007/s10653-025-02723-2.
18. Domínguez-González, M. R., M. C. Barciela-Alonso, V. G. Calvo-Millán, P. Herbello-Hermelo, and P. Bermejo-Barrera. 2020. The Bioavailability of Arsenic Species in Rice. *Analytical and Bioanalytical Chemistry* 412 (13): 3253–3259. https://doi.org/10.1007/s00216-020-02589-6.
19. Oziegbe, O., E. J. Oziegbe, and O. Ojo-Omoniyi. 2025. Bioremediation of Heavy Metals in Aquatic Environment: A Review. *Cleaner Chemical Engineering* 11: 100193 (1–18). https://doi.org/10.1016/j.clce.2025.100193.
20. Raab, A., D. B. Stengel, and J. Feldmann. 2025. Differential Prevalence of Arsenic Speciation in the Kelps Laminaria Digitata and L. Hyperborea. *Marine Pollution Bulletin* 211: 117473 (1–6). https://doi.org/10.1016/j.marpolbul.2024.117473.
21. Ma, Z., L. Lin, M. Wu, H. Yu, T. Shang, T. Zhang, and M. Zhao. 2018. Total and Inorganic Arsenic Contents in Seaweeds: Absorption, Accumulation, Transformation and Toxicity. *Aquaculture* 497: 49–55. https://doi.org/10.1016/j.aquaculture.2018.07.040.
22. Luvonga, C., C. A. Rimmer, L. L. Yu, and S. B. Lee. 2020. Organoarsenicals in Seafood: Occurrence, Dietary Exposure, Toxicity, and Risk Assessment Considerations: A Review. *Journal of Agricultural and Food Chemistry* 68: 943–60. https://doi.org/10.1021/acs.jafc.9b07532.
23. Garcia-Sartal, C., S. Taebunpakul, E. Stokes, M. C. Barciela-Alonso, P. Bermejo-Barrera, and H. Goenaga-Infante. 2012. Two-Dimensional HPLC Coupled to ICP-MS and Electrospray Ionisation (ESI)-MS/MS for Investigating the Bioavailability in Vitro of Arsenic Species from Edible Seaweed. *Analytical and Bioanalytical Chemistry* 402 (10): 3359–3369. https://doi.org/10.1007/s00216-011-5483-4.
24. Heitland, P., A. Raab, and S. Krell. 2025. Unexpected Arsenic Compound in Routine Determination of Arsenic Species in Urine. *Journal of Trace Elements in Medicine and Biology* 92: 127745 (1–3). https://doi.org/10.1016/j.jtemb.2025.127745.
25. Hameed, S., S. Ashraf, S. Ashraf, N. Nazish, M. Sughra, and A. Ijaz. 2025. Blood, Hair, and Nail as Biomarkers of Arsenic Exposure among Leather Industry Workers in Sialkot. *Journal of Health, Wellness and Community Research* 3 (12) (1–5). https://doi.org/10.61919/avenwy68.
26. József, P., N. Dávid, K. Sándor, and B. Áron. 2019. A Comparison Study of Analytical Performance of Chromium Speciation Methods. *Microchemical Journal* 149: 103958 (1–8). https://doi.org/10.1016/j.microc.2019.05.058.

27. Pechancová, R., R. J. N. Bettencourt da Silva, D. Milde, and T. Pluháček. 2025. Bottom-up Measurement Uncertainty Evaluation for Cr(III) and Cr(VI) Speciation in Natural Water by HPLC-ICP-MS. *Microchemical Journal* 212: 113190 (1–7). https://doi.org/10.1016/j.microc.2025.113190.
28. Song, G., H. Tan, C. Cheng, P. Li, X. Sun, Y. Zhou, and Y. Fang. 2024. Development of a Fast Method Using Inductively Coupled Plasma Mass Spectrometry Coupled with High-Performance Liquid Chromatography and Exploration of the Reduction Mechanism of Cr(VI) in Foods. *Toxics* 12 (5): 325 (1–12). https://doi.org/10.3390/toxics12050325.
29. *IARC*. 2025. "IARC List of Classifications". Accessed October 27, 2025. https://monographs.iarc.who.int/list-of-classifications/.
30. Spinazzè, A., D. Spanu, P. Della-Bella, C. Corti, F. Borghi, G. Fanti, A. Cattaneo, et al. 2022. On the Determination of Cr(VI) in Cr(III)-Rich Particulates: From the Failure of Official Methods to the Development of an Alternative Protocol. *International Journal of Environmental Research and Public Health* 19 (19): 12111 (1–15). https://doi.org/10.3390/ijerph191912111.
31. Tumolo, M., V. Ancona, D. Paola, D. Losacco, C. Campanale, C. Massarelli, and V. F. Uricchio. 2020. Chromium Pollution in European Water, Sources, Health Risk, and Remediation Strategies: An Overview. *International Journal of Environmental Research and Public Health* 17 (15): 5438 (1–24). https://doi.org/10.3390/ijerph17155438.
32. European Commission. 2000. Commission Decision of 17 July 2000 on the Implementation of a European Pollutant Emission Register (EPER) According to Article 15 of Council Directive 96/61/EC Concerning Integrated Pollution Prevention and Control (IPPC). *Official Journal of the European Communities* L 192, 28 July 2000.
33. European Commission. 2020. Commission Regulation (EU) 2020/1245 of 2 September 2020 amending and correcting Regulation (EU) No 10/2011 on plastic materials and articles intended to come into contact with food. *Official Journal of the European Union* L 288, 3 September 2020.
34. European Commission. 2018a. Commission Directive (EU) 2018/725 of 16 May 2018 Amending, for the Purpose of Adaptation to Technical and Scientific Developments, Point 13 of Part III of Annex II t Directive 2009/48/EC of the European Parliament and of the Council on the Safety of Toys, as Regards Chromium VI. *Official Journal of the European Union* L 122, 17 May 2018.
35. European Chemicals Agency (ECHA). 2025. ECHA Proposes Restrictions on Chromium VI Substances to Protect Health. Accessed October 13, 2025. https://echa.europa.eu/-/echa-proposes-restrictions-on-chromium-vi-substances-to-protect-health.
36. European Parliament and Council. 2019a. Regulation (EU) 2019/1009 of the European Parliament and of the Council of 5 June 2019 Laying Down Rules on the Making Available on the Market of EU Fertilising Products and Amending Regulations (EC) No 1069/2009 and (EC) No 1107/2009 and repealing Regulation (EC) No 2003/2003. *Official Journal of the European Union* L 170, 25 June 2019.
37. Zhu, N., Y. Tian, S. Tao, Z. Qiao, Z. Yang, L. Hu, J. Liu, and Z. Zhang. 2025. Chromium Speciation Monitoring Platform for Drinking Water: Machine Learning-Assisted Dual-Emission Fluorescence Sensor Array. *Environmental Science and Technology* 12 (7): 792–798. https://doi.org/10.1021/acs.estlett.5c00506.
38. Łożyńska, M., M. Trojanowska, A. Molik, and R. Świetlik. 2025. Measurement Report: Seasonal Trends and Chemical Speciation of Chromium(III/VI) in Different Fractions of Urban Particulate Matter: A Case Study of Radom, Poland. *Atmospheric Chemistry and Physics* 25 (18): 10499–10510. https://doi.org/10.5194/acp-25-10499-2025.
39. Arellano-Sánchez, M. G., J. Vievard, L. Moufarrej, C. Devouge-Boyer, M. Hubert-Roux, C. Afonso, and M. Mignot. 2023. Separation, Speciation and Quantification of Both Chromium (VI) and Chromium (III) in Tanned Leather Samples: A Comparative Study and Validation of Analytical Methods. *Collagen and Leather* 5 (1): 1–16. https://doi.org/10.1186/s42825-023-00125-7.

40. Kashyap, S., and T. K. Mandal. 2025. Prototype of a POCT Device for Colorimetric Sensing of Chromium(VI) for Healthcare Applications. *Journal of Electronic Materials* 54: 9470–9477. https://doi.org/10.1007/s11664-025-11768-z.
41. Pechancová, R., T. Pluháček, and D. Milde. 2019. Recent Advances in Chromium Speciation in Biological Samples. *Spectrochimica Acta – Part B: Atomic Spectroscopy* 152:109–22. https://doi.org/10.1016/j.sab.2018.12.008.
42. Shirkhanloo, H., M. Ghazaghi, and H. Z. Mousavi. 2016. Chromium Speciation in Human Blood Samples Based on Acetyl Cysteine by Dispersive Liquid-Liquid Biomicroextraction and in-vitro Evaluation of Acetyl Cysteine/Cysteine for Decreasing of Hexavalent Chromium Concentration. *Journal of Pharmaceutical and Biomedical Analysis* 118: 1–8. https://doi.org/10.1016/j.jpba.2015.10.018.
43. Charkiewicz, A. E., W. J. Omeljaniuk, M. Garley, and J. Nikliński. 2025. Mercury Exposure and Health Effects: What Do We Really Know? *International Journal of Molecular Sciences* 26 (5): 232. https://doi.org/10.3390/ijms26052326.
44. EFSA. 2012. Scientific Opinion on the Risk for Public Health Related to the Presence of Mercury and Methylmercury in Food. *EFSA Journal* 10 (12): 2985 (1–241). https://doi.org/10.2903/j.efsa.2012.2985.
45. Manceau, A., J. P. Bourdineaud, R. B. Oliveira, S. L. F. Sarrazin, D. P. Krabbenhoft, C. A. Eagles-Smith, J. T. Ackerman, et al. 2021. Demethylation of Methylmercury in Bird, Fish, and Earthworm. *Environmental Science and Technology* 55 (3): 1527–1534. https://doi.org/10.1021/acs.est.0c04948.
46. Bosch, A. C., B. O'Neill, G. O. Sigge, S. E. Kerwath, and L. C. Hoffman. 2016. Heavy Metals in Marine Fish Meat and Consumer Health: A Review. *Journal of the Science of Food and Agriculture* 96: 32–48. https://doi.org/10.1002/jsfa.7360.
47. European Commission. 2003. Commission Directive 2003/40/EC of 16 May 2003 Establishing the List, Concentration Limits and Labelling Requirements for the Constituents of Natural Mineral Waters and the Conditions for Using Ozone-Enriched Air for the Treatment of Natural Mineral Waters and Spring Waters. *Official Journal of the European Union* L 126, 22 May 2003.
48. European Commission. 2011. Commission Regulation (EU) No 836/2011 of 19 August 2011 Amending Regulation (EC) No 333/2007 Laying Down the Methods of Sampling and Analysis for the Official Control of the Levels of Lead, Cadmium, Mercury, Inorganic Tin, 3-MCPD and Benzo(a)pyrene in Foodstuffs. *Official Journal of the European Union* L 215, 20 August 2011.
49. FAO/WHO Codex Committee on Food Additives and Contaminants. 2005. Discussion Paper on the Guideline Levels for Methylmercury in Fish: Comments. Thirty-seventh Session, The Hague, Netherlands, 25–29 April 2005.
50. Médieu, A., D. Point, V. Allain, N. Bodin, M. Lemire, P. Ayotte, Z. Dhurmeea, et al. 2025. Species-Specific Mercury Speciation in Billfishes and Its Implications for Food Safety Monitoring and Dietary Advice. *Environment International* 195: 109252 (1–10). https://doi.org/10.1016/j.envint.2025.109252.
51. Iwai, K., M. Iwai-Shimada, N. Tatsuta, Y. Kobayashi, K. Asato, M. Nishimoto, K. Nakai, and S. F. Nakayama. 2025. High-Throughput Mercury Speciation Analysis of Breast Milk Using HPLC-ICP-MS. *Microchemical Journal* 211: 113142 (1–9). https://doi.org/10.1016/j.microc.2025.113142.
52. Pacer, E. J., C. D. Palmer, and P. J. Parsons. 2025. A Rapid Method for the Determination of Methylmercury and Inorganic Mercury Species in Whole Blood by Liquid Chromatography with Detection Using Vapor Generation ICP-MS/MS. *Analytical Methods* 17 (8): 1840–1849. https://doi.org/10.1039/D4AY02116A.
53. Wang, Y., L. Li, C. Yao, X. Tian, Y. Wu, Q. Xie, and D. Wang. 2021. Mercury in Human Hair and Its Implications for Health Investigation. *Current Opinion in Environmental Science and Health* 22: 100271 (1–6). https://doi.org/10.1016/j.coesh.2021.100271.

54. Zheng, C., J. Tang, X. Pan, H. Shen, Z. Hu, J. Zhang, L. Wang, P. Wu, and Y. Tan. 2025. Development and Validation of a High-Performance Liquid Chromatography-Inductively Coupled Plasma Mass Spectrometry Method for the Simultaneous Determination of Arsenic and Mercury Species in Human Urine. *Chemosensors* 13 (3): 78 (1–14). https://doi.org/10.3390/chemosensors13030078.
55. Wang, M., Y. Fan, Q. Hong, W. Huang, X. Han, Z. Qu, N. Yan, and H. Xu. 2025. Speciation Detection Deciphering $C_2H_2$/Cl-Driven Mercury Escape Mechanisms in PVC Production. *Environmental Science and Technology* 59 (46): 25103–25112. https://doi.org/10.1021/acs.est.5c13769.
56. Guo, F., P. Zeng, J. Liu, H. Hu, W. Zhu, Y. Wang, and H. Cheng. 2024. Simultaneous Quantification of Tin and Lead Species in Antarctic Krill and Fish by Interfacing High-Performance Liquid Chromatography with Inductively Coupled Plasma Mass Spectrometry Based on Strong Cation-Exchange and Amphion Columns. *Food Chemistry* 443: 138552 (1–9). https://doi.org/10.1016/j.foodchem.2024.138552.
57. Kushwaha, A., N. Hans, S. Kumar, and R. Rani. 2018. A Critical Review on Speciation, Mobilization and Toxicity of Lead in Soil-Microbe-Plant System and Bioremediation Strategies. *Ecotoxicology and Environmental Safety* 147: 1035–1045. https://doi.org/10.1016/j.ecoenv.2017.09.049.
58. Xia, J., Y. Fang, Y. Chen, Y. Pan, P. Li, M. Xue, and Q. Hu. 2017. Lead Speciation Analysis in Rice by Reversed Phase Chromatography with Inductively Coupled Plasma Mass Spectrometry. *Journal of Food Composition and Analysis* 60: 74–80. https://doi.org/10.1016/j.jfca.2017.03.002.
59. European Parliament and Council. 2024. Directive (EU) 2024/869 of 13 March 2024 Amending Directive 2004/37/EC of the European Parliament and of the Council and Council Directive 98/24/EC as Regards the Limit Values for Lead and Its Inorganic Compounds and for Diisocyanates. *Official Journal of the European Union* L 869, 19 March 2024.
60. European Parliament and Council. 2011. Directive 2011/65/EU of the European Parliament and of the Council of 8 June 2011 on the Restriction of the Use of Certain Hazardous Substances in Electrical and Electronic Equipment (recast). *Official Journal of the European Union* L 174, 1 July 2011.
61. Song, Y., F. Guo, P. Zeng, J. Liu, Y. Wang, and H. Cheng. 2022. Simultaneous Measurements of Cr, Cd, Hg and Pb Species in ng $L^{-1}$ Levels by Interfacing High Performance Liquid Chromatography and Inductively Coupled Plasma Mass Spectrometry. *Analytica Chimica Acta* 1212: 339935 (1–9). https://doi.org/10.1016/j.aca.2022.339935.
62. Lwin, C. S., H. Jung, M. S. Kim, E. J. Lee, and T. G. Lee. 2025. Geochemical Speciation, Uptake, and Transportation Mechanisms of Arsenic, Cadmium, and Lead in Soil–Rice Systems: Additional Aspects and Challenges. *Antioxidants* 14 (5): 607 (1–18). https://doi.org/10.3390/antiox14050607.
63. Cläsgens, C., T. Schwank, and K. Pitzke. 2024. Occupational Exposure to Organotin Substances: Development of a Liquid Chromatographic Separation Method for 11 Organotin Compounds in Workplace Air Samples via HPLC-ICP-MS. *Journal of Chromatography A* 1718: 464695. https://doi.org/10.1016/j.chroma.2024.464695.
64. European Parliament and Council. 2003. Regulation (EC) No 782/2003 of 14 April 2003 on the Prohibition of Organotin Compounds on Ships. *Official Journal of the European Union L* 115, 9 May 2003.
65. European Parliament and Council. 2006. Regulation (EC) No 1907/2006 of 18 December 2006 concerning the Registration, Evaluation, Authorisation and Restriction of Chemicals (REACH), Establishing a European Chemicals Agency, amending Directive 1999/45/EC and Repealing Council Regulation (EEC) No 793/93 and Commission Regulation (EC) No 1488/94 as Well as Council Directive 76/769/EEC and Commission Directives 91/155/EEC, 93/67/EEC, 93/105/EC and 2000/21/EC. *Official Journal of the European Union* L 396, 30 December 2006.

66. Gui, W., C. Tian, Q. Sun, S. Li, W. Zhang, J. Tang, and G. Zhu. 2016. Simultaneous Determination of Organotin Pesticides by HPLC-ICP-MS and Their Sorption, Desorption, and Transformation in Freshwater Sediments. *Water Research* 95: 185–194. https://doi.org/10.1016/j.watres.2016.02.056.
67. Chung, S. W. C., A. H. T. Wu, and J. S. Y. Lau. 2020. Development of a Robust and Sensitive Method for Analyzing Butyltins, Phenyltins and Octyltins in Seafood. *Food Analytical Methods* 13 (2): 337–347. https://doi.org/10.1007/s12161-019-01659-5.
68. Tian, Z., L. Xing, D. Li, L. Ma, R. Wu, N. Tian, and Ch. Zhang. 2020. Speciation of Tin in Tobacco by High-Performance Liquid Chromatography–Inductively Coupled Plasma–Mass Spectrometry (HPLC-ICP-MS). *Analytical Letters* 53 (15): 2501–2516. https://doi.org/10.1080/00032719.2020.1746325.
69. Lin, P. H., M. A. Edwards, and R. E. Richardson. 2025. Investigation of Organotin Compounds in Drinking Water Contacting Aged PVC Pipes. *ACS ES and T Water* 5: 5069–5078. https://doi.org/10.1021/acsestwater.5c00161.
70. European Parliament and Council. 2019b. Regulation (EU) 2019/1021 of 20 June 2019 on Persistent Organic Pollutants. *Official Journal of the European Union* L 169, 25 June 2019.
71. French Republic. 2025. Loi no 2025-188 du 27 février 2025 visant à protéger la population des risques liés aux substances perfluoroalkylées et polyfluoroalkylées. *Journal officiel de la République française, 28* février 2025.
72. Dobrzyńska, E., P. Wasilewski, and M. Pośniak. 2025. Per- and Polyfluoroalkyl Substances (PFASs): A Comprehensive Review of Environmental Distribution, Health Impacts, and Regulatory Landscape. *Applied Sciences* 15 (22): 11884 (1–28). https://doi.org/10.3390/app152211884.
73. Coggan, T. L., T. Anumol, J. Pyke, J. Shimeta, and B. O. Clarke. 2019. A Single Analytical Method for the Determination of 53 Legacy and Emerging Per- and Polyfluoroalkyl Substances (PFAS) in Aqueous Matrices. *Analytical and Bioanalytical Chemistry* 411 (16): 3507–3520. https://doi.org/10.1007/s00216-019-01829-8.
74. Scher, D. P., J. E. Kelly, C. A. Huset, K. M. Barry, R. W. Hoffbeck, V. L. Yingling, and R. B. Messing. 2018. Occurrence of Perfluoroalkyl Substances (PFAS) in Garden Produce at Homes with a History of PFAS-Contaminated Drinking Water. *Chemosphere* 196: 548–555. https://doi.org/10.1016/j.chemosphere.2017.12.179.
75. Li, W. L., and K. Kannan. 2024. Determination of Legacy and Emerging Per- and Polyfluoroalkyl Substances (PFAS) in Indoor and Outdoor Air. *ACS ES and T Air* 1 (9): 1147–55. https://doi.org/10.1021/acsestair.4c00095.
76. Dueñas-Mas, M. J., A. Ballesteros-Gómez, and J. de Boer. 2023. Determination of Several PFAS Groups in Food Packaging Material from Fast-Food Restaurants in France. *Chemosphere* 339: 139734 (1–7). https://doi.org/10.1016/j.chemosphere.2023.139734.
77. Drage, D. S., M. Sharkey, H. Berresheim, M. Coggins, and S. Harrad. 2023. Rapid Determination of Selected PFAS in Textiles Entering the Waste Stream. *Toxics* 11 (1): 55 (1–13). https://doi.org/10.3390/toxics11010055.
78. Groffen, T., L. Bervoets, Y. Jeong, T. Willems, M. Eens, and E. Prinsen. 2021. A Rapid Method for the Detection and Quantification of Legacy and Emerging Per- and Polyfluoroalkyl Substances (PFAS) in Bird Feathers Using UPLC-MS/MS. *Journal of Chromatography B* 1172: 122653 (1–8). https://doi.org/10.1016/j.jchromb.2021.122653.
79. Piva, E., A. Giorgetti, P. Ioime, L. Morini, F. Freni, F. Lo Faro, F. Pirani, M. Montisci, P. Fais, and J. P. Pascali. 2021. Hair Determination of Per- and Polyfluoroalkyl Substances (PFAS) in the Italian Population. *Toxicology* 458: 152849 (1–7). https://doi.org/10.1016/j.tox.2021.152849.
80. Qian, S., H. Lu, T. Xiong, Y. Zhi, G. Munoz, C. Zhang, Z. Li, et al. 2023. Bioaccumulation of Per- and Polyfluoroalkyl Substances (PFAS) in Ferns: Effect of PFAS Molecular Structure and Plant Root Characteristics. *Environmental Science and Technology* 57 (11): 4443–4453. https://doi.org/10.1021/acs.est.2c06883.

81. Khan, B., R. M. Burgess, and M. G. Cantwell. 2023. Occurrence and Bioaccumulation Patterns of Per- and Polyfluoroalkyl Substances (PFAS) in the Marine Environment. *ACS ES and T Water* 3: 1243–1259. https://doi.org/10.1021/acsestwater.2c00296.
82. JRC. 2011. Towards a Review of the EC Recommendation for a Definition of the Term "Nanomaterial," Part 1: Compilation of Information Concerning the Experience with the Definition. JRC Scientific and Policy Report. https://doi.org/10.2788/36237
83. Ali, F., K. Neha, and S. Parveen. 2023. Current Regulatory Landscape of Nanomaterials and Nanomedicines: A Global Perspective. *Journal of Drug Delivery Science and Technology* 80: 104118. https://doi.org/10.1016/j.jddst.2022.104118.
84. Eker, F., H. Duman, E. Akdaşçi, E. Bolat, S. Sarıtaş, S. Karav, and A. M. Witkowska. 2024. A Comprehensive Review of Nanoparticles: From Classification to Application and Toxicity. *Molecules* 29 (15): 3482 (1–83). https://doi.org/10.3390/molecules29153482.
85. Nanotechnology Products Database. 2025. Nanotechnology Products Database. Accessed December 26, 2025. https://product.statnano.com/.
86. European Commission. 2018b. Commission Regulation (EU) 2018/1881 of 3 December 2018 Amending Regulation (EC) No 1907/2006 of the European Parliament and of the Council on the Registration, Evaluation, Authorisation and Restriction of Chemicals (REACH) as Regards Annexes I, III, VI, VII, VIII, IX, X, XI and XII to Address Nanoforms of Substances. *Official Journal of the European Union* L 308, 4 December 2018.
87. European Parliament and Council. 2017. Regulation (EU) 2017/745 of 5 April 2017 on Medical Devices, Amending Directive 2001/83/EC, Regulation (EC) No 178/2002 and Regulation (EC) No 1223/2009 and Repealing Council Directives 90/385/EEC and 93/42/EEC. *Official Journal of the European Union* L 117, 5 May 2017.
88. European Parliament and Council. 2012. Regulation (EU) No 528/2012 of 22 May 2012 Concerning the Making Available on the Market and Use of Biocidal Products. *Official Journal of the European Union* L 167, 27 June 2012.
89. European Parliament and Council. 2015. Regulation (EU) 2015/2283 of 25 November 2015 on Novel Foods, Amending Regulation (EU) No 1169/2011 of the European Parliament and of the Council and Repealing Regulation (EC) No 258/97 of the European Parliament and of the Council and Commission Regulation (EC) No 1852/2001. *Official Journal of the European Union* L 327, 11 December 2015.
90. European Parliament and Council. 2009. Regulation (EC) No 1223/2009 of 30 November 2009 on cosmetic products. *Official Journal of the European Union L* 342, 22 December 2009.
91. More, S., V. Bampidis, D. Benford, C. Bragard, T. Halldorsson, A. Hernández-Jerez, S. Hougaard-Bennekou, et al. 2021a. Guidance on Technical Requirements for Regulated Food and Feed Product Applications to Establish the Presence of Small Particles Including Nanoparticles. *EFSA Journal* 19: 6769 (1–48). https://doi.org/10.2903/j.efsa.2021.6769.
92. More, S., V. Bampidis, D. Benford, C. Bragard, T. Halldorsson, A. Hernández-Jerez, S. Hougaard-Bennekou, et al. 2021b. Guidance on Risk Assessment of Nanomaterials to Be Applied in the Food and Feed Chain: Human and Animal Health. *EFSA Journal* 19 (8) e6768. https://doi.org/10.2903/j.efsa.2021.6768.
93. Andreassen, M., G. Aquilina, M. L. Bastos, P. Boon, L. Castle, B. Fallico, R. FitzGerald, et al. 2025. Follow-up of the Re-Evaluation of Silver (E 174) as a Food Additive (EFSA-Q-2023-00169). *EFSA Journal* 23 (4): e9316. https://doi.org/10.2903/j.efsa.2025.9316.
94. EFSA. 2016. Scientific Opinion on the Re-evaluation of Gold (E 175) as a Food Additive. *EFSA Journal* 14 (1): 4362 (1–43). https://doi.org/10.2903/j.efsa.2016.4362.
95. EFSA. 2015. Scientific Opinion on the Re-Evaluation of Iron Oxides and Hydroxides (E 172) as Food Additives. *EFSA Journal* 13 (12): 4317 (1–57). https://doi.org/10.2903/j.efsa.2015.4317.

96. Younes, M., P. Aggett, F. Aguilar, R. Crebelli, B. Dusemund, M. Filipič, M. J. Frutos, et al. 2018. Re-Evaluation of Silicon Dioxide (E551) as a Food Additive. *EFSA Journal* 16 (1): e5088. https://doi.org/10.2903/j.efsa.2018.5088.
97. Younes, M., G. Aquilina, L. Castle, K. H. Engel, P. Fowler, M. J. Frutos-Fernández, P. Fürst, et al. 2021. Safety Assessment of Titanium Dioxide (E171) as a Food Additive. *EFSA Journal* 19 (5): e6585. https://doi.org/10.2903/j.efsa.2021.6585.
98. Tschiche, H. R., F. S. Bierkandt, O. Creutzenberg, V. Fessard, R. Franz, B. Giese, R. Greiner, et al. 2022. Environmental Considerations and Current Status of Grouping and Regulation of Engineered Nanomaterials. *Environmental Nanotechnology, Monitoring and Management* 18: 100707 (1–11). https://doi.org/10.1016/j.enmm.2022.100707.
99. Gregar, F., D. Baron, and T. Pluháček. 2025. Advances in ICP-MS-Based Nanoparticle Characterization: Techniques and Challenges in Biological Sample Analysis. *Journal of Separation Science* 48 (9): e70259 (1–23). https://doi.org/10.1002/jssc.70259.
100. Zhang, W., J. Long, J. Li, M. Zhang, X. Ye, W. Chang, and H. Zeng. 2020. Effect of Metal Oxide Nanoparticles on the Chemical Speciation of Heavy Metals and Micronutrient Bioavailability in Paddy Soil. *International Journal of Environmental Research and Public Health* 17 (7): 2482 (1–13). https://doi.org/10.3390/ijerph17072482.
101. Johnston, B. D., T. M. Scown, J. Moger, S. A. Cumberland, M. Baalousha, K. Linge, R. Van-Aerle, K. Jarvis, J. R. Lead, and C. R. Tyler. 2010. Bioavailability of Nanoscale Metal Oxides $TiO_2$, $CeO_2$, and ZnO to Fish. *Environmental Science and Technology* 44 (3): 1144–1151. https://doi.org/10.1021/es901971a.
102. Gosens, I., J. Minnema, A. J. F. Boere, E. Duistermaat, P. Fokkens, J. Vidmar, K. Löschner, et al. 2024. Biodistribution of Cerium Dioxide and Titanium Dioxide Nanomaterials in Rats after Single and Repeated Inhalation Exposures. *Particle and Fibre Toxicology* 21 (1): 33 (1–23). https://doi.org/10.1186/s12989-024-00588-4.
103. López-Mayán, J. J., R. Domínguez-González, M. C. Barciela-Alonso, E. Peña-Vázquez, A. Moreda-Piñeiro, P. Taboada-Antelo, and P. Bermejo-Barrera. 2025. Bioaccessibility and Cellular Transport Study of Silver and Titanium Dioxide Nanoparticles from Exposed Seaweed and Mussels Using Caco-2 Cells. *Microchimica Acta* 192: 216 (1–12). https://doi.org/10.1007/s00604-025-07066-4.
104. Abou-Okada, M., M. El-Matbouli, and M. Saleh. 2025. Silver and Zinc Oxide Nanoparticles for Effective Aquaculture Wastewater Treatment. *Nanomaterials* 15 (7): 559–583. https://doi.org/10.3390/nano15070559.

# 7 Mesoporous Nanoparticles and Polymers for Entrapping Toxic Metal Ions and Antimicrobial Studies

*Elisabete Oliveira, Joana Galhano, José Luis Capelo-Martínez, and Javier Fernández-Lodeiro*

## 7.1 INTRODUCTION

The rapid growth of industrial activities, urbanization, and intensive agricultural practices has led to a significant increase in environmental contamination by toxic metal ions and pathogenic microorganisms [1, 2]. Heavy metals, such as lead, cadmium, mercury, chromium, and arsenic, are among the most persistent pollutants in natural waters and soils due to their non-biodegradable nature, high toxicity, and tendency to bioaccumulate in living organisms [3, 4]. Even at low concentrations, these metallic species pose severe risks to ecosystems and human health, being associated with neurological disorders, organ damage, carcinogenic effects, and antimicrobial resistance [3, 5]. Consequently, the development of effective, sustainable, and multifunctional remediation technologies has become a global priority.

Conventional methods for toxic metal removal, including chemical precipitation, ion exchange, membrane filtration, and adsorption using bulk materials, have been widely implemented in water and wastewater treatment (Figure 7.1) [4]. However, these approaches often have critical limitations such as low selectivity, incomplete removal leaving trace concentrations of contaminants, high energy consumption, secondary waste generation, and limited reusability [6]. In parallel, the increasing prevalence of microbial contamination and antibiotic-resistant bacteria has intensified the need for materials capable of simultaneously addressing chemical and biological pollutants within integrated treatment strategies.

Despite their widespread usage in metal contamination sites, traditional adsorbents and other conventional remediation strategies, such as those referred to previously, present some significant limitations. These limitations are commonly associated with difficulties in following strict regulatory limits, particularly when

DOI: 10.1201/9781003632092-8 

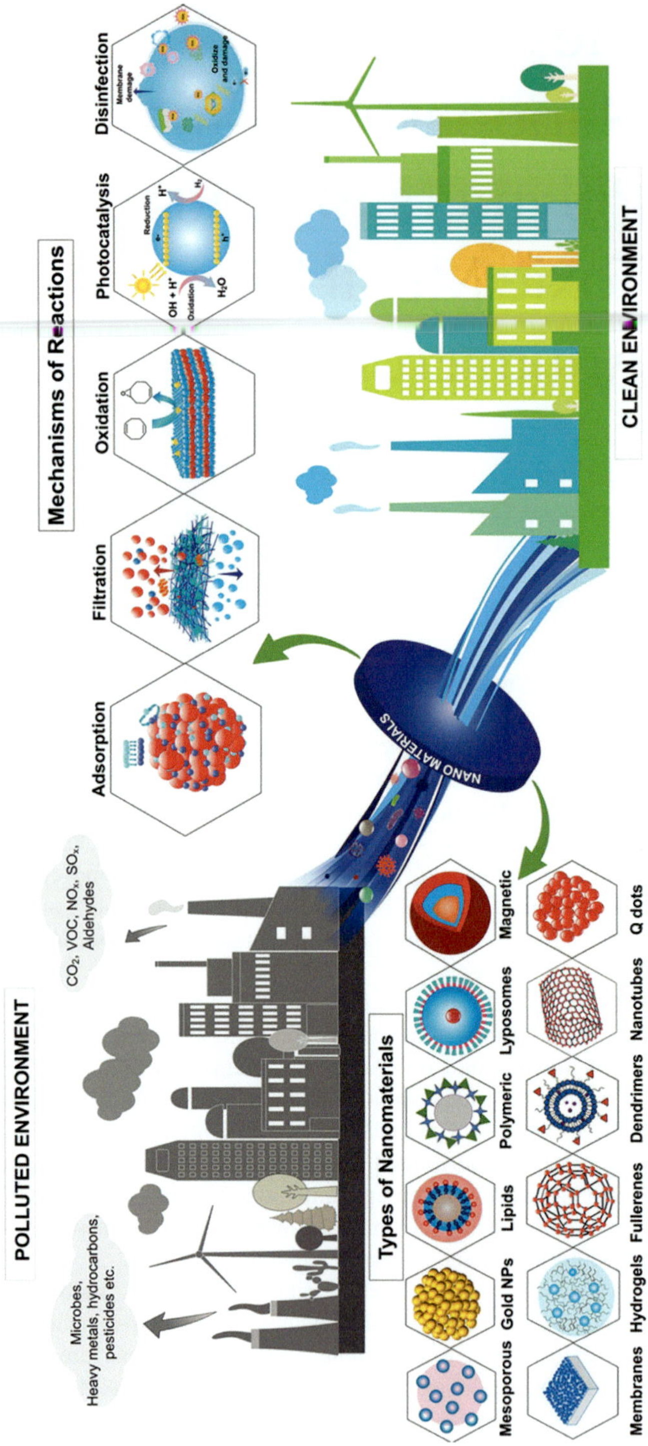


**FIGURE 7.1** General overview of the types of materials and reaction mechanisms used for pollutant removal. (Copyright © 2024, Nosheen Asghar et al. [3]).

considering the presence of the analytes at trace-level concentrations, which becomes ever more difficult when considering the complex mixture of analytes present in field samples of contaminated aqueous or soil samples. Regardless of their relatively low cost and ease of implementation, limitations regarding their limits of detection (LOD) can become concerning because some metallic species can induce severe health risks even at extremely low concentrations. As such, nanotechnology-based remediation methodologies, particularly mesoporous nanomaterials, have arisen as a promising alternative mostly due to their specific characteristics. Although the scalability of these nanomaterials is not yet optimized for large-scale applications and might present higher costs than those caused by common strategies, with continuous research, these materials might have a significant impact on commonly employed remediation strategies. Rather than replacing already existing methods, these might complement each other, particularly in cases where the sample matrixes are composed of a variety of analytes and selectivity is crucial. In this context, mesoporous nanoparticles (MNPs) and functional polymers have emerged as highly promising platforms for environmental remediation and antimicrobial applications (Figure 7.1) [7, 8]. Mesoporous nanomaterials offer exceptionally high surface areas, tunable pore architectures, and versatile surface chemistry. These features enable efficient adsorption, confinement, and transformation of toxic metal ions, while facilitating rapid mass transfer and strong interactions between contaminants and active sites. Moreover, the nanoscale dimensions of these materials allow for enhanced reactivity and improved performance compared to their bulk counterparts.

MNPs based on silica, metal oxides, and carbon frameworks represent distinct but complementary material classes. Silica-based MNPs are valued for their chemical stability, biocompatibility, and easy functionalization. Metal oxide mesoporous systems introduce catalytic, redox, and magnetic functionalities that enhance metal detoxification and material recovery; while carbon-based mesoporous materials provide high electrical conductivity, large pore volumes, and rich surface chemistry, enabling synergistic adsorption–reduction mechanisms [9]. When combined with functional polymers, these materials can be further engineered to improve selectivity, antimicrobial activity, mechanical stability, and reusability, opening pathways for multifunctional hybrid systems [10].

Beyond metal ion sequestration, MNPs and functional polymers have demonstrated significant antimicrobial activity through direct interactions with microbial membranes, generation of reactive species, or controlled release of antimicrobial agents [11]. The ability to design materials that entrap toxic metal ions as well as inhibit microbial growth is particularly attractive for water treatment, biomedical, and environmental protection applications, where complex contamination scenarios are increasingly common.

This chapter provides a comprehensive overview of MNPs and polymer-based materials for toxic metal entrapment and antimicrobial studies. It discusses the environmental impact of major toxic metal ions, critically examines conventional remediation strategies and their limitations, and highlights recent advances in mesoporous silica, metal oxide, and carbon-based nanoparticles. Emphasis is placed on structure–property–function relationships, synthesis strategies, removal mechanisms, and performance metrics. Finally, the chapter explores how polymer integration and surface functionalization enhance material efficiency, contributing to next-generation multifunctional remediation systems.

## 7.2 TOXIC METAL IONS: ENVIRONMENTAL IMPACT AND REMEDIATION STRATEGIES

The contamination of aquatic systems by toxic metal ions is a serious global environmental and public health concern. Rapid industrialization, mining, agricultural activities, and improper waste disposal have led to elevated concentrations of non-biodegradable metals in surface and groundwater resources. Unlike organic pollutants, toxic metals persist in the environment, accumulate in living organisms, and often transfer through food chains, resulting in long-term ecological and human health risks [12, 13].

Among the most hazardous metal ions commonly detected in water and wastewater are lead ($Pb^{2+}$), cadmium ($Cd^{2+}$), mercury ($Hg^{2+}$), hexavalent chromium ($Cr^{6+}$), and arsenic ($As^{3+}$). Lead contamination originates mainly from battery manufacturing, metal processing, pigments, and aging water distribution systems. It is strongly associated with neurological damage, renal dysfunction, and developmental disorders, particularly in children [13, 14]. Cadmium, released from electroplating, fertilizers, and battery waste, is highly toxic and has been linked to bone demineralization, kidney failure, and carcinogenic effects [15]. Mercury pollution, caused due to coal combustion, mining, and chemical industries, is especially problematic due to its conversion into methylmercury, a highly toxic and bioaccumulative species that affects the central nervous system and biomagnifies in aquatic food webs [16]. Hexavalent chromium is extensively used in leather tanning, electroplating, and pigment production; it is highly soluble, mobile, and carcinogenic, even at trace concentrations, making it one of the most dangerous inorganic water pollutants [17]. Arsenic contamination occurs due to both natural geological sources and anthropogenic activities and it is associated with skin lesions, cardiovascular diseases, and various cancers following chronic exposure [18].

To address metal pollution, several conventional remediation strategies have been widely implemented. Chemical precipitation is one of the oldest methods in which metal ions are converted into insoluble hydroxides or sulfides and subsequently removed as sludge [19]. Adsorption processes using activated carbon, clays, metal oxides, or biosorbents are extensively applied due to their simplicity and effectiveness [20]. Membrane-based technologies, including reverse osmosis, nanofiltration, and ultrafiltration, offer high removal efficiency and are increasingly being used in industrial wastewater treatment [19]. Additional approaches include ion exchange, electrochemical treatments, and coagulation–flocculation processes [19].

Despite their broad use, these traditional methods exhibit important limitations [19, 21]. Chemical precipitation and coagulation–flocculation generate large volumes of hazardous sludge that require costly handling and disposal. Adsorption processes often suffer from limited selectivity, reduced efficiency in complex water matrices containing competing ions, and difficulties in adsorbent regeneration. Membrane technologies, although efficient, are associated with high energy consumption, membrane fouling, and elevated operational costs, which limit their large-scale application. Moreover, many conventional methods only transfer metal contaminants from the aqueous phase to another phase rather than permanently immobilizing or detoxifying them, raising concerns about secondary pollution and long-term sustainability.

These challenges highlight the need for advanced and more selective remediation strategies, capable of combining high removal efficiency, stability, reusability, and reduced environmental impact. In this context, nanostructured materials, particularly MNPs and functional polymer-based systems, have emerged as promising alternatives for the effective remediation of toxic metal ions and are discussed in detail in the following sections.

## 7.3 MESOPOROUS NANOPARTICLES FOR METAL REMEDIATION

Human history is closely linked to the evolution of materials science and, consequently, to the discovery of new classes of materials. Among the various categories, metals, polymers, and inorganic non-metallic materials are those that exert the greatest influence on industry, science, technology, and the global economy in the present scenario [22]. Within these broad classes lies an important subgroup—porous materials—which possess structural features and functional properties that distinguish them from their bulk, non-porous counterparts.

According to the International Union of Pure and Applied Chemistry (IUPAC), porous materials are divided into three families based on pore diameter: (i) microporous materials, with pores smaller than 2 nm; (ii) mesoporous materials, characterized by pores between 2 and 50 nm; (iii) and macroporous materials, which contain pores larger than 50 nm [23]. They can also be categorized by their chemical composition into organic, inorganic, or hybrid organic-inorganic systems. This combined structural and compositional diversity gives porous materials a remarkable degree of tunability, enabling their application in catalysis, adsorption and separation processes, biomedicine, drug delivery, and environmental remediation.

Within this broader context, MNPs constitute a versatile class of nanomaterials characterized by different pore sizes, high surface areas, tunable pore structures, and well-defined morphologies. Their "nature" is typically defined by composition (silica-based, metal oxides, carbonaceous, or hybrid systems) and structural organization (ordered hexagonal, cubic, disordered, hollow, or core–shell architectures). These parameters influence their chemical reactivity, adsorption behavior, diffusion kinetics, stability, and suitability for environmental and antimicrobial applications.

According to the literature provided, silica-based MNPs offer superior chemical stability and functionalization potential; metal oxide MNPs exhibit catalytic, magnetic, or redox-responsive behavior; and carbon-based MNPs have excellent electrical conductivity and large pore volumes suitable for pollutant capture and charge transfer processes. Together, these material families form a starting point for advanced systems designed for toxic metal remediation and antimicrobial action.

### 7.3.1 Silica-Based Mesoporous Nanoparticles

Silica-based materials are the most extensively studied and widely applied inorganic porous systems due to their chemical inertness, thermal and hydrothermal stability, nontoxicity, highly tunable structure, expanding their applicability in adsorption,

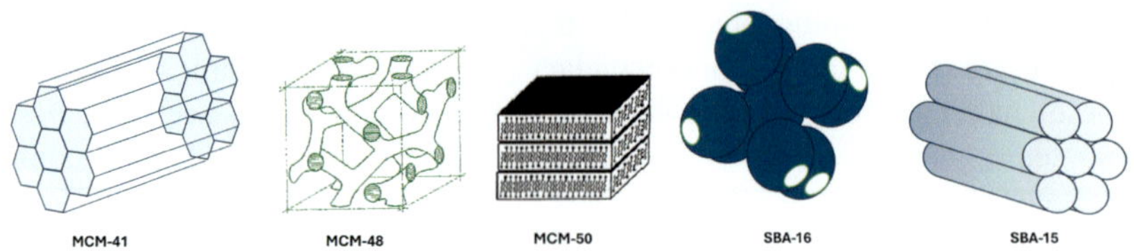


| | MCM-41 | SBA-15 | SBA-16 |
|---|---|---|---|
| **Surfactant type** | Cationic / Anionic | Pluronic P123 (Ion assisted) | Pluronic F127 (Ion assisted) |
| **Medium** | Basic | Acidic | Acidic |
| **Pore-size** | 2-10 nm | 5-30 nm | 5-30 nm |
| **Stability** | Thermal | Thermal & Hydrothermal | Thermal & Hydrothermal |
| **Mesophase structures** | 2D hexagonal | 2D hexagonal | 3D cubic |
| **Main application** | Zeolites, Catalysts, Delivery systems | Zeolites, Catalysts, Delivery systems | Zeolites, Catalysts, Delivery & Extraction systems |

**FIGURE 7.2** Chemical structure and main properties of the most used silica mesoporous nanoparticles.

separation, drug delivery, and antimicrobial systems (Figure 7.2). They are composed of repetitive $SiO_4$ tetrahedral units assembled through siloxane (Si–O–Si) linkages, while surface silanol groups (Si–OH) determine reactivity, hydrophilicity, and interfacial properties [24, 25]. These structural features allow silica frameworks to adopt either crystalline or amorphous morphologies, the latter being particularly relevant for mesoporous materials because their disordered networks generate very high surface areas ranging from 100 to 1000 $m^2/g$ [24].

The history of silica-based MNPs began in 1992 with the production of MCM-41 nanoparticles, developed by researchers at Mobil Oil Co. [26]. This discovery marked the birth of the M41S family which includes three archetypal structures: MCM-41 (2D hexagonal arrangement), MCM-48 (3D cubic arrangement), and MCM-50 (lamellar layered structure; Figure 7.2) [27, 28].

These materials represented a breakthrough because, unlike previous amorphous mesoporous solids with irregular, poorly controlled porosity, M41S frameworks offered uniform pore sizes (2–10 nm), ordered arrays of channels, and highly reproducible synthetic procedures. MCM-41 quickly became one of the most investigated mesoporous materials due to its ease of synthesis, high specific surface area, and straightforward functionalization. Its micelle-templated formation mechanism, based on interactions between silica precursors and cationic surfactants such as cetyltrimethylammonium bromide (CTAB), enabled predictable tailoring of morphology and pore architecture.

Encouraged by the success of the M41S family, researchers soon developed new classes of silica mesostructures by varying surfactants, templating strategies, and synthesis conditions. From this, the Santa Barbara Amorphous (SBA) family was developed in the late 1990s, with SBA materials expanding the range of achievable pore sizes and structural symmetries [29–31].

SBA-15 has large mesopores (5–30 nm), thick pore walls (3.5–6.4 nm), and a 2D hexagonal symmetry similar to MCM-41 but with significantly enhanced mechanical, thermal, and hydrothermal stability. SBA-16 presents a unique 3D cubic cage-like mesostructure composed of interconnected spherical cavities and it favors improved mass transport and encapsulation of large biomolecules.

SBA materials rely mainly on soft-templating strategies, particularly using non-ionic triblock copolymers such as Pluronic P123 and F127. The resulting frameworks provide larger pore diameters and enhanced structural robustness compared to MCM-41, making SBA-type MNPs highly suitable for drug delivery, enzyme immobilization, adsorption of heavy metals, and biomedical imaging.

MNPs synthesis is typically carried out through sol–gel chemistry, using silicon alkoxide precursors such as tetraethyl orthosilicate (TEOS) or tetramethyl orthosilicate (TMOS) [32]. After assembly, the surfactant template is removed by calcination or, preferably for nanoparticle systems, solvent extraction, which maintains better pore integrity and preserves surface silanol groups.

The structural features of silica-based MNPs, such as pore size, symmetry, connectivity, wall thickness, and surface chemistry are highly tunable. This tunability determines their behavior in adsorption, catalysis, diffusion processes, biological interactions, and metal-ion binding. MCM-41 is ideal for applications requiring high surface area and small, uniform channels; SBA-15, with larger pores and thicker walls, offers improved loading capacity for biomolecules and enhanced stability; and SBA-16 provides a 3D interconnected network advantageous for fast diffusion, guest encapsulation, and extraction of large species.

Beyond M41S and SBA materials, numerous other silica-based mesoporous families, such as FSM, MSU, and KIT, have been developed, each contributing to the diversity of structures and functionalities available.

#### 7.3.1.1 Synthesis Approaches

The growing interest in mesoporous materials has led to the development of several approaches aimed at producing different mesostructures with tunable properties. Mesoporous silica nanoparticles are mainly synthesized using a limited number of well-established approaches [33–35].

The most widely used method is the sol–gel process combined with soft templating, where silica precursors condense around self-assembled surfactant or block-copolymer micelles to generate ordered mesoporous structures. A second important method is the hydrothermal synthesis which involves carrying out the sol–gel reaction under elevated temperature and pressure to improve pore ordering and structural stability. In addition, microwave-assisted synthesis is employed as a rapid alternative, providing faster and more homogeneous heating. It significantly reduces synthesis time while maintaining mesostructural order. These methods constitute the core synthetic strategies for producing silica-based MNPs.

#### 7.3.1.2 Metal Remediation

MNPs have emerged as one of the most effective inorganic platforms for toxic metal remediation due to their high specific surface area, uniform pore size distribution, chemical stability, and exceptional versatility for surface functionalization. The ordered mesoporous architecture, typically found in materials such as MCM-41, SBA-15, and SBA-16, provides a large number of accessible adsorption sites and facilitates a quick mass transfer of metal ions from aqueous media into the mesopores [9, 36].

The remediation performance of mesoporous silica is primarily governed by adsorption processes occurring at surface silanol groups which act as coordination sites for metal ions through electrostatic interactions, hydrogen bonding, and surface complexation [9]. The open and interconnected mesoporous framework enhances diffusion kinetics, allowing efficient uptake even at low metal concentrations.

While silica-based MNPs already exhibit moderate affinity toward divalent and trivalent metal ions, surface functionalization significantly improves adsorption capacity and selectivity. Incorporation of functional groups such as amines, thiols, carboxylates, and chelating ligands introduces specific binding sites that promote strong coordination with targeted metal ions [37]. Amino-functionalized mesoporous silica has demonstrated enhanced uptake of $Cu^{2+}$, $Ni^{2+}$, and $Fe^{3+}$ (harder metals), whereas thiol-modified materials show particularly strong affinity for soft metal ions such as $Hg^{2+}$, $Cd^{2+}$, and $Pb^{2+}$ (soft metals) [37].

Quantitative adsorption studies reported in the literature emphasize the efficiency of mesoporous silica systems in metal remediation. Representative adsorption capacities further illustrate the influence of surface chemistry and pore architecture on metal ion uptake. For instance, pristine MCM-41 typically exhibits moderate adsorption capacities, such as ~9.7 mg/g for $Cu^{2+}$ and ~18.8 mg/g for $Pb^{2+}$ under mildly acidic conditions [38]. Upon surface functionalization, however, adsorption efficiencies increase significantly (Table 7.1). Amine- or chelator-modified MCM-41 materials have reported maximum capacities exceeding 250 mg/g for $Pb^{2+}$ and $Cu^{2+}$, while $Cr^{6+}$ uptake values as high as ~390 mg/g have been achieved. SBA-15-based materials generally show higher and more stable adsorption performances, with $Pb^{2+}$ capacities up to ~240 mg/g and $Cd^{2+}$ removal reaching ~190 mg/g depending on the chelating ligand employed. Especially, SBA-16 systems often surpass their counterparts, benefitting from their 3D pore connectivity; adsorption capacities up to ~290 mg/g for $Pb^{2+}$, ~260 mg/g for $Hg^{2+}$, and ~230 mg/g for $Cd^{2+}$ have been reported, clearly demonstrating the strong influence of pore architecture and surface chemistry on metal ion sequestration efficiency [39].

Another interesting result is the application of SBA-type mesoporous silica for mercury remediation. Functionalized SBA-16 systems incorporating organic ligands or chromophoric groups have demonstrated $Hg^{2+}$ extraction efficiencies exceeding 70%, with adsorption capacities as high as 69 mg/g, highlighting their potential for both detection and remediation of mercury-contaminated waters (Figure 7.3) [40].

An important advantage of mesoporous silica nanoparticles is their structural robustness under aqueous conditions and their capacity for regeneration. Studies have shown that adsorption–desorption cycles can be performed with minimal loss of efficiency, particularly when mild eluents are used to recover metal ions without

**TABLE 7.1**
**Mesoporous Silica Nanoparticles for Toxic Metal Remediation**

| Mesoporous Silica Type | Surface Functionalization | Main Adsorption Mechanism | Target Metal Ions | Maximum Reported Adsorption Capacity (mg/g) | Experimental Conditions | Main Advantages |
|---|---|---|---|---|---|---|
| **MCM-41** | Pristine | Electrostatic interaction with silanol groups | $Pb^{2+}$, $Cu^{2+}$ | $Pb^{2+}$: ~18.8<br>$Cu^{2+}$: ~9.7 | pH 5<br>Initial metal concentration: 10 mg/L<br>Incubation time: 2 h | Simple synthesis; high surface area |
| **MCM-41** | Amine-based ligands, thiol groups, chelating agents (EDTA, molybdophosphate), polymeric modifiers (PEI, cyclodextrin), metal oxides ($TiO_2$) | Electrostatic adsorption, metal–ligand chelation, chemisorption (S–metal), redox-assisted adsorption ($Cr^{6+}$), synergistic polymer effects | $Pb^{2+}$, $Cu^{2+}$, $Cd^{2+}$, $Cr^{3+}/Cr^{6+}$, $Hg^{2+}$ | $Pb^{2+}$: 267<br>$Cu^{2+}$: 259.2<br>$Cd^{2+}$: 191.1<br>$Cr^{6+}$: 392<br>$Hg^{2+}$: 1343.9 | pH 5<br>Initial metal concentration: 100 mg/L<br>Nanoparticle mass: 0.1 g | Easy synthesis; high surface area; wide functionalization chemistry |
| | | | | $Cr^{6+}$: 392 | pH 7<br>Initial metal concentration: 50 mg/L<br>Nanoparticle mass: 0.1 g | |
| | | | | $Hg^{2+}$: 1343.9 | Nanoparticle mass: 0.05 g | |
| **SBA-15** | Amine, thiol, Schiff bases, EDTA, dendrimers, redox-active ligands | Surface complexation, metal–ligand chelation, chemisorption (Hg–S), redox-assisted adsorption | $Pb^{2+}$, $Cd^{2+}$, $Cu^{2+}$, $Ni^{2+}$, $Co^{2+}$, $Zn^{2+}$, $Cr^{3+}/Cr^{6+}$, $Hg^{2+}$ | $Pb^{2+}$: 242.5<br>$Cd^{2+}$: 187.9 | pH 5<br>Initial metal concentration: 45 mg/L<br>Nanoparticle mass: 0.02 g | High stability; tunable surface chemistry; good regeneration |
| | | | | $Cu^{2+}$: 128 | pH 5<br>Initial metal concentration: 100 mg/L<br>Nanoparticle: 1 g/L | |
| | | | | $Cr^{6+}$: 113 | pH 5<br>Initial metal concentration: 100 mg/L<br>Nanoparticle mass: 0.05 g | |
| | | | | $Hg^{2+}$: 7.59 | pH 2.5<br>Initial metal concentration: 10 mg/L<br>Nanoparticle mass: 0.04 g | |

*(Continued)*

**TABLE 7.1** *(Continued)*
**Mesoporous Silica Nanoparticles for Toxic Metal Remediation**

| Mesoporous Silica Type | Surface Functionalization | Main Adsorption Mechanism | Target Metal Ions | Maximum Reported Adsorption Capacity (mg/g) | Experimental Conditions | Main Advantages |
|---|---|---|---|---|---|---|
| **SBA-16** | Organic ligands/dyes | Coordination confinement | $Hg^{2+}$ | $Hg^{2+}$: ~70 | Nanoparticle mass: 0.001 g<br>Exposure time: 2 h | 3D cage structure; fast diffusion |
| **SBA-16** | Biomolecule-derived ligands (guanine, maltodextrin), amine, thiol, EDTA | Multidentate chelation combined with 3D confinement effects and enhanced diffusion | $Pb^{2+}$, $Hg^{2+}$, $Cd^{2+}$, $Zn^{2+}$, $Cu^{2+}$ | $Pb^{2+}$: 289.9<br>$Hg^{2+}$: 259.9 | pH 5<br>Initial metal concentration: 100 mg/L<br>Nanoparticle mass: 0.02 g | Fast diffusion; very high adsorption capacities; efficient for large ions |
| | | | | $Cd^{2+}$: 228.8 | pH 5<br>Initial metal concentration: 100 mg/L<br>Nanoparticle mass: 0.025 g | |
| | | | | $Zn^{2+}$: 564.7 | pH 6 | |

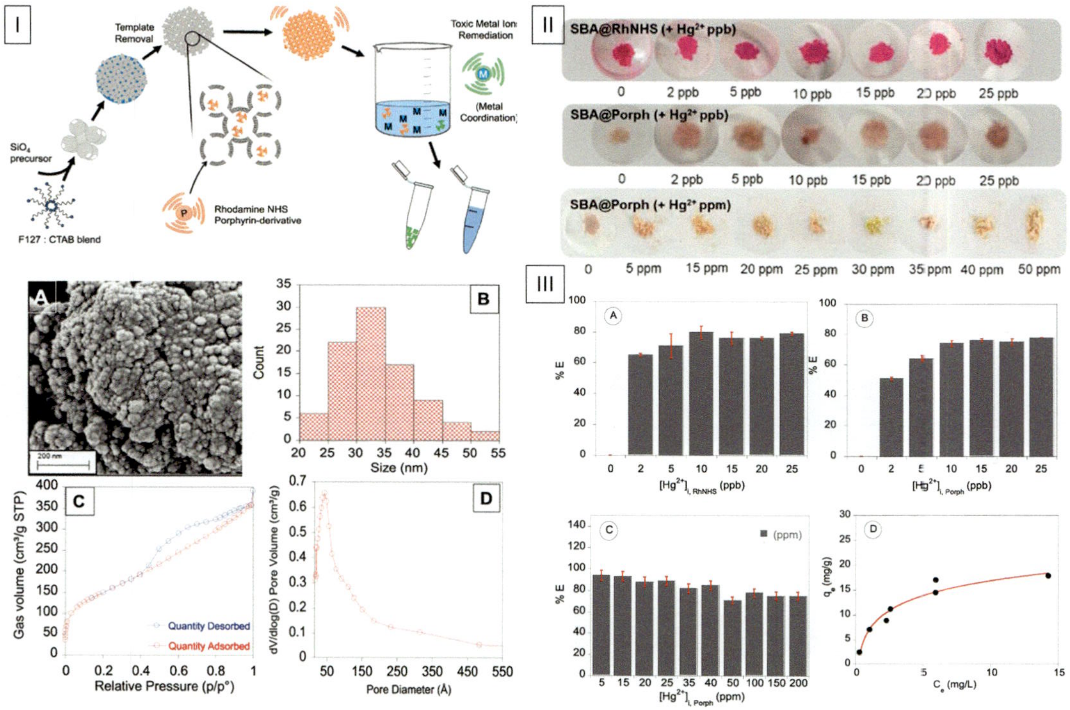


**FIGURE 7.3** (I) (Top) General schematic synthetic pathway of SBA-16 silica mesoporous nanoparticles, chromophores adsorption and metal extraction. (Bottom) SEM image (A) and size histogram (B), nitrogen adsorption–desorption isotherms (C), and pore size (D) of SBA-16 mesoporous nanoparticles (MNs). (II) Colorimetric and fluorimetric images of SBA-16@RhNHS and SBA-16@Porph in the presence of different $Hg^{2-}$ concentrations. (III) Extraction efficiency (%E) of $Hg^{2+}$ into SBA-16@RhNHS (A), SBA-16@Porph in the ppb (B), and ppm (C) range in water. (D) Adsorption isotherm of $Hg^{2+}$ into SBA-16@Porph mesoporous nanosystem in water. (Adapted with permission from [41] 2018 Elsevier Ltd. All rights reserved.)

damaging the silica framework. This reusability significantly enhances the sustainability and economic feasibility of mesoporous silica-based remediation systems.

Mesoporous silica nanoparticles represent a highly adaptable and efficient class of adsorbents for toxic metal remediation. Their tunable pore structure, ease of functionalization, and strong adsorption performance across a broad range of metal ions make them suitable for integration into advanced water treatment technologies. Ongoing developments focus on hybrid silica–polymer systems and multifunctional materials capable of simultaneous metal removal and antimicrobial activity, further expanding the applicability of mesoporous silica in complex environmental remediation scenarios.

### 7.3.2 Metal Oxide Mesoporous Nanoparticles

Mesoporous metal oxide nanoparticles have attracted significant attention in environmental remediation due to their high surface area, tunable porosity, chemical stability, and intrinsic catalytic or magnetic properties. Compared with bulk oxides, the mesoporous structure improves accessibility of active sites and enhances mass transport, leading to improved removal and transformation of toxic metal ions [41].

Titanium dioxide ($TiO_2$) is one of the most widely studied metal oxides because of its strong photocatalytic activity, chemical inertness, and environmental compatibility. In mesoporous form, $TiO_2$ exhibits enhanced adsorption capacity and improved photoinduced redox behavior. Several studies report its effectiveness in the reduction of hexavalent chromium ($Cr^{6+}$) to the less toxic $Cr^{3+}$ species through combined adsorption–photocatalytic mechanisms [41, 42]. The presence of mesopores facilitates diffusion of chromium ions to photoactive sites, increasing reaction efficiency.

Zinc oxide (ZnO) presents similar photocatalytic properties as $TiO_2$ and has also been investigated for $Cr^{6+}$ remediation. Mesoporous ZnO materials benefit from increased surface area and improved charge separation, enhancing their catalytic performance under light irradiation [42]. In addition, ZnO exhibits intrinsic antimicrobial activity, making it suitable for multifunctional water treatment systems.

Iron oxide ($Fe_3O_4$) MNPs are particularly attractive due to their magnetic properties, which allow easy separation from treated water using an external magnetic field. This feature significantly improves material recovery and reusability, reducing secondary pollution. $Fe_3O_4$-based mesoporous systems also provide abundant surface hydroxyl groups that promote metal ion adsorption and, in some cases, redox transformations [41, 43].

Aluminum oxide ($Al_2O_3$) mesoporous materials are mainly employed as adsorbents due to their high surface area, surface acidity, and strong affinity toward metal ions. Adsorption occurs through electrostatic interactions and coordination with surface hydroxyl groups, while the mesoporous architecture enhances uptake and diffusion. $Al_2O_3$ is also frequently used as a stable support for other active species [41].

Mesoporous metal oxide nanoparticles combine adsorption, catalytic activity, magnetic separation, and redox functionality, making them promising candidates for advanced remediation technologies. However, issues such as particle aggregation, sensitivity to water chemistry, and challenges in large-scale applications remain, encouraging the development of hybrid and surface-modified systems.

#### 7.3.2.1 Synthesis

Mesoporous metal oxide nanoparticles are generally synthesized using soft-templated sol–gel methods, nanocasting approaches, or metal incorporation into preformed mesoporous frameworks [41]. These strategies are widely reported for oxides such as $TiO_2$, ZnO, $Fe_3O_4$, and $Al_2O_3$, enabling control over pore structure, surface area, and functional properties.

#### 7.3.2.2 Metal Remediation

Metal oxide MNPs are highly effective for heavy metal remediation due to their vast surface area and porous structure, offering high removal capacities (e.g., >500 mg/g for $As^{5+}$ with Mg-Al-Ti oxides) for various toxic metals like lead, copper, arsenic, and chromium, primarily through adsorption via surface functional groups (hydroxyl, carboxyl) and magnetic recovery, with performance depending on specific oxides (Fe, Ti, Zn, Mg) and functionalization.

Quantitative studies reported in the literature indicate that mesoporous metal oxide nanoparticles exhibit significant metal removal capacities, typically ranging from tens to several hundreds of milligrams per gram, depending on oxide composition and experimental conditions. Li et al. developed amine-functionalized mesoporous $Fe_3O_4$ nanoparticles for toxic metal removal from aqueous media, achieving maximum adsorption capacities for $Pb^{2+}$, $Cd^{2+}$, and $Cu^{2+}$ in the range of 369.0–523.6 mg/g, values that exceed those reported for many conventional adsorbents [44].

Mesoporous magnesium-aluminum-titanium oxide nanoparticles show exceptional arsenic removal, up to 1215 mg/g for $As^{5+}$ due to high surface area and tunable properties [44]. These values highlight the strong potential of mesoporous metal oxides as efficient and reusable adsorbents for toxic metal remediation.

### 7.3.3 Carbon-Based Mesoporous Nanoparticles

Carbon-based MNPs constitute a versatile class of porous materials, combining large surface areas, interconnected mesoporous networks, chemical stability, and intrinsic electrical conductivity. These characteristics make them particularly attractive for environmental remediation, where efficient mass transport, surface reactivity, and robustness under harsh aqueous conditions are required [7, 45].

An important property of carbon-based MNPs is their high electrical conductivity, which enables redox-assisted remediation pathways not accessible to insulating materials such as silica. Moreover, their large pore volume and interconnected mesoporous structure facilitate the rapid diffusion of metal ions and enhance access to adsorption sites. Surface chemistry can be readily tuned via oxidation, heteroatom doping (e.g., N or S), or incorporation of metal or metal oxide nanoparticles, significantly improving selectivity and binding affinity toward toxic metal species.

#### 7.3.3.1 Synthesis

The synthesis of ordered mesoporous carbons has been extensively reported and generally follows three main approaches [45]. The hard-templating (nanocasting) method is one of the most established strategies in which a carbon precursor is infiltrated into a preformed mesoporous silica template (e.g., SBA-15 or KIT-6), followed by carbonization and subsequent removal of the silica framework. This route enables the formation of highly ordered structures such as CMK-3 and CMK-5, characterized by uniform pore sizes, high surface areas, and interconnected mesoporous channels [7, 45].

Soft-templating strategies rely on the cooperative self-assembly between carbon precursors and amphiphilic surfactants or block copolymers. During carbonization, the organic template decomposes, yielding ordered mesoporous carbon frameworks. Compared to nanocasting, this approach offers greater flexibility in tuning pore size, morphology, and surface chemistry while simplifying synthesis and improving scalability.

More recently, polymer-derived and direct pyrolysis approaches have been developed to overcome cost and scalability limitations. These methods involve the controlled carbonization of self-assembled block copolymers or polymeric frameworks, enabling the production of ordered mesoporous carbons without external hard templates. Such strategies allow the incorporation of heteroatoms and the generation of large-pore architectures suitable for advanced environmental applications.

#### 7.3.3.2 Metal Remediation

Carbon-based mesoporous materials have demonstrated excellent performance in the removal of toxic metal ions from water [46]. In particular, several studies report efficient remediation of hexavalent chromium ($Cr^{6+}$), where mesoporous carbons and carbon-based composites combine strong adsorption with redox-mediated reduction of $Cr^{6+}$ to the less toxic $Cr^{3+}$. These systems often exhibit fast kinetics, high removal efficiency, and good reusability, especially when magnetic or redox-active phases are incorporated.

Beyond chromium, functionalized mesoporous carbons have shown strong affinity toward $Pb^{2+}$, $Cd^{2+}$, $Hg^{2+}$, and $As^{3+}$, benefiting from specific interactions between metal ions and oxygen-, nitrogen-, or sulfur-containing surface groups. Their ordered pore architecture ensures efficient mass transfer, while the chemical stability of carbon frameworks allows operation over a wide pH range and in complex water matrixes.

The combination of ordered mesoporosity, high conductivity, tunable surface chemistry, and structural robustness positions carbon-based MNPs as powerful platforms for toxic metal remediation, either as unconnected adsorbents or as components of hybrid oxide–carbon systems.

From a mechanistic viewpoint, the specific mechanism of action for remediation differs from the type of material presented. For example, silica-based nanoparticles tend to promote the occurrence of adsorption or surface electrostatic-mediated interactions, whereas carbon-based mesoporous materials also integrate redox mechanisms in their metal–nanoparticle interfaces. Regardless, it is not possible to indicate only one specific mechanism of action for a specific material, but the most probable mechanism to occur. The following sections of this chapter will explore in more detail this synergistic interaction of mechanisms, particularly in hybrid nanoparticle-polymer systems. Once again, it is not possible to pinpoint unique and specific mechanisms that might contribute to metal–nanoparticle/polymer interactions but is important to retain that these interactions occur due to a combination of mechanistic phenomena.

## 7.4 FUNCTIONAL POLYMERS FOR METAL ION ENTRAPMENT

Another approach for the production of solid supported materials for environmental applications, particularly in terms of metal ion entrapment, involves the production of functional polymeric materials. Functional polymers are defined as materials that can (i) be composed of a specific chemical group or be (ii) a polymer that has a specific activity such as physical, chemical, biological or pharmacological [47, 48]. This specific activity can be produced by the addition of active components, which can be, among others, an organic compound that is added to the matrix and produces a colorimetric alteration in response to the presence of a specific analyte, or an organic molecule or nanomaterial with chelation activity for a specific set of heavy metal ions [49, 50]. The addition of such components to a non-functionalized basal polymeric matrix allows the production of materials which can be easily modulated to fulfill a specific goal, with almost infinite combination possibilities.

The production of these complex polymeric materials, especially when considering the entrapment of organic molecules onto a solid supported material allows for their use in other contexts other than a laboratorial one, allowing for in situ applications. One major advantage of functional polymers is the inherent variability they present in terms of the polymeric mesh, and the possibility to modulate the composition, size and shape of the produced material accordingly to the intended objective [51]. However, regardless of their varied applications, considering the purposes of this chapter, the focus will be mostly on their use as environmental monitoring and remediation tools.

### 7.4.1 Synthesis

Several techniques can be employed to produce functional polymers. Solvent casting, electrospinning, electrospray, are some of the most employed, however several more can be employed, always considering that the choice of production technique depends on the intended size and shape that the polymeric matrix will present [52–54]. This section will provide a brief overview of some of the most relevant methods for polymer production. It is also important to consider that a combination of these methods might also be useful for the production of such materials, for example in a layer-by-layer approach [55]. Figure 7.4 provides microscopy images demonstrating the possible distribution of nanoparticles within a polymeric material and the effects of solvent choice on polymer production.

Several matrixes can be used as a basis for the production of these materials, both petrochemical derived, and/or biocompatible, or even blended mixtures of two or more monomeric units. However, considering a more directed environmental application, as is the scope of this chapter, the choice of the basal polymeric material must be a conscientious choice and only performed after monitorization of the possible ecotoxicological profile effects, especially considering the existent issue with microplastics, oftentimes originated from a continuous exposure of polymeric materials and its subsequent continuous lixiviation [56–58].

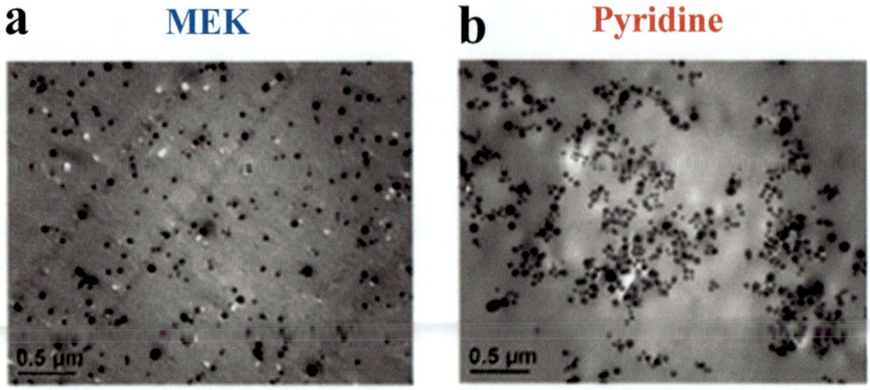


**FIGURE 7.4** TEM imaging of solvent-casted polymers with the incorporation of silica nanoparticles in (a) methylethylketone (MEK) and (b) pyridine. (Reprinted (adapted) with permission from [54]. Copyright 2014 American Chemical Society.)

#### 7.4.1.1 Solvent Casting

Solvent casting methodologies are based on the solubilization of the polymeric material and the additives in compatible solvents followed by temperature variations that promote solvent evaporation and the formation of a thin, homogeneous film [59, 60]. Several aspects must be considered with this technique, with the most important being the compatibility between the solvents in which the mesh and the additives are solubilized. Ideally, the same solvent should be applied, however, if this is not possible, it is necessary to test the compatibility of the mesh and the additive in the two solvents to ensure the formation of a homogeneous polymer and prevent the precipitation of the additives out of the polymer matrix.

In addition, temperature cycles might also be necessary, especially when considering water-soluble polymers. Several polymers such as starch or chitosan require higher temperatures, ca. 80–90°C, during the process of polymer solubilization and additive entrapment in the mesh to promote the formation of the 3D arrangement and architecture of the fibers inside the mesh [49, 61]. However, the specifications of the process are dependent on the type of mesh and additive.

#### 7.4.1.2 Electrospray

Electrospray, also known as electrodynamic atomization, methodologies are based on the formation of a polymeric membrane through application of an electric field between a nozzle, which projects the solubilized polymer, and a surface or collector [62]. In this process, the stabilized polymer solution, with or without additives, is projected onto the surface, and as the projected droplets enter in contact with the intended target, the solvent evaporates and promotes polymerization of the material onto the surface onto which was sprayed. Several parameters can be modulated for this method such as voltage, flow rate, or distance to target, to better fit the intended thickness of the formed membrane [62, 63]. This technique is mostly used to produce surface coating and one of its major advantages is that it allows for the formation of non-flat surfaces, leading to rugged coatings or more detailed surfaces.

#### 7.4.1.3 Spin Coating

Spin coating methodologies are based on the deposition of a solubilized polymeric matrix, with or without additives, to a flat surface followed by the application of a high rotation centrifugal force, to homogeneously disperse the solution [64, 65]. As the solution is sped to high rotations, the excess solvent evaporates, forming a thin membrane layer. The modulation of aspects such as membrane thickness is dependent on the speed at which the solution is spun and can be modulated from nanometers to micrometers. This technique is usually employed to produce petrochemical-based polymers, as these usually do not require temperature cycles to promote polymerization. However, one of its major disadvantages is the solution waste associated with this process that occurs as the spinning procedure occurs, as some is thrown off the edges of the spinning plate [66].

### 7.4.2 Polymer–Metal Interactions

Polymer–metal interactions can be of various types, differing in strength, specificity and reversibility [67]. Coordination/chelation interactions are commonly the strongest interactions that might occur and are based on the formation of stable mono- or multidentate bonds between the metal ions to electron-donating groups present in the polymeric material, such as carboxyl, hydroxyl, amino, or thiol groups [68, 69]. These groups might be directly present on the matrix or provided by the presence of additives present in the mesh such as organic molecules or functionalized nanomaterials. The immobilization of these groups on the polymeric material and their spatial arrangement inside the polymeric matrix promotes the occurrence of sorption phenomena from the metal ions to the matrix and their subsequent removal from the aqueous medium.

Other interactions might also occur, such as ion exchange, electrostatic, hydrogen bonding or physical entrapment, however these are weaker interactions, leading to a lower recovery rate of the metal ions toward the polymeric material [58, 70].

Another alternative to induce a specific activity on a polymeric matrix is through the direct functionalization or enrichment of the polymeric matrix with carboxy, hydroxy or amino groups increasing the overall reactivity of the polymer [71–73]. The high versatility of these processes allows for a broad range of functional polymeric materials to be produced to fit a specific application.

#### 7.4.2.1 Molecular-Imprinted Polymers

Molecular-imprinted polymers (MIPs) are a class of polymers that are designed precisely to recognize a specific analyte, such as a metal ion [74, 75]. During the synthesis process of these materials, the polymer is synthesized in the presence of the intended analyte to provide a "template" around which the polymeric matrix can form. Several strategies can be employed to produce such polymers such as those described in sections above and also others [75]. After the removal of the template, the polymer then selectively binds to the template analyte when in its presence. This is a strategy that is employed for a variety of techniques and provides high recovery rates. In the specific case of metal ion removal from aqueous environments, these polymers promote the occurrence of selective adsorption mechanisms, with a preferential uptake of that templated ion towards the matrix and have been proven to be reused after several adsorption cycles.

## 7.5 HYBRID MESOPOROUS NANOPARTICLE-POLYMER SYSTEMS

Combining the two materials described in the above sections allows for the production of hybrid MNP–polymer systems. These materials can derive synergistic effects from the combination of both counterparts, increasing their overall efficiency [76, 77].

By integrating MNPs into a polymeric matrix, several advantages can arise such as the significant increase of the overall available surface area of the material, allowing for higher recovery rates. Additionally, the insertion of nanoparticles onto a polymeric structure tends to increase their mechanical properties. Further considering functionalizations that can be performed in one or both individual counterparts, nanomaterials and/or polymeric materials, an almost infinite set of combinations can be derived from these hybrid materials, tailored to fit the intended goal. One example might be the use of these materials with a multi-purpose application, considering their increased remediation efficiency and antimicrobial activity.

### 7.5.1 Integration Strategies of Mesoporous Nanoparticles in Polymeric Matrixes

One of the easiest and most reported ways to integrate nanoparticles into a polymeric matrix is through the stabilization of the nanoparticles into a solvent-stabilized polymeric solution, and as the material polymerizes, the nanoparticles are physically entrapped and stabilized on the polymeric material [78]. This methodology promotes homogeneous integration of the particles onto the material, preserving their internal porous structure, and subsequently their increased surface area [54]. This approach can be applied to all solvent-based techniques for the production of polymeric materials. Regardless, it is crucial to evaluate the compatibility of the solvents towards the nanoparticles and the polymeric matrix, as poor solvent compatibility might precipitate the particles to the outer side of the membrane, forming a non-homogeneous coating of the membrane instead of physical entrapment.

Other mechanisms might also be employed such as in situ formation of nanoparticles inside a polymeric matrix, for example through a simple sol–gel reaction, metal precursor salt reduction, among others, however, this process does not allow for the formation of entrapped hollow MNPs [79–82].

One of the major advantages of this integration is the synergistic effects that are derived from the combination of these two counterparts, not only in terms of functionality, but also in terms of the increased mechanical stability of the material. Even though powdered nanoparticles can still be used for applications such as metal ion removal, their recovery and reuse, if possible, oftentimes require additional instrumentation and sample handling. With their integration on a macro-scale polymeric material, homogeneous dispersion is ensured as well as an easier handling of the material itself.

### 7.5.2 Metal Remediation

These hybrid materials can act as strong remediation tools due to the increased surface area and availability to capture specific or non-specific metal ions. The mechanisms by which these polymer–metal interactions occur are similar to those reported for the single counterparts, however with increased recovery percentages. These are related to the stabilization of the nanoparticles into the solid supported material and subsequent increased availability of the interaction sites. Table 7.2 demonstrates a summarized version of some of the hybrid systems reported in literature with their respective solid supported materials and recovery percentages. These materials demonstrate a high recovery rate of metal ions from solutions, with varied results in terms of metal ion selectivity. These results further confirm the impactful application of these hybrid materials for real life applications in terms of metal ion contamination in aqueous environments. However, further studies are necessary to fully employ such materials at a larger scale.

A crucial point to consider in terms of the applicability and sustainability of these materials is also their capacity to be recycled and reused for remediation. Several cycles of metal capturing, washing, and reuse should be incorporated into the initial development of these materials, in order to determine how their efficacy varies throughout various cycles. If these materials are able to retain high effectiveness throughout various cycles, it improves their overall applicability, particularly in cases where high metal concentrations are verified. A complete study of these properties must be performed to ensure the long-term applicability and sustainability of the material.

**TABLE 7.2**
**Summary of Metal Ion Efficiency of Hybrid Nanoparticle–Polymer Materials**

| Polymer Matrix | Nanoparticle | Metal Ion | Recovery | Ref |
|---|---|---|---|---|
| **Alginate** | Mesoporous halloysite | $Pb^{2+}$ | 99.97% | [77] |
| | | $Cu^{2+}$ | 95.01% | |
| | | $Cd^{2+}$ | 92.5% | |
| | | $Ni^{2+}$ | 80.85% | |
| **Chitosan** | MCM-41 (mesoporous silica) | $Cr^{6+}$ | 61.78–85.71% | [83] |
| **Polyacrylamide** | $MoS_2$ | $Cu^{2+}$ | 333.3 mg/g | [84] |
| | | $Cr^{6+}$ | 200.02 mg/g | |
| **Alginate** | Mesoporous hydrogel beads with FeNPs core structure | $Pb^{2+}$ | 21.09 mg/g | [85] |
| | | $Cu^{2+}$ | 15.54 mg/g | |
| | | $Cd^{2+}$ | 2.47 mg/g | |
| | | $Ni^{2+}$ | 2.68 mg/g | |

## 7.6 ANTIMICROBIAL PROPERTIES OF MESOPOROUS AND POLYMERIC MATERIALS

Recent studies have demonstrated that combining MNPs with polymeric matrices enables the development of antimicrobial materials that simultaneously offer mechanical stability, controlled release, and multi-mechanistic antimicrobial activity, particularly for applications in biomedical devices, food packaging, and environmental protection. Recent work has demonstrated that polymeric matrices doped with MNP systems can achieve strong and durable antimicrobial activity by combining the intrinsic bactericidal properties of metallic cores with the structural advantages of mesoporous coatings [86]. In particular, mesoporous silica-coated silver nanoparticles ($Ag@mSiO_2$) embedded into thermoplastic polyurethane films exhibited broad-spectrum antimicrobial activity against both Gram-negative and Gram-positive bacteria, highlighting then their potential for applications in antimicrobial coatings, food packaging, and biomedical materials (Figure 7.5).

Hybrid MNP–polymer systems also allow for antimicrobial applications due to the presence of the mesoporous structure inside the polymeric matrix. These antimicrobial applications are derived from the antimicrobial effects presented by the additives present in the polymeric mesh, whether compounds or nanomaterials that are functionalized or loaded with those bioactive compounds [87–89]. These agents can produce antimicrobial activity through a variety of mechanisms; however, the following two are some of the most reported: (i) reactive oxygen species (ROS) generation; and (ii) membrane disruption [53].

In terms of ROS generation, several compounds can produce species such as superoxide, hydrogen peroxide or hydroxyl radicals that increase cellular oxidative stress, damaging crucial cellular components, whether being cytosolic or membranal, disrupting the normal functioning of the bacteria and triggering subsequent cell death [90–92]. These species can also impact the normal functioning of certain crucial metabolic enzymes or pathways in bacterial central metabolism, also increasing oxidative stress levels to critical levels [88].

Other compounds might also produce membranal disruption, particularly compounds which present a large variety of aromatic moieties, increasing their lipophilicity and hindering the physical structure of the bacterial membrane, causing the expulsion of cytosolic content and triggering cell death mechanisms [93]. However, antimicrobial action occurs in response to not only one of these mechanisms, but to a combination of mechanisms that ultimately trigger cell death [89].

These hybrid MNP–polymer systems can also act as controlled release vessels for antimicrobial agents from their release from the mesoporous and polymeric matrixes in which they are inserted into [87]. These can be antibiotics or other classes of organic molecules with intrinsic antimicrobial activity such as the insertion in DNA chains, hindrance to normal enzyme functioning, antioxidant defense systems, or even membrane disruption.

### 7.6.1 Silica MNPs

In terms of silica MNPs, one of the most reported mechanisms is through loading or surface functionalization of antimicrobial agents to the nanoparticles themselves [94].

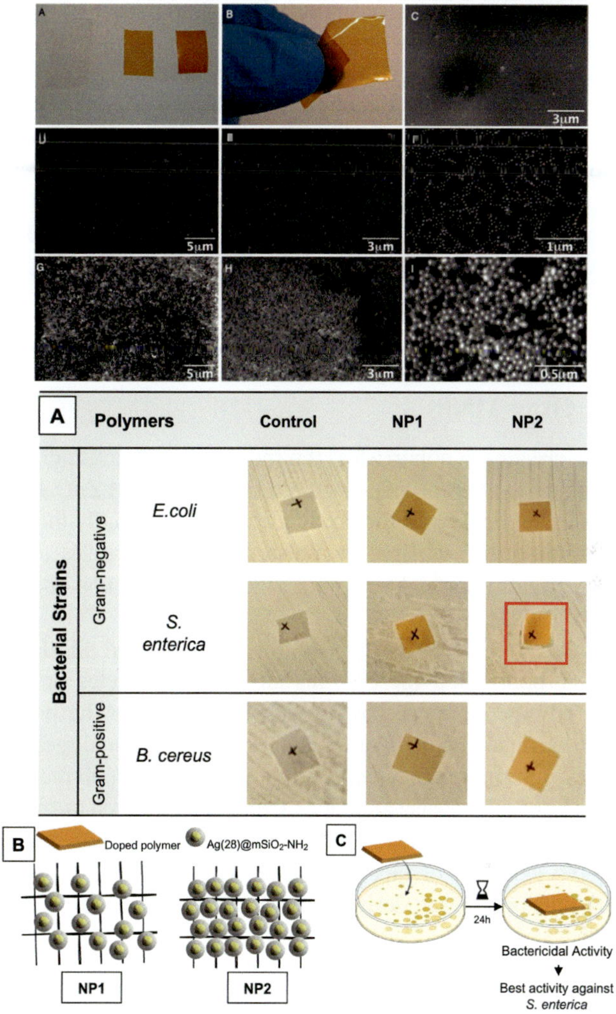


**FIGURE 7.5** (Top) Undoped (clear) and doped film with different quantities of nanoparticles (yellow films). (B) Flexibility of the doped film. (C–I) Scanning electron microscope (SEM) images of the films. (Down) (A) Antimicrobial assays against *E. coli*, *S. enterica* and *B. cereus*. (B) Schematic representation of the doped polymers. (C) Bactericidal activity against *S. enterica*. (Reproduced with permission under the Creative Commons CC BY 4.0 license from [87] copyright © 2024 MDPI.)

These nanoparticles can be loaded with the antimicrobial agents, prior to their incorporation on the polymeric matrix, producing subsequently bioactive polymeric materials.

These loading processes can be mediated by several mechanisms, such as surface coating and functionalization of the particles with the antimicrobial moiety, or through simple electrostatic interactions that promote the loading of the compounds to the internal high-surface interface of the mesoporous structure [94]. Regardless of the method, these systems tend to promote a controlled release of the bioactive molecules toward the medium, however, through a physical entrapment of the molecules inside the polymeric material, two major phenomena can occur: (i) the controlled release will be prolonged due to the physical hindrance provided by the membrane fibers, thus reducing the release kinetics of the bioactive compounds; or (ii) if the released compounds also present affinity toward the polymeric material will also become entrapped in the matrix, conferring intrinsic antimicrobial properties to a polymer that otherwise might not have presented any [87, 88, 91]. These materials combine the functionalities of both solid counterparts, nanoparticles and polymers, to produce a multi-purpose material, which can be easily modulated through modulation of each individual part or the material as a whole.

### 7.6.2 Metal Oxide MNPs

In terms of this class of nanoparticles, the most reported antimicrobial effects are derived from the generation of ROS mediated by $TiO_2$– and ZnO-based nanoparticles. These materials can trigger the production of ROS, usually modulated through an external stimulus, such as light or moisture. These particles tend to present a broad spectrum of antimicrobial activity, as they produce a non-specific antimicrobial activity [90, 92, 95].

A similar analysis can be performed for carbon-based MNPs, whether carbon-coated or carbon-doped mesoporous silica nanoparticles or even nanotube/graphene-type hybrids. As these particles are embedded in the polymeric matrix, their disposition can also directly disrupt cell membranes upon direct contact with the bacteria or release antimicrobial agents present in the hybrid polymeric material. The kinetic parameters of this release are dependent on factors such as the tridimensional arrangement of the nanoparticles inside the polymer and can be modulated accordingly through the variation of factors such as polymer thickness and nanoparticle mass per area of the polymer to derive optimized synergistic antimicrobial effects.

## 7.7 CHALLENGES, LIMITATIONS, AND FUTURE PROSPECTS

Despite the remarkable progress achieved in the development of MNPs, functional polymers, and hybrid nanoparticle–polymer systems for toxic metal remediation and antimicrobial applications, several challenges and limitations must still be addressed before their implementation in real-world environmental applications.

One of the primary challenges concerns material stability and performance under realistic conditions. Most studies are conducted using aqueous solution models under controlled laboratory conditions, whereas real wastewaters are complex matrices containing competing ions, organic matter, fluctuating pH, and variable ionic strength. These factors can significantly reduce adsorption efficiency, modify antimicrobial performance, or promote fouling of active sites. Maintaining high selectivity and removal efficiency in such complex environments remains a critical problem.

Scalability and cost also represent important limitations. While mesoporous materials exhibit excellent performance at the laboratory scale, their synthesis often involves surfactants, templates, organic solvents, or multi-step procedures that increase production costs and limit large-scale feasibility. Similarly, some functional polymers and hybrid systems rely on sophisticated fabrication techniques or expensive additives. Developing low-cost, scalable, and environmentally benign synthesis routes is therefore essential for industrial water treatment applications.

Another major apprehension concerns material recovery, regeneration, and long term durability. Although many mesoporous and hybrid systems demonstrate good reusability over several adsorption–desorption cycles, gradual loss of performance can occur due to pore blockage, surface degradation, or leaching of functional components. In antimicrobial systems, sustained release of active agents may lead to depletion over time, reducing long-term efficacy. Moreover, improper recovery of nanoscale materials raises concerns about secondary contamination and potential ecotoxicological effects.

From an environmental and regulatory perspective, the ecotoxicity and life cycle impact of nanomaterials and polymer-based systems require careful evaluation. The release of nanoparticles, degradation products, or microplastic fragments into aquatic ecosystems could lead to new environmental risks if not properly controlled. Future research must therefore integrate life-cycle assessment, toxicity testing, and safe-by-design principles to ensure that remediation technologies do not introduce new forms of pollution.

Despite these challenges, the future prospects for MNPs and polymer-based remediation systems are highly promising. Advances in surface chemistry, green synthesis, and computational material design are expected to enable more selective, robust, and multifunctional materials. The development of hybrid systems capable of simultaneous metal ion removal and antimicrobial action represents a particularly attractive strategy for addressing complex contamination scenarios, such as those found in wastewater treatment, healthcare settings, and resource-limited environments.

Emerging directions also include stimuli-responsive and smart materials, capable of modulating adsorption or antimicrobial activity in response to pH, light, redox conditions, or the presence of specific contaminants. In addition, the integration of mesoporous materials into macroscopic supports, membranes, or reusable polymeric matrices offers a pathway toward safer handling, easier recovery, and enhanced operational stability.

To sum up, MNPs, polymeric materials and their hybrid material combination are a versatile tool against toxic metal contamination and microbial pollution in complex environments. Some possible research directions might include the development of low-cost and easily scalable synthesis protocols for all materials herein described, and the alignment of their performance with existing regulatory limits, to allow for a real-world application.

Regardless of their still existing limitations such as scalability, long-term stability and regulatory approval, these materials demonstrate exceptional potential for toxic metal entrapment and antimicrobial applications. Regardless, their successful transition from a laboratorial setting to a real-life application will depend on continued interdisciplinary research combining materials science, environmental engineering, toxicology, and sustainability assessment will be essential to unlock the full potential of these advanced multifunctional systems.

## ACKNOWLEDGEMENTS

The authors thank Fundação para a Ciência e a Tecnologia (FCT), I.P., and the Associate Laboratory for Green Chemistry LAQV REQUIMTE (project UID/50006/2023); as well as the PROTEOMASS Scientific Society (Portugal). J.G. thanks FCT/MCTES for doctoral support (Grant 2022.09495.BD; https://doi.org/10.54499/2022.09495.BD). EO thanks NOVA FCT (Universidade Nova de Lisboa) and FCT for the assistant professor contract (2023.11076.TENURE.116).

## REFERENCES

1. Ji, B. Towards Environment-Sustainable Wastewater Treatment and Reclamation by the Non-aerated Microalgal-Bacterial Granular Sludge Process: Recent Advances and Future Directions. *Sci Total Environ*, 2022, *806*, 150707. https://doi.org/10.1016/j.scitotenv.2021.150707.
2. Del Prado-Audelo, M. L.; García Kerdan, I.; Escutia-Guadarrama, L.; Reyna-González, J. M.; Magaña, J. J.; Leyva-Gómez, G. Nanoremediation: Nanomaterials and Nanotechnologies for Environmental Cleanup. *Front Environ Sci*, 2021, *9*, 793765. https://doi.org/10.3389/fenvs.2021.793765.
3. Asghar, N.; Hussain, A.; Nguyen, D. A.; Ali, S.; Hussain, I.; Junejo, A.; Ali, A. Advancement in Nanomaterials for Environmental Pollutants Remediation: A Systematic Review on Bibliometrics Analysis, Material Types, Synthesis Pathways, and Related Mechanisms. *J Nanobiotechnol*, 2024, *22* (1), 26. https://doi.org/10.1186/s12951-023-02151-3.
4. Dagdag, O.; Quadri, T. W.; Haldhar, R.; Kim, S.-C.; Daoudi, W.; Berdimurodov, E.; Akpan, E. D.; Ebenso, E. E. An Overview of Heavy Metal Pollution and Control. In *Heavy Metals in the Environment*; ACS Publication, 2023; pp 3–24. https://doi.org/10.1021/bk-2023-1456.ch001.
5. Hama Aziz, K. H.; Mustafa, F. S.; Omer, K. M.; Hama, S.; Hamarawf, R. F.; Rahman, K. O. Heavy Metal Pollution in the Aquatic Environment: Efficient and Low-Cost Removal Approaches to Eliminate Their Toxicity: A Review. *RSC Adv*, 2023, *13* (26), 17595–17610. https://doi.org/10.1039/d3ra00723e
6. Vardhan, K. H.; Kumar, P. S.; Panda, R. C. A Review on Heavy Metal Pollution, Toxicity and Remedial Measures: Current Trends and Future Perspectives. *J Mol Liq*, 2019, *290*, 111197. https://doi.org/10.1016/j.molliq.2019.111197.
7. Grozdov, D.; Zinicovscaia, I. Mesoporous Materials for Metal-Laden Wastewater Treatment. *Materials*, 2023, *16* (17), 5864. https://doi.org/10.3390/ma16175864.
8. Helmi, M. Polymeric Adsorbents for Heavy Metal Removal. In *Polymeric Adsorbents*; Elsevier, 2024; pp 259–296. https://doi.org/10.1016/B978-0-323-99746-1.00004-5.
9. Mane, P. V.; Rego, R. M.; Yap, P. L.; Losic, D.; Kurkuri, M. D. Unveiling Cutting-Edge Advances in High Surface Area Porous Materials for the Efficient Removal of Toxic Metal Ions from Water. *Prog Mater Sci*, 2024, *146*, 101314. https://doi.org/10.1016/j.pmatsci.2024.101314.
10. Fatima, R.; Katiyar, P.; Kushwaha, K. Recent Advances in Mesoporous Silica Nanoparticle: Synthesis, Drug Loading, Release Mechanisms, and Diverse Applications. *Front Nanotechnol*, 2025, *7*. https://doi.org/10.3389/fnano.2025.1564188.
11. Manzano, M.; Vallet-Regí, M. Mesoporous Silica Nanoparticles in Biomedicine: Advances and Prospects. *Adv Mater*, 2025, e12433. https://doi.org/10.1002/adma.202512433.
12. Tchounwou, P. B.; Yedjou, C. G.; Patlolla, A. K.; Sutton, D. J. Heavy Metal Toxicity and the Environment. In *Molecular, Clinical and Environmental Toxicology*; Springer, 2012; pp 133–164. https://doi.org/10.1007/978-3-7643-8340-4_6.

13. Jaishankar, M.; Tseten, T.; Anbalagan, N.; Mathew, B. B.; Beeregowda, K. N. Toxicity, Mechanism and Health Effects of Some Heavy Metals. *Interdiscip Toxicol*, 2014, *7* (2), 60–72. https://doi.org/10.2478/intox-2014-0009.
14. Sharma, M.; Kant, R.; Sharma, A. K.; Sharma, A. K. Exploring the Impact of Heavy Metals Toxicity in the Aquatic Ecosystem. *Int J Energy Water Resources*, 2025, *9* (1), 267–280. https://doi.org/10.1007/s42108-024-00284-1.
15. El Rasafi, T.; Oukarroum, A.; Haddioui, A.; Song, H.; Kwon, E. E.; Bolan, N.; Tack, F. M. G.; Sebastian, A.; Prasad, M. N. V.; Rinklebe, J. Cadmium Stress in Plants: A Critical Review of the Effects, Mechanisms, and Tolerance Strategies. *Crit Rev Environ Sci Technol*, 2022, *52* (5), 675–726. https://doi.org/10.1080/10643389.2020.1835435.
16. Ullrich, S. M.; Tanton, T. W.; Abdrashitova, S. A. Mercury in the Aquatic Environment: A Review of Factors Affecting Methylation. *Crit Rev Environ. Sci. Technol.*, 2001, *31* (3), 241–293. https://doi.org/10.1080/20016491089226.
17. Barrera-Díaz, C. E.; Lugo-Lugo, V.; Bilyeu, B. A Review of Chemical, Electrochemical and Biological Methods for Aqueous Cr(VI) Reduction. *J. Hazard. Mater.*, 2012, *223–224*, 1–12. https://doi.org/10.1016/J.JHAZMAT.2012.04.054.
18. Bhat, A.; Ravi, K.; Tian, F.; Singh, B. Arsenic Contamination Needs Serious Attention: An Opinion and Global Scenario. *Pollutants*, 2024, *4* (2), 196–211. https://doi.org/10.3390/pollutants4020013.
19. Shofia, S. I.; Vickram, A. S.; Saravanan, A.; Deivayanai, V. C.; Yaashikaa, P. R. Sustainable Separation Technologies for Heavy Metal Removal from Wastewater: An Upgraded Review of Physicochemical Methods and Its Advancements. *Sustain Chem Environ*, 2025, *10*, 100264. https://doi.org/10.1016/j.scenv.2025.100264.
20. Qasem, N. A. A.; Mohammed, R. H.; Lawal, D. U. Removal of Heavy Metal Ions from Wastewater: A Comprehensive and Critical Review. *NPJ Clean Water*, 2021, *4* (1), 36. https://doi.org/10.1038/s41545-021-00127-0.
21. Vidu, R.; Matei, E.; Predescu, A. M.; Alhalaili, B.; Pantilimon, C.; Tarcea, C.; Predescu, C. Removal of Heavy Metals from Wastewaters: A Challenge from Current Treatment Methods to Nanotechnology Applications. *Toxics*, 2020, *8* (4), 101. https://doi.org/10.3390/toxics8040101.
22. Chen, Y. *Design, Synthesis, Multifunctionalization and Biomedical Applications of Multifunctional Mesoporous Silica-Based Drug Delivery Nanosystems*; Springer Berlin Heidelberg: Berlin, Heidelberg, 2016. https://doi.org/10.1007/978-3-662-48622-1.
23. Everett, D. H. Manual of Symbols and Terminology for Physicochemical Quantities and Units, Appendix II: Definitions, Terminology and Symbols in Colloid and Surface Chemistry. *Pure Appl Chem*, 1972, *31* (4), 577–638. https://doi.org/10.1351/pac197231040577.
24. Zhao, D.; Feng, J.; Huo, Q.; Melosh, N.; Fredrickson, G. H.; Chmelka, B. F.; Stucky, G. D. Triblock Copolymer Syntheses of Mesoporous Silica with Periodic 50 to 300 Angstrom Pores. *Science (1979).*, 1998, *279* (5350), 548–552. https://doi.org/10.1126/science.279.5350.548.
25. Fripiat, J. J. Silanol Groups and Properties of Silica Surfaces. In *Soluble Silicates*; ACS Publication, 1982; pp 165–184. https://doi.org/10.1021/bk-1982-0194.ch011.
26. Kresge, C. T.; Leonowicz, M. E.; Roth, W. J.; Vartuli, J. C.; Beck, J. S. Ordered Mesoporous Molecular Sieves Synthesized by a Liquid-Crystal Template Mechanism. *Nature*, 1992, *359* (6397), 710–712. https://doi.org/10.1038/359710a0.
27. Siefker, J.; Karande, P.; Coppens, M.-O. Packaging Biological Cargoes in Mesoporous Materials: Opportunities for Drug Delivery. *Expert Opin Drug Deliv*, 2014, *11* (11), 1781–1793. https://doi.org/10.1517/17425247.2014.938636.
28. Feng, Y.; Panwar, N.; Tng, D. J. H.; Tjin, S. C.; Wang, K.; Yong, K.-T. The Application of Mesoporous Silica Nanoparticle Family in Cancer Theranostics. *Coord Chem Rev*, 2016, *319*, 86–109. https://doi.org/10.1016/j.ccr.2016.04.019.

29. dos Santos, S. M. L.; Nogueira, K. A. B.; de Souza Gama, M.; Lima, J. D. F.; da Silva Júnior, I. J.; de Azevedo, D. C. S. Synthesis and Characterization of Ordered Mesoporous Silica (SBA-15 and SBA-16) for Adsorption of Biomolecules. *Micropor Mesopor Mater*, 2013, *180*, 284–292. https://doi.org/10.1016/j.micromeso.2013.06.043.
30. Sierra, C.; Mesa, M.; Ramirez, A.; Lopez, B.; Gut, J.L. Synthesis of Micron-Sized Particles of Mesoporous Silica from Tri-Block Surfactants in Presence of Fluoride, Usable as Stationary Phase in HPLC. *Stud Surface Sci Catalysis*, 2004, *154*, 573–580. https://doi.org/10.1016/S0167-2991(04)80853-X.
31. Lin, C.-L.; Pang, Y.-S.; Chao, M.-C.; Chen, B.-C.; Lin, H.-P.; Tang, C.-Y.; Lin, C.-Y. Synthesis of SBA-16 and SBA-15 Mesoporous Silica Crystals Templated with Neutral Block Copolymer Surfactants. *J Phys Chem Solids*, 2008, *69* (2–3), 415–419. https://doi.org/10.1016/j.jpcs.2007.07.006.
32. Wan, Y., Zhao, D. On the Controllable Soft-Templating Approach to Mesoporous Silicates. *Chem Rev*, 2007, *107* (7), 2821–2860. https://doi.org/10.1021/cr068020s.
33. Zhao, D.; Sun, J.; Li, Q.; Stucky, G. D. Morphological Control of Highly Ordered Mesoporous Silica SBA-15. *Chem Mater*, 2000, *12* (2), 275–279. https://doi.org/10.1021/cm9911363.
34. Rivera-Muñoz, E. M.; Huirache-Acuña, R. Sol Gel-Derived SBA-16 Mesoporous Material. *Int J Mol Sci*, 2010, *11* (9), 3069–3086. https://doi.org/10.3390/ijms11093069.
35. Lovingood, D. D.; Owens, J. R.; Seeber, M.; Kornev, K. G.; Luzinov, I. Controlled Microwave-Assisted Growth of Silica Nanoparticles under Acid Catalysis. *ACS Appl Mater Interfaces*, 2012, *4* (12), 6875–6883. https://doi.org/10.1021/am3020247.
36. Rani, L.; Srivastav, A. L.; Kaushal, J.; Nguyen, X. C. Recent Advances in Nanomaterial Developments for Efficient Removal of Hg(II) from Water. *Environ Sci Poll Res*, 2022, *29* (42), 62851–62869. https://doi.org/10.1007/s11356-022-21869-5.
37. Mane, S. M.; Raorane, C. J.; Shin, J. C. Synthesis of Mesoporous Silica Adsorbent Modified with Mercapto–Amine Groups for Selective Adsorption of Cu2+ Ion from Aqueous Solution. *Nanomaterials*, 2022, *12* (18), 3232. https://doi.org/10.3390/nano12183232.
38. Putz, A.-M.; Ivankov, O. I.; Kuklin, A. I.; Ryukhtin, V.; Ianăşi, C.; Ciopec, M.; Negrea, A.; Trif, L.; Horváth, Z. E.; Almásy, L. Ordered Mesoporous Silica Prepared in Different Solvent Conditions: Application for Cu(II) and Pb(II) Adsorption. *Gels*, 2022, *8* (7), 443. https://doi.org/10.3390/gels8070443.
39. Ezzeddine, Z.; Batonneau-Gener, I.; Ghssein, G.; Pouilloux, Y. Recent Advances in Heavy Metal Adsorption via Organically Modified Mesoporous Silica: A Review. *Water (Basel).*, 2025, *17* (5), 669. https://doi.org/10.3390/w17050669.
40. Marcelo, G. A.; Pires, S. M. G.; Faustino, M. A. F.; Simões, M. M. Q.; Neves, M. G. P. M. S.; Santos, H. M.; Capelo, J. L.; Mota, J. P.; Lodeiro, C.; Oliveira, E. New Dual Colorimetric/Fluorimetric Probes for $Hg^{2+}$ Detection & Extraction Based on Mesoporous SBA-16 Nanoparticles Containing Porphyrin or Rhodamine Chromophores. *Dyes Pigment*, 2019, *161*, 427–437. https://doi.org/10.1016/j.dyepig.2018.09.068.
41. Bastakoti, B. P.; Kuila, D.; Salomon, C.; Konarova, M.; Eguchi, M.; Na, J.; Yamauchi, Y. Metal-Incorporated Mesoporous Oxides: Synthesis and Applications. *J Hazard Mater*, 2021, *401*, 123348. https://doi.org/10.1016/j.jhazmat.2020.123348.
42. Islam, Md. M.; Mohana, A. A.; Rahman, Md. A.; Rahman, M.; Naidu, R.; Rahman, M. M. A Comprehensive Review of the Current Progress of Chromium Removal Methods from Aqueous Solution. *Toxics*, 2023, *11* (3), 252. https://doi.org/10.3390/toxics11030252.
43. Yuan, S.; Wang, M.; Liu, J.; Guo, B. Recent Advances of SBA-15-Based Composites as the Heterogeneous Catalysts in Water Decontamination: A Mini-Review. *J Environ Manage*, 2020, *254*, 109787. https://doi.org/10.1016/j.jenvman.2019.109787.

44. Xin, X.; Wei, Q.; Yang, J.; Yan, L.; Feng, R.; Chen, G.; Du, B.; Li, H. Highly Efficient Removal of Heavy Metal Ions by Amine-Functionalized Mesoporous $Fe_3O_4$ Nanoparticles. *Chem Eng J*, 2012, *184*, 132–140. https://doi.org/10.1016/j.cej.2012.01.016.
45. Li, Z.; Liu, L.; Wang, Z.; Gao, P.; Li, G. K. Synthesis and Application of Mesoporous Materials: Process Status, Technical Problems, and Development Prospects: A Mini-Review. *Energy Fuel*, 2023, *37* (5), 3413–3427. https://doi.org/10.1021/acs.energyfuels.2c03882.
46. Liu, Y.; Liu, Y.; Tang, L.; Zhou, Y.; Yang, G.; Zeng, G.; Liu, H. Mesoporous Carbon-Based Composites for Adsorption of Heavy Metals. In *Nanohybrid and Nanoporous Materials for Aquatic Pollution Control*; Elsevier, 2019; pp 63–102. https://doi.org/10.1016/B978-0-12-814154-0.00003-7.
47. Rivas, B. L.; Pereira, E. D.; Palencia, M.; Sánchez, J. Water-Soluble Functional Polymers in Conjunction with Membranes to Remove Pollutant Ions from Aqueous Solutions. *Prog Polym Sci*, 2011, *36* (2), 294–322. https://doi.org/10.1016/j.progpolymsci.2010.11.001.
48. Horie, K.; Barón, M.; Fox, R. B.; He, J.; Hess, M.; Kahovec, J.; Kitayama, T.; Kubisa, P.; Maréchal, E.; Mormann, W.; et al. Definitions of Terms Relating to Reactions of Polymers and to Functional Polymeric Materials (IUPAC Recommendations 2003). *Pure Appl Chem*, 2004, *76* (4), 889–906. https://doi.org/10.1351/pac200476040889.
49. Galhano, J.; Kurutos, A.; Dobrikov, G. M.; Duarte, M. P.; Santos, H. M.; Capelo-Martínez, J. L.; Lodeiro, C.; Oliveira, E. Fluorescent Polymers for Environmental Monitoring: Targeting Pathogens and Metal Contaminants with Naphthalimide Derivatives. *Hazard Mater*, 2024, *480*, 136107. https://doi.org/10.1016/J.JHAZMAT.2024.136107.
50. Yuan, R.; He, X.; Zhu, C.; Tao, L. Recent Developments in Functional Polymers via the Kabachnik–Fields Reaction: The State of the Art. *Molecules*, 2024, *29* (3), 727. https://doi.org/10.3390/molecules29030727.
51. EL-Ghoul, Y.; Alminderej, F. M.; Alsubaie, F. M.; Alrasheed, R.; Almousa, N. H. Recent Advances in Functional Polymer Materials for Energy, Water, and Biomedical Applications: A Review. *Polymers (Basel).*, 2021, *13* (24), 4327. https://doi.org/10.3390/polym13244327.
52. Elsayed, R.; Teow, Y. H. Advanced Functional Polymer Materials for Biomedical Applications. *J Appl Polym Sci*, 2025, *142* (3), e56391. https://doi.org/10.1002/APP.56391.
53. Galhano, J.; Capelo-Martinez, J. L.; Lorenzo, J.; Lodeiro, C.; Oliveira, E. Advances in Antimicrobial Applications of Ag, Cu, and AgCu Nanoparticle-Doped Polymeric Composite Materials: A Comprehensive Review. *ACS Nano*, 2025, *19* (35), 31301–31330. https://doi.org/10.1021/acsnano.5c08822.
54. Jouault, N.; Zhao, D.; Kumar, S. K. Role of Casting Solvent on Nanoparticle Dispersion in Polymer Nanocomposites. *Macromolecules*, 2014, *47* (15), 5246–5255. https://doi.org/10.1021/ma500619g.
55. Alotaibi, K. M.; Shukla, A. K.; Alshahrani, A. A.; Alzhrani, E. M.; Alotaibi, S. S.; Alswieleh, A. Layer-by-Layer Assembled Nanocomposite Membranes with Functionalized Dendritic Mesoporous Silica Nanoparticles for Selective Heavy Metal Removal. *ACS Omega*, 2025, *10* (26), 28112–28127. https://doi.org/10.1021/acsomega.5c02486.
56. Piao, Z.; Agyei Boakye, A. A.; Yao, Y. Environmental Impacts of Biodegradable Microplastics. *Nature Chem Eng*, 2024, *1* (10), 661–669. https://doi.org/10.1038/s44286-024-00127-0.
57. Schwarz, A. E.; Herlaar, S.; Cohen, Q. M.; Quik, J. T. K.; Golkaram, M.; Urbanus, J. H.; van Emmerik, T. H. M.; Huijbregts, M. A. J. Microplastic Aquatic Impacts Included in Life Cycle Assessment. *Resour Conserv Recycl*, 2024, *209*, 107787. https://doi.org/10.1016/j.resconrec.2024.107787.

58. Groh, K. J.; Arp, H. P. H.; MacLeod, M.; Wang, Z. Assessing and Managing Environmental Hazards of Polymers: Historical Development, Science Advances and Policy Options. *Environ Sci Process Impacts*, 2023, *25* (1), 10–25. https://doi.org/10.1039/D2EM00386D.
59. Hossain, K. M. Z.; Felfel, R. M.; Ogbilikana, P. S.; Thakker, D.; Grant, D. M.; Scotchford, C. A.; Ahmed, I. Single Solvent-Based Film Casting Method for the Production of Porous Polymer Films. *Macromol Mater Eng*, 2018, *303* (4). https://doi.org/10.1002/mame.201700628.
60. Felton, L. A. Mechanisms of Polymeric Film Formation. *Int J Pharm*, 2013, *457* (2), 423–427. https://doi.org/10.1016/j.ijpharm.2012.12.027.
61. Nguyen, K. D. Temperature Effect of Water Coagulation Bath on Chitin Fiber Prepared through Wet-Spinning Process. *Polymers (Basel).*, 2021, *13* (12), 1909. https://doi.org/10.3390/polym13121909.
62. Boda, S. K.; Li, X.; Xie, J. Electrospraying an Enabling Technology for Pharmaceutical and Biomedical Applications: A Review. *J Aerosol Sci*, 2018, *125*, 164–181. https://doi.org/10.1016/j.jaerosci.2018.04.002.
63. Lai, W.; Di, L.; Zhao, C.; Tian, Y.; Duan, Y.; Pan, Y.; Ye, D.; Jiang, L.; Guo, Y.; He, G.; et al. Electrospray Deposition for Electronic Thin Films on 3D Freeform Surfaces: From Mechanisms to Applications. *Adv Mater Technol*, 2024, *9* (22). https://doi.org/10.1002/admt.202400192.
64. Allaf, R. M. Influence of Spin Coating Parameters on the Fabrication of Free Standing Porous and Nonporous Poly(ε-Caprolactone) Based Films. *Sci Rep*, 2025, *15* (1), 25119. https://doi.org/10.1038/s41598-025-10425-z.
65. Jiang, Y.; Minett, M.; Hazen, E.; Wang, W.; Alvarez, C.; Griffin, J.; Jiang, N.; Chen, W. New Insights into Spin Coating of Polymer Thin Films in Both Wetting and Nonwetting Regimes. *Langmuir*, 2022, *38* (41), 12702–12710. https://doi.org/10.1021/acs.langmuir.2c02206.
66. Aziz, F.; Ismail, A. F. Spray Coating Methods for Polymer Solar Cells Fabrication: A Review. *Mater Sci Semicond Process*, 2015, *39*, 416–425. https://doi.org/10.1016/j.mssp.2015.05.019.
67. Abtahi, S.; Hendeniya, N.; Mahmud, S. T.; Mogbojuri, G.; Iheme, C. L.; Chang, B. Metal-Coordinated Polymer–Inorganic Hybrids: Synthesis, Properties, and Application. *Polymers (Basel).*, 2025, *17* (2), 136. https://doi.org/10.3390/polym17020136.
68. Archer, R. D.; Hardiman, C. J.; Kim, K. S.; Grandbois, E. R.; Goldstein, M. Metal-Chelate Polymers: Structural/Property Relationships as a Function of the Metal Ion. In *Metal-Containing Polymeric Systems*; Springer US: Boston, MA, 1985; pp 355–366. https://doi.org/10.1007/978-1-4615-9415-4_20.
69. Guibal, E. Interactions of Metal Ions with Chitosan-Based Sorbents: A Review. *Sep Purif Technol*, 2004, *38* (1), 43–74. https://doi.org/10.1016/j.seppur.2003.10.004.
70. Cyganowski, P.; Dzimitrowicz, A. A Mini-Review on Anion Exchange and Chelating Polymers for Applications in Hydrometallurgy, Environmental Protection, and Biomedicine. *Polymers (Basel).*, 2020, *12* (4), 784. https://doi.org/10.3390/polym12040784.
71. Curran, J. M.; Fawcett, S.; Hamilton, L.; Rhodes, N. P.; Rahman, C. V.; Alexander, M.; Shakesheff, K.; Hunt, J. A. The Osteogenic Response of Mesenchymal Stem Cells to an Injectable PLGA Bone Regeneration System. *Biomaterials*, 2013, *34* (37), 9352–9364. https://doi.org/10.1016/j.biomaterials.2013.08.044.
72. Drobota, M.; Ursache, S.; Aflori, M. Surface Functionalities of Polymers for Biomaterial Applications. *Polymers (Basel).*, 2022, *14* (12), 2307. https://doi.org/10.3390/polym14122307.
73. Mountaki, S. A.; Kaliva, M.; Loukelis, K.; Chatzinikolaidou, M.; Vamvakaki, M. Responsive Polyesters with Alkene and Carboxylic Acid Side-Groups for Tissue Engineering Applications. *Polymers (Basel).*, 2021, *13* (10), 1636. https://doi.org/10.3390/polym13101636.

74. Sharma, G.; Kandasubramanian, B. Molecularly Imprinted Polymers for Selective Recognition and Extraction of Heavy Metal Ions and Toxic Dyes. *J Chem Eng Data*, 2020, *65* (2), 396–418. https://doi.org/10.1021/acs.jced.9b00953.
75. Metwally, M.; Benhawy, A.; Khalifa, R.; El Nashar, R.; Trojanowicz, M. Application of Molecularly Imprinted Polymers in the Analysis of Waters and Wastewaters. *Molecules*, 2021, *26* (21), 6515. https://doi.org/10.3390/molecules26216515.
76. Santhamoorthy, M.; Asaithambi, P.; Ramkumar, V.; Elangovan, N.; Perumal, I.; Kim, S. C. A Review on the Recent Advancements of Polymer-Modified Mesoporous Silica Nanoparticles for Drug Delivery Under Stimuli-Trigger. *Polymers (Basel).*, 2025, *17* (12), 1640. https://doi.org/10.3390/polym17121640.
77. Hassan, M.; Liu, Y.; Naidu, R.; Du, J.; Qi, F.; Donne, S. W.; Islam, M. M. Mesoporous Biopolymer Architecture Enhanced the Adsorption and Selectivity of Aqueous Heavy Metal Ions. *ACS Omega*, 2021, *6* (23), 15316–15331. https://doi.org/10.1021/acsomega.1c01642.
78. Domingues, J. M.; Miranda, C. S.; Homem, N. C.; Felgueiras, H. P.; Antunes, J. C. Nanoparticle Synthesis and Their Integration into Polymer-Based Fibers for Biomedical Applications. *Biomedicines*, 2023, *11* (7), 1862. https://doi.org/10.3390/biomedicines11071862.
79. Kelarakis, A. In Situ Generation of Nanoparticles on and within Polymeric Materials. *Polymers (Basel).*, 2024, *16* (11), 1611. https://doi.org/10.3390/polym16111611.
80. Aldakheel, F.; Ntetsikas, K.; Yudhanto, A.; Lubineau, G.; Hadjichristidis, N. In Situ Formation of Silica Nanoparticles Decorated with Well-Defined Homopolymers and Block Copolymers. *ACS Appl Polym Mater*, 2023, *5* (6), 4244–4255. https://doi.org/10.1021/acsapm.3c00440.
81. Muraviev, D. N.; Macanás, J.; Farre, M.; Muñoz, M.; Alegret, S. Novel Routes for Inter-Matrix Synthesis and Characterization of Polymer Stabilized Metal Nanoparticles for Molecular Recognition Devices. *Sens Actuators B Chem*, 2006, *118* (1–2), 408–417. https://doi.org/10.1016/j.snb.2006.04.047.
82. Madkour, M.; Bumajdad, A.; Al-Sagheer, F. To What Extent Do Polymeric Stabilizers Affect Nanoparticles Characteristics? *Adv Colloid Interface Sci*, 2019, *270*, 38–53. https://doi.org/10.1016/j.cis.2019.05.004.
83. Dinari, M.; Soltani, R.; Mohammadnezhad, G. Kinetics and Thermodynamic Study on Novel Modified–Mesoporous Silica MCM-41/Polymer Matrix Nanocomposites: Effective Adsorbents for Trace $Cr^{VI}$ Removal. *J Chem Eng Data*, 2017, *62* (8), 2316–2329. https://doi.org/10.1021/acs.jced.7b00197.
84. Shwetha Shree, H. N.; Arpitha, H. J.; Priyanka, H. D.; Banu, N.; Gayathri, B. H.; Srinivasan, R.; Al-Sadoon, M. K.; Durai, M.; Gnanasekaran, L.; Ramasundaram, S.; et al. Effective Removal of Metal Ions Using MoS2 Functionalized Chitosan Schiff Base Incorporated with C3N4 Nanoparticle from Aqueous Solutions. *Int J Biol Macromol*, 2025, *307*, 141402. https://doi.org/10.1016/j.ijbiomac.2025.141402.
85. Hassan, M.; Naidu, R.; Du, J.; Qi, F.; Ahsan, M. A.; Liu, Y. Magnetic Responsive Mesoporous Alginate/β-Cyclodextrin Polymer Beads Enhance Selectivity and Adsorption of Heavy Metal Ions. *Int J Biol Macromol*, 2022, *207*, 826–840. https://doi.org/10.1016/j.ijbiomac.2022.03.159.
86. Nuti, S.; Fernández-Lodeiro, A.; Galhano, J.; Oliveira, E.; Duarte, M. P.; Capelo-Martínez, J. L.; Lodeiro, C.; Fernández-Lodeiro, J. Tailoring Mesoporous Silica-Coated Silver Nanoparticles and Polyurethane-Doped Films for Enhanced Antimicrobial Applications. *Nanomaterials*, 2024, *14* (5), 462. https://doi.org/10.3390/nano14050462.
87. Presentato, A.; Armetta, F.; Spinella, A.; Chillura Martino, D. F.; Alduina, R.; Saladino, M. L. Formulation of Mesoporous Silica Nanoparticles for Controlled Release of Antimicrobials for Stone Preventive Conservation. *Front Chemi*, 2020, *8*. https://doi.org/10.3389/fchem.2020.00699.

88. Laskowska, M.; Doskocz, M.; Karczmarska, A.; Pogoda, K.; Olender, A.; Gumbarewicz, E.; Bogut, A.; Gagoś, M.; Stepulak, A.; Dąbrowski, W.; et al. Mesoporous Silica-Based Nanocomposite Surfaces with ROS-Generating Copper Phosphonate Functional Groups in the Fight against Dangerous Microorganisms. *Appl Surf Sci*, 2026, *720*, 165228. https://doi.org/10.1016/j.apsusc.2025.165228.
89. Li, B.; Liao, Y.; Su, X.; Chen, S.; Wang, X.; Shen, B.; Song, H.; Yue, P. Powering Mesoporous Silica Nanoparticles into Bioactive Nanoplatforms for Antibacterial Therapies: Strategies and Challenges. *J Nanobiotechnol*, 2023, *21* (1), 325. https://doi.org/10.1186/s12951-023-02093-w.
90. Pandey, M.; Singh, M.; Wasnik, K.; Gupta, S.; Patra, S.; Gupta, P. S.; Pareek, D.; Chaitanya, N. S. N.; Maity, S.; Reddy, A. B. M.; et al. Targeted and Enhanced Antimicrobial Inhibition of Mesoporous ZnO–$Ag_2$ O/Ag, ZnO–CuO, and ZnO–$SnO_2$ Composite Nanoparticles. *ACS Omega*, 2021, *6* (47), 31615–31631. https://doi.org/10.1021/acsomega.1c04139.
91. Laskowska, M.; Kowalczyk, P.; Karczmarska, A.; Pogoda, K.; Zubko, M.; Laskowski, Ł. Functionalised Mesoporous Silica Thin Films as ROS-Generating Antimicrobial Coatings. *Int J Mol Sci*, 2025, *26* (15), 7154. https://doi.org/10.3390/ijms26157154.
92. Bai, Q.; Zhang, Y.; Cai, R.; Wu, H.; Fu, H.; Zhou, X.; Chai, J.; Teng, X.; Liu, T. AMP-Coated TiO2 Doped ZnO Nanomaterials Enhanced Antimicrobial Activity and Efficacy in Otitis Media Treatment by Elevating Hydroxyl Radical Levels. *Int J Nanomed*, 2024, *19*, 2995–3007. https://doi.org/10.2147/IJN.S449888.
93. Häffner, S. M.; Parra-Ortiz, E.; Browning, K. L.; Jørgensen, E.; Skoda, M. W. A.; Montis, C.; Li, X.; Berti, D.; Zhao, D.; Malmsten, M. Membrane Interactions of Virus-like Mesoporous Silica Nanoparticles. *ACS Nano*, 2021, *15* (4), 6787–6800. https://doi.org/10.1021/acsnano.0c10378.
94. Galhano, J.; Marcelo, G. A.; Duarte, M. P.; Oliveira, E. Ofloxacin@Doxorubicin-Epirubicin Functionalized MCM-41 Mesoporous Silica-Based Nanocarriers as Synergistic Drug Delivery Tools for Cancer Related Bacterial Infections. *Bioorg Chem*, 2022, *118*, 105470. https://doi.org/10.1016/j.bioorg.2021.105470.
95. Al-Shaeri, M.; Satar, R.; Ahmed, S. I.; Oves, M.; Ansari, S. A.; Chibber, S. Utilization of Doped Nanoparticles of ZnO and $TiO_2$ As Antimicrobial Agent. *Oriental J Chem*, 2019, *35* (3), 1235–1243. https://doi.org/10.13005/ojc/350346.

# 8 An Insight into the Environmental Impact of Inorganic Nanoparticles
## *Upsides and Downsides*

*Ana Justo-Vega, Juan José López-Mayán, Antonio Moreda-Piñeiro, and Pilar Bermejo-Barrera*

## 8.1 INTRODUCTION

The rapid development and industrialization of modern society have contributed to unprecedented environmental contamination. Human activities pose risks to the balance of ecosystems in terms of pollution and environmental degradation, leading to long-term adverse effects on ecology. Pollution is one of the leading causes of human mortality, directly and indirectly contributing to nine million deaths each year, according to Lancet Commission on Pollution and Health [1]. It can be classified as air, water, or soil pollution depending on where the contaminant is released. Environmental contaminants include heavy metals, chemical waste, pesticides and herbicides, pharmaceuticals, and, more recently, emerging pollutants such as microplastics and nanomaterials (NMs) [2]. This call for action – both to combat climate change and to ensure the sustainable management of availability of water resources – is expressed in the United Nations 2030 Development Agenda [3].

As of 2000, nanotechnology has emerged as a revolutionary discipline to foster development in a wide variety of sectors: textile and food industries, medicine and healthcare, electronics and computing, energy, agriculture, materials science, automotive and aerospace, and construction and infrastructure, among others. As a result of such widespread use, NMs have ended up being released into the environment and are now considered contaminants of emerging concerns (CECs) [4]. The deterioration of ecosystems issues an urgent need for environmental protection. In this sense, analytical efforts have been made to develop new strategies for the analysis, understanding, and remediation of environmental systems. This chapter covers the environmental implications of inorganic nanoparticles (NPs), addressing both the challenges and concern they pose as emerging contaminants and the opportunities they offer for cutting-edge remediation strategies.

DOI: 10.1201/9781003632092-9

## 8.2 INORGANIC NANOPARTICLES: DEFINITION, ORIGINS, OCCURRENCE, REGULATORY FRAMEWORK, AND ENVIRONMENTAL DISTRIBUTION

According to the International Organization for Standardization (ISO), ISO/TR 18401:2017, nanoparticle is defined as a "nano-object with all external dimensions in the nanoscale where the lengths of the longest and the shortest axes of the nano-object do not differ significantly", where nanoscale is specified as the "length range approximately from 1 nm to 100 nm" [5]. More recently, the European Commission (EC) Recommendation 2022/C 229/01 defines NM as "a natural, incidental or manufactured material consisting of solid particles that are present, either on their own or as identifiable constituent particles in aggregates or agglomerates, and where 50% or more of these particles in the number-based size distribution fulfil at least one of the following conditions: (a) one or more external dimensions of the particle are in the size range 1 nm to 100 nm; (b) the particle has an elongated shape, such as a rod, fibre or tube, where two external dimensions are smaller than 1 nm and the other dimension is larger than 100 nm; (c) the particle has a plate-like shape, where on external dimension is smaller than 1 nm and the other dimensions are larger than 100 nm" [6]. Under this framework, nanoparticles are considered a specific subset of NMs.

Based on their origin, NMs can be classified as natural, incidental, or engineered [7]. Natural NMs arise from (bio)geochemical or mechanical processes occurring in nature, such as volcanic ash and ocean spray. Incidental NMs are unintentionally generated as by-products of human activities, such as vehicle exhaust gases. Engineered NMs (ENMs), in contrast, are deliberately designed and manufactured by humans for specific applications.

As previously mentioned, nanotechnology has revolutionised numerous industrial domains, and emerging applications are expected to further contribute to the expansion of the global nanotechnology market. The global market for NMs was valued at USD 10.3 billion in 2020 and is projected to reach USD 38.1 billion by 2029, reflecting a compound annual growth rate (CAGR) of 17.8%. In terms of volume, the global NMs market size was estimated at 11 million tonnes [8]. According to Nanotechnology Products Database (NPD), there are currently available more than 11,100 nanotechnology-based consumer products, being electronics the leading sector (NPD) [9]. However, real figures are difficult to obtain, as comprehensive quantitative information on NMs manufacturing volumes in Europe is lacking. Existing estimates reported in the literature are typically obtained through surveys to producers and extrapolated from regional to global scales.

The exponential global increase in manufacture, application, and disposal of NMs has led to their unavoidable release into environmental systems. To such an extent that NMs are now considered CECs. Among these, titanium dioxide nanoparticles ($TiO_2$NPs), zinc oxide nanoparticles (ZnONPs), and silver nanoparticles (AgNPs) are recognized by the Organization for Economic Cooperation and Development (OECD) as priority materials requiring focused assessment [10]. NMs originated from a variety of anthropogenic sources subsequently disperse into atmospheric, terrestrial, and aquatic compartments. Despite increasing research attention, their ecological and environmental consequences remain uncertain, and potential hazards and the mechanisms underpinning their effects are not fully understood.

NMs are not static entities and once released into the environment they undergo a suite of physicochemical and biological transformations. Consequently, NM ecotoxicology involves the study of their environmental fate and transport, bioavailability, and uptake by ecological receptors [11].

Entry of NMs into the environment typically occurs through three major emission pathways [4]: (i) unintentional release during the production of nano-enabled materials, (ii) emissions during product use, and (iii) direct or indirect discharge associated with the disposal of NMs-containing wastes. Their environmental fate includes the assessment of both physical and chemical alterations occurring within natural systems, as well as their transport dynamics, and distribution patterns across environmental matrices. These processes are strongly influenced by the intrinsic physicochemical characteristics of NMs, particularly surface properties, and by external environmental conditions such as pH, ionic strength, temperature, organic matter content, oxygen availability, and the presence of co-existing chemicals.

Because multiple transformation processes may occur simultaneously, interactions among them can modify the extent or direction of NMs behaviour. Three major transformation processes can explain the environmental fate of NMs [10]: (i) Physical, including homo- and hetero-aggregation, agglomeration, sedimentation, and deposition; (ii) chemical transformations, such as dissolution, redox reactions, photochemical reactions, and corona formation; and (iii) biological, including biodegradation and biomodification.

Following these transformations, the transport characteristics and environmental persistence of NMs are altered, enabling their accumulation in compartments such as surface waters, sediments, soils, wastewater sludge, and the atmosphere. In addition, aquatic and terrestrial ecosystems are frequently subjected to a diverse array of chemical stressors, raising the possibility that NMs may facilitate the transport of conventional pollutants by acting as vectors [12]. This underscores the need to evaluate not only the isolated effects of NMs but also their potential interactions with co-occurring environmental stressors and the cumulative effects of such interactions.

Environmental risk refers to the likelihood and severity of adverse effects on ecosystems or human health resulting from the exposure to environmental contaminants. In its 2024 Global Risks Report, the World Economic Forum identifies environmental risk as one of the most critical global threats [13]. As NMs are increasingly recognized as emerging pollutants, elucidating their environmental fate and associated ecological impact becomes critical for the development of informed risk assessment and management strategies.

Growing scientific evidence has confirmed the presence, mobility, and persistence of NMs within multiple environmental compartments. Building on these observations, initial modelling efforts have been made to estimate environmental NPs concentrations. However, these estimates remain highly uncertain due to the limited availability of robust data on NP production [14].

In the context of the European Union (EU), legislation is gradually being updated to include specific requirements for NMs (accessible online at https://eur-lex.europa.eu/). However, current NMs regulations largely focus on cosmetic formulations, medical devices, and food-related applications (Commission Regulation (EC) **450/2009**, Regulation (EC) **1223/2009,** Regulation (EU) **1169/2011**, Regulation (EU) **528/2012**, Regulation (EU) **2015/2283**, Regulation (EU) **2017/745**, Commission Regulation (EU) **2022/63**), while environmental protection considerations receive comparatively little attention. This regulatory deficiency may contribute to undesirable ecological alterations, including potential trophic transfer of NMs through food chain, thereby risking ecosystem balance and functioning.

### 8.2.1 NPs Occurrence in Water

NPs can enter aquatic environments through several pathways: (i) effluents from wastewater treatment facilities, (ii) direct release associated with product use (e.g., NMs-based paintings and coatings for boats, sunscreen residues washed off during swimming, or nanofibers from swimwear textiles), and (iii) indirect inputs from atmospheric deposition or surface rainwater runoff [14].

Once introduced into aquatic systems, dissolution represents a major transformation route, particularly for metallic NPs. Dissolution processes can be driven by (i) the formation of partially soluble metal oxides [15], (ii) oxidative reactions of the particle constituents [16], and (iii) complexation of constituent elements with ligands present in the surrounding environment [17]. Under certain conditions, sulfidation may also occur, which slows down oxidative reactions and thus reduces dissolution rates [18]. Within aquatic media, NPs may remain suspended in the water column, settle into benthic sediments, or bind to dissolve or particulate organic matter (OM) [12]. Aggregation represents a prevalent phenomenon in aquatic medium. It increases the effective particle size while simultaneously reducing the available surface area, thereby decreasing the overall reactivity. Because concentrations of natural OM are typically much higher than those of NPs, heteroaggregation processes are generally favored.

Size of NP size plays a crucial role because it influences NP bioavailability to exposed organisms. Excessive aggregation can produce particle clusters too large to cross cellular membranes, limiting NP uptake. Conversely, dissolution or partial dissolution generates smaller particles that more readily penetrate biological barriers. Mobility within aquatic environments is likewise size-dependent: large aggregates tend to settle rapidly, restricting both their transport potential and bioavailability, whereas small aggregates remain suspended for longer periods, disperse more widely, and interact more frequently with biota [19].

Pourzahedi et al. applied life cycle assessment to AgNPs incorporated into 15 different consumer products and found that although certain products exhibited relatively high leaching rates, the environmental burden from AgNPs release was still minor compared with the impacts associated with product manufacture [20]. In another LCA-based investigation, Mitrano et al. assessed AgNPs emissions from textiles and determined that the majority of releases occur during product use, especially during laundering, rather than at the end-of-life disposal [21]. According to Corsi et al. [4] textile products represent the dominant source of AgNPs discharge into aquatic environments. Similarly, Markus et al. modelled the transport of ZnO, $TiO_2$, and AgNPs along the Rhine River and demonstrated that these NPs can be transported over substantial distances, behaving in a manner comparable to suspended particulate matter.

Model-based predictions by Gottschalk et al. [22] estimated AgNPs concentrations in surface waters to reach 0.764 ng $L^{-1}$ in Europe and 0.1166 ng $L^{-1}$ in North America. In the same study, predicted $TiO_2$NPs concentrations were 0.015 μg $L^{-1}$ for European waters and 0.002 μg $L^{-1}$ for North American waters. Furthermore, Neal et al. [23] projected average $TiO_2$NPs concentrations in surface water from the United Kingdom to be approximately 2.1 μg $L^{-1}$, with values ranging from 0.55 to 6.48 μg $L^{-1}$.

### 8.2.2 NPs Occurrence in Soil and Sediments

NPs can migrate through aquatic systems or undergo settling processes that ultimately deposit them within sediment layers. Fine suspended particulates, such as clays, OM, iron oxides, and other minerals, serve as vectors for pollutants transport in surface waters. Once settled to the sediment bed, cohesive sediments gradually compact, diminishing their suspendability over time [24]. Comparable to colloidal particles, NPs settling and diffusion are governed by gravitational forces and Brownian motion, respectively [25]. Within porous substrates, Brownian motion drives NP mobility. However, as NPs aggregate, their increased mass enhances gravitational settling and fosters interactions with soil surfaces. The nature and magnitude of these interactions depend largely on the surface charges of the NPs and the surrounding soil matrix. Electrostatic repulsion (like-charged interfaces) promotes NPs transport, whereas electrostatic attraction (oppositely charged interfaces) enhances sorption and limits mobility.

Both soils and sediments are widely regarded as final accumulating sinks for NPs, with the potential to act as secondary sources of groundwater contamination. Experimental evidence indicates that up to 99% of $TiO_2$NPs entering the wastewater treatment plants (WWTPs) are retained in the sludge. Also, the dominant pathway for NPs entry into terrestrial environments is the land application of sewage sludge [25].

Dynamic material flow modelling estimates that nanoscale $TiO_2$ inputs into marine sediments may reach approximately 44,000 tonnes, while AgNPs may accumulate roughly 30 tonnes [4]. Reported environmental concentrations of AgNPs include values of 30.1 mg $kg^{-1}$ in sediments and 2.3 mg $kg^{-1}$ in sludge-amended soils [26], with broader sediment estimates ranging from 0.19 to 470.6 μg $kg^{-1}$ [27] and biosolid concentrations near 1 μg $g^{-1}$ [28]. In Lake Ontario, Canada, Metcalfe et al. [29] monitored AgNPs across water, suspended particulates, and bottom sediments near a WWTP outfall. Their results showed that more than 75% of the measured silver was associated with suspended and benthic sediments, whereas less than 2% of total silver (97–509 μg $kg^{-1}$) was present in nanoparticulate form. Cong et al. similarly reported AgNPs concentrations of 0.27 0.01 μg $kg^{-1}$ in estuarine sediments in Denmark [30].

Concentrations of $TiO_2$NPs in sediments have been estimated to range from below detection to 2.7 mg $kg^{-1}$ [24] and an average ≤ 2.74 mg $kg^{-1}$ [31]. Measurements by Kiser et al. [32] in biosolids from WWTPs across several US states (Arizona,

Colorado, California, Maryland, Iowa, and New York) indicated that $TiO_2$NPs concentrations between 1.8 and 6.4 g $kg^{-1}$, Johnson et al. [33] and Khosravi et al. [34] observed mean levels of 305 and 317.4 mg $kg^{-1}$, respectively, in biosolids from facilities in the United Kingdom and Canada. In addition, sediment samples from Xiamen Bay, China, contained $TiO_2$NPs concentrations of approximately 2.74 g $kg^{-1}$ [35].

Adam et al. [36] evaluated the fate of different NMs forms throughout entire environmental compartments, categorizing them into pristine, dissolved, transformed, matrix-embedded, and product-embedded states. Their modelling indicates that AgNPs are predominantly released into surface waters and soils as transformed silver species (46% and 79%, respectively), while emissions to air occur mainly as pristine and matrix-embedded forms (42% and 40%, respectively). In contrast, $TiO_2$NPs are discharged largely in their pristine state across water, soil, and air (97%, 91%, and 80%, respectively). Sediments continue to be recognized as the principal environmental reservoir for AgNPs [37].

### 8.2.3 Occurrence of NPs in Air

There are multiple pathways that can introduce NPs into the atmosphere. Some NPs are emitted directly during industrial activities such as grinding, blending, packaging, or through the use of disinfectant spays, and as a result of emissions from waste-incineration facilities [38]. However, in urban settings, one of dominant contributors is road traffic powered by diesel or gasoline engines [12]. Mobile sources together with fixed combustion installations account for more than one-third of the NPs load detected in ambient air [19]. Research has demonstrated that CNTs and related fibrous structures may form unintentionally during diesel and petrol combustion. Notably, diesel exhaust alone is responsible for more than 90% of the carbon-based nanoparticulate fraction identified in the atmosphere.

Respiratory conditions ranked among the top global causes of mortality in 2021 (World Health Organization [WHO]). The WHO has previously highlighted the burden of disease linked to exposure to outdoor air pollution. Among the various pollutants, particulate matter ($PM_{10}$, $PM_{2.5}$) remains a major focus of air-quality evaluations [39]. According to Li et al. [40], ambient $PM_{2.5}$ leads to several million premature deaths each year and shortens average human life expectancy by roughly one year.

Once in the atmosphere, NPs can disperse widely due to their small size and low gravitational settling rates that allow them to remain airborne for long periods. They may also cluster into larger aggregates and ultimately be transferred to water or soil systems through both wet and dry deposition.

Nowack and Mueller [41] developed models to estimate environmental releases of ENPs under two conditions: a realistic emissions scenario and a higher-emissions alternative. Their projections yielded atmospheric concentrations for nano-Ag of 0.0017 μg $m^{-3}$ (realistic) and 0.0044 μg $m^{-3}$ (high-emission), and for nano-$TiO_2$ of 0.0015 μg $m^{-3}$ (realistic) and 0.042 μg $m^{-3}$ (high-emission). Another modelling study by Gottschalk et al. [42] provided predicted environmental concentrations (PECs) for Europe, the United Stated, and Switzerland. For Europe territory, estimated values ranged from 0-0.066 to 0.02 ng $m^{-3}$ for nano-Ag, and bellow 0.0005 ug $m^{-3}$ for nano-$TiO_2$. Giese et al. [27] reported predicted atmospheric PECs for silver-based ENMs within the range 0.00 to 0.49 ng $m^{-3}$. Variations among model predictions primarily come from differences in assumed production quantities of NMs and the residence times incorporated into each modelling framework.

## 8.3 ADDRESSING ECOLOGICAL CHALLENGES AND TOXICITY ASSESSMENT

Toxicology has traditionally been linked to the work of Paracelsus, whose articulation of the principles of dose and dose-response laid the foundation for the discipline and gave rise to the famous "the dose makes the poison" [43]. Nevertheless, these classical toxicological parameters are not always directly transferable to the evaluation of nanoscale materials. The investigation of harmful outcomes associated with ENPs has emerged as the specialized field of nanotoxicology, a domain that gained prominence after the rapid expansion of nanotechnology [44]. As the production and application of NMs continue to grow, concerns have intensified regarding their potential impacts on human health and ecological systems. Consequently, to mitigate possible risks and ensure the development of safer nanotechnology, the scientific community and regulatory bodies have been called upon to confront knowledge gaps.

Conventional toxicology typically relates adverse effects to variables such as exposure route, dose, concentration, time, and frequency. In contrast, nanotoxicological assessment must incorporate additional parameters, including particle composition, size, surface characteristics, and morphology [44]. For instance, optical studies have demonstrated that even minor alterations in NP size or geometry can shift the colour of a particulate suspension. Analogously, subtle changes in surface chemistry can alter biological behaviour, influencing toxicity profiles, biodistribution, clearance and elimination. These complexities hinder broad generalizations regarding NP toxicity, as biological responses depend on intricate interactions between NMs physicochemical properties and their biological surroundings.

Despite the challenges, a growing number of experimental evidence demonstrates that numerous NPs exhibit measurable toxicity in both *in vitro* and *in vivo* systems. Reported effects include oxidative stress, inflammatory responses, and secondary DNA damage [45]. A central mechanism implicated in nanotoxicity is the excessive generation of reactive oxygen species (ROS), such as superoxide anions and hydroxyl radicals [46]. Overproduction of ROS disrupts cellular redox balance and may initiate a self-amplifying cascade, commonly referred to as ROS-induced ROS release, though mitochondrial signalling pathways. Escalating ROS levels can subsequently damage intracellular structures, nucleic acids, lipid membranes, and ion channels. When such cellular injury accumulates, tissue function may be impaired, potentially progressing to organ dysfunction or inflammatory pathology. Moreover, cell death pathways including apoptosis, necrosis, and pyroptosis can be activated, releasing intracellular components that provoke localized inflammation or systemic immune activation [47].

### 8.3.1 Adverse Consequences of NPs on Fauna

NPs may enter organisms *via* inhalation [48, 60], ingestion [49], or direct ocular and dermal contact [50, 51]. Evidence from the scientific literature indicates that ENPs have the ability to migrate up to trophic levels, thereby increasing hazards for species at higher positions in the food web and, eventually, for human populations. A number of adverse effects have been illustrated on several organisms, including bacteria, algae, aquatic invertebrates, fishes, mammals, and even humans [12].

There is no single mechanism governing NP–organism interactions; therefore, resulting disturbances differ across biological systems. At the microbial level, NPs presence can alter microbial community structure and interfere with essential metabolic functions. In higher organisms, exposure to NPs often results in physiological alterations that, although not immediately lethal, can propagate through the food chain and modify ecosystem dynamics [12, 14]. Plant species may also internalize NPs, enabling their movement into aquatic and terrestrial food webs. Such processes, uptake, bioaccumulation, and trophic transfer, highlight the broader toxicological significance of NPs within natural environments. Bioadverse effects are NP-dependent and also specie-dependent, and are generally attributed to the contribution of: (i) the presence of NPs alone, and (ii) the presence of both NPs and their dissolved ionic species due to ion release.

Selected literature examples illustrating the toxicity of various NMs in multiple organisms are available elsewhere [12, 14, 52]. Nonetheless, the bioaccumulation of NPs within organisms and their biomagnification across successive trophic levels raise serious concerns. As NPs are assimilated by lower-level species and subsequently transferred upward the food web, exposure intensifies for higher trophic organisms, including humans.

### 8.3.2 Adverse Consequences of NPs on Flora

The growth of the world population is driving global crop demand to unprecedented levels. Agriculture plays a critical role to meet these demands, and thus sustainable and more efficient agricultural practices are needed. In this sense, nanotechnology offers significant potential to enhance the sustainability of modern agriculture. Notably, nanoscale approaches are emerging as promising tools for improving nutrient-use efficiency (nano-fertilizers), enabling precise genetic modification, and advancing targeted pest-management strategies [53]. Based on recent developments, it is suggested that nanotechnology may be crucial for resilient and environmentally responsible agriculture. Nonetheless, potential negative effects should not be neglected, as it represents one of the major channels for NP release into the environment. Overall, interactions between plants and NPs typically proceed through three sequential stages: (i) NP deposition onto plant surfaces such as leaves, roots, or stems; (ii) subsequent penetration through the cuticular and epidermal layers; and three (iii) internal transport and potential transformation within plant tissues [53]. Although leaves share common anatomical elements (an epidermis containing stomata, underlying mesophyll tissue, and a vascular network) their specific rearrangement can vary among plant species [54]. Stomata are microscopic pores on leaf surfaces that regulate water and gas exchange with surroundings. With typical pore sizes of the order of 10 μm, open stomata can be sufficiently large to allow NP penetration. Their open/closed response is dynamically controlled by factors such as $CO_2$ concentration, humidity, temperature, and light intensity. Alternatively, plant roots constitute another portal of entry for NPs. The outermost layer of a primary root is the epidermis, which protects the root. Just beneath the epidermis is the cortex, which stores nutrients and helps transport water inside. Below the cortex lies the endodermis, a selective barrier that regulates movement of substances into the vascular system. Attached to the endodermis is the pericycle, a layer involved in the formation of lateral roots. And, at the centre of the root, is the vascular system, consisting of xylem and phloem, which transports water and nutrients throughout the plant [55]. Once NPs penetrate the cuticle and epidermis (whether in leaves or roots), they gain relatively easy access to the plant's vascular system. The xylem and phloem are critical for the long-distance transport of NPs in plants. Two key physiological factors influence this transport: (i) sap composition and (ii) sap flow rate. The first one can affect the degree of NP aggregation, sedimentation, and dissolution, while the second one determines the transport velocity of NPs [53, 56].

Reported negative effects of NPs on plants manifest as inhibited seed germination, reduced root and shoot development, disturbances in transpiration, weakened photosynthesis performance and nutrient absorption, altered hormone regulation, and cell damage or death resulting from genetic disruption [12]. For example, exposure of wheat (*Triticum aestivum*) to AgNPs at 20, 200, and 2000 mg $kg^{-1}$ concentrations, resulted in significant reductions in plant weight, plant height, and overall biomass [57]. Similar effects were observed in *Capsicum annuum* after AgNPs exposure at 1 mg $L^{-1}$ [58]. Exposure of *Lolium perenne* and *Hordeum vulgare* to zero-valent iron NPs (nZVI) led to a reduction in seed germination at 500 mg $L^{-1}$ [59]. Exposure to $TiO_2$NPs at 0.25 mM has been shown to trigger genotoxic responses in plants (*Allium cepa* and *Nicotiana tabacum*), with more severe DNA damage occurring at increasing concentrations [60]. Exposure of *Panicum maximum* to ZnONPs at 500 mg $kg^{-1}$ soil led to adverse effects at the initial growth of plants and leaf biomass and productivity [61]. *Glycine max* grown in soil amended with either nano-$CeO_2$ or nano ZnO experienced foliar stress and damage [62].

## 8.4 ENVIRONMENTAL APPLICATIONS OF NPS: NANOREMEDIATION

Beyond their classification as potential contaminants, NMs are increasingly recognized for their capacity to mitigate environmental stressors. Due to their nanoscale features – small size, large surface area-to-volume ratio, morphology, porosity, and chemical composition – NMs possess high environmental remediation capabilities. These physicochemical properties enable the design of advanced materials that enhance the efficiency of clean-up practices helping in the mitigation and management of environmental pollution. Consequently, NMs are integrated into membrane technologies, adsorbent media, nanocatalytic systems, as well as engineered surfaces and protective coatings. This targeted application of NMs and nanotechnology-based

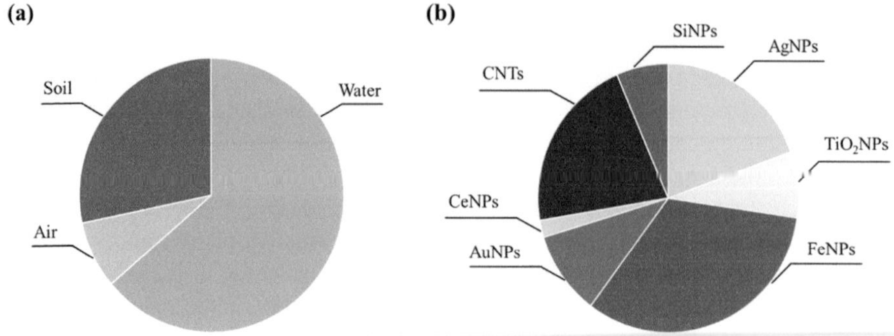


**FIGURE 8.1** Publication metrics on environmental nanoremediation based on (a) environmental compartment, and (b) type of nanomaterial employed. (Data extracted from Web of Science database).

solutions to prevent, mitigate, or remediate environmental contamination is collectively referred to as nanoremediation [2, 63].

Figure 8.1a shows the distribution of nanotechnology-related environmental remediation research across the three environmental compartments – water, air, and soil – based on Web of Science search results for "Nanotechnology" combined with each remediation term. Based on the figure, the distribution of publications indicates a strong dominance of research focused on water systems. Approximately two-thirds (63.8%) of the identified studies address nanotechnology-based approaches for water remediation, reflecting the major global emphasis on improving drinking water quality, wastewater treatment, and contaminant removal in aquatic environments. Soil remediation represents a smaller but still substantial share of the literature (28.2%), suggesting sustained interest in the application of NMs for the stabilization or removal of pollutants in terrestrial systems. In contrast, air remediation constitutes the smallest proportion of research outputs (7.9%), indicating that the application of nanotechnology for air pollution control remains comparatively underexplored. Figure 8.1b presents the NM types most frequently employed in environmental remediation, based on Web of Science data retrieved using the search terms "environmental remediation" in combination with the name of each NM.

In general terms, nanoremediation strategies for contaminated water, soil, and air rely on several key mechanisms [64]: pollutant adsorption onto nanoscale material surfaces; catalytic reactions that lead to pollutant transformation and degradation; and separation of contaminants through filtration processes employing nanoporous membranes. In addition, sensing and monitoring platforms based on nanotechnology (nanosensors and nanoprobes) enable sensitive real-time detection of environmental contaminants. Such capabilities support proactive decision-making and facilitate timely implementation of environmental management and remediation strategies.

### 8.4.1 Nanoscale Sorbents and Adsorbents

Nanoscale sorbents and adsorbents efficiently absorb and adsorb contaminants due to their small size, which offers a large specific surface area together with reactive sites for interaction. Numerous studies have demonstrated the capacity of metallic NPs to capture organic pollutants from aqueous systems. For instance, AgNPs and CuNPs have been shown to efficiently remove Congo red and crystal violet dyes, respectively, though adsorption processes [65, 66]. Similarly, metal oxide NPs of Zn, Cu, and Ni exhibit strong affinity toward methyl blue [67]. ZnO quantum dots-based absorbents have also proven effective in the uptake of anionic azo dyes, including methyl orange, Congo red, and Eriochrome black-T from water [68]. In addition, nanocellulose-based materials have been reported to adsorb both positively and negatively charged dyes [69].

Over recent years, nanoadsorbents have emerged as one of the most promising technologies for wastewater purification. Their use is largely attributed to intrinsic physicochemical characteristics such as high porosity, tunable surface functional groups, relatively low production cost, and straightforward synthesis routes [70]. Wastewater derived from animal farms frequently contains mixtures of antibiotics (e.g., tetracycline) and heavy metals (e.g., cadmium), both of which contribute significantly to environmental contamination. Graphene-based NMs have been successfully applied as adsorbents to remove these co-occurring pollutants [71].

Beyond aqueous systems, NMs have also been explored as sorbents for the remediation of crude-oil-contaminated soil and water, underscoring their versatility in environmental clean-up applications [72].

### 8.4.2 NANOCATALYSTS

A substantial body of research has examined the catalytic activity of metal-based NMs, particularly palladium, platinum, and gold NPs, in accelerating redox processes associated with pollutant degradation. Semiconductor NMs such as $TiO_2$ and ZnO are especially well known for their photocatalytic behaviour, whereby incident light activates the catalyst surface to drive the breakdown of organic contaminants [73]. The performance of these nanocatalysts can be further enhanced through surface modification, compositional tuning, or immobilization onto suitable support matrices, collectively improving their selectivity and overall degradation efficiency.

Several literature examples can be cited on the use of nanocatalysis for pollutant removal. For instance, Mishra and Jain [74] developed a nano-composite of multiwalled carbon nanotubes manganese oxide/polypyrrole (MWCNT-$MnO_2$/PPy) for sewage sludge treatment. Carbon-based NMs have also shown considerable potential in this domain. For example, nanodiamonds have demonstrated the capacity to eliminate a wide range of emerging organic pollutants via electro-oxidation catalysis, photocatalytic pathways, oxidative reactions, and adsorption mechanisms [75]. Other photocatalytic systems include silica-supported tin oxide ($SnO_2/SiO_2$) NPs for the degradation of orange II dye [76] and free-standing cellulose film impregnated with $MnO_2$ NPs for efficiently discoloring Indigo Carmine dye [77].

NMs have also been engineered to target gaseous pollutants. One strategy involves developing nanocomposites capable of adsorbing and catalytically transforming species such as nitrogen oxides and carbon monoxide. For instance, Dai et al. [78]

fabricated an ohmic junction of copper@cuprous oxide nanocomposite ($Cu@Cu_2O$ photocatalyst) for the catalytic oxidation and reduction of nitric oxide (NO) and $CO_2$, respectively. Likewise, Kamaei et al. [79] reported the use of $TiO_2$NPs-coated glass beads to promote the photocatalytic breakdown of the volatile organic compound ethylbenzene.

Together, these studies illustrate the versatility of NMs as catalytic platforms across aqueous and gaseous environments.

### 8.4.3 Nano-enabled Filtration Membranes

The incorporation of nanotechnology into membrane-based filtration systems has enabled the development of advanced treatment technologies with markedly improved separation performance compared with conventional filtration approaches. Nanostructured membranes are engineered with tightly controlled pore size and tailored surface properties, allowing for the selective rejection of target contaminants while maintaining efficient permeate flow.

Nanofiltration membranes are commonly fabricated from ceramic and polymeric materials. More recently, bio-based systems incorporating cellulose and chitosan have been developed for dye removal from aqueous media. In many cases, these membranes are manufactured by embedding nanofillers within a polymeric matrix to enhance rejection efficiency. Such nano-enabled membranes have demonstrated the effective removal of a broad spectrum of synthetic dyes, including direct red 16, methyl blue, Congo red, acid fuchsin, Eriochome black T, and acid orange II [80, 81].

Beyond liquid-phase treatment, NMs have also been integrated into membranes designed for air purification. Carbon- and silica-based NMs are particularly attractive for air filtration because of their high surface area and strong adsorption capacity [82]. For example, graphene oxide (GO)-modified polyacrylonitrile nanofibrous membranes, as well as reduced GO (rGO)-based nanoair filters, have shown excellent removal efficiencies of 2.5 μm particulate matter [83].

### 8.4.4 Nanosensors and Monitoring Devices

Electrochemical sensors and biosensing platforms incorporating NMs have shown considerable potential for the rapid and sensitive detection of environmental pollutants, including pesticides, heavy metals, and microbial contaminants. These devices frequently employ nanoscale receptor elements such as quantum dots, carbon nanotubes, graphene and related derivatives, as well as metal and metal oxide NPs, which enhance signal transduction and analytical performance [84]. Significant research efforts have focused on the development of electrochemical immunosensors, aptasensors, and genosensors capable of detecting a broad spectrum of target analytes with high specificity [85].

Recent applications illustrate the diversity of nano-enabled sensing strategies. For example, electrochemical sensors made of rGO and copper NPs have been designed to monitor hazardous hormone concentrations in aquatic environments [86]. Silver NP-based electrochemical sensors have likewise been employed for the detection of organic contaminants in water [87].

## 8.5 CONCLUSION

NMs represent a dual-faced presence in the environment: they are simultaneously powerful technological tools and CECs. Their unique size-dependent properties drive innovation in environmental management strategies and sensing technologies, yet these same characteristics govern their environmental mobility, transformation, and potential toxicity. Evidence indicates that NPs can accumulate in environmental compartments and interact with organisms at multiple biological levels, raising concerns regarding long-term ecological and human-health impacts. Moving forward, the responsible development of nanotechnology will require continued refinement of analytical methods, realistic exposure and risk assessment strategies, and regulatory frameworks that accurately reflect the complexity of NMs in real-world environments. Ultimately, a balanced perspective recognizing both the potentials and the risks of NMs is essential to ensure that their environmental applications progress in a safe, sustainable, and scientifically informed manner.

## REFERENCES


1. Fuller, Richard, Philip J. Landrigan, Kalpana Balakrishnan, Glynda Bathan, Stephan Bose-O'Reilly, Michael Brauer, Jack Caravanos, et al. 2022. "Pollution and Health: A Progress Update." *The Lancet Planetary Health* 6 (6): e535–47. https://doi.org/10.1016/S2542-5196(22)00090-0.
2. Dabas, Seema. 2025. "Metal and Metal Oxide Based Nanoremediation : A Sustainable Alternative." *Discover Materials* 5:35 https://doi.org/10.1007/s43939-025-00242-6.
3. United Nations. 2015. "Transforming our World: The 2030 Agenda for Sustainable Development." https://sdgs.un.org/publications/transforming-our-world-2030-agenda-sustainable-development-17981.
4. Corsi, Ilaria, Arianna Bellingeri, Maria Concetta Eliso, Giacomo Grassi, Giulia Liberatori, Carola Murano, Lucrezia Sturba, Maria Luisa Vannuccini, and Elisa Bergami. 2021. "Eco-Interactions of Engineered Nanomaterials in the Marine Environment: Towards an Eco-Design Framework." *Nanomaterials* 11 (8): 1903. https://doi.org/10.3390/nano11081903.
5. International Organization of Standards. 2017. "Nanotechnologies—Plain Language Explanation of Selected Terms from the ISO/IEC 80004 Series" ISO/TR 18401:2017." https://www.iso.org/obp/ui/#iso:std:iso:tr:18401:ed-1:v1:en (access May 2026).
6. European Commssion. 2022. "Commission Recommendation 2022/C 229/01 of 10.6.2022 on the Definition of Nanomaterial." https://doi.org/10.2788/36237.
7. Mekuye, Bawoke, and Birhanu Abera. 2023. "Nanomaterials: An Overview of Synthesis, Classification, Characterization, and Applications." *Nano Select* 4 (8): 486–501. https://doi.org/10.1002/nano.202300038.
8. European Union Observatory for Nanomaterials (EUON). https://euon.echa.europa.eu/es/ (access May 2023).
9. Nanotechnology Products Database (NPD). n.d. https://product.statnano.com/. (access May 2025).
10. Devasena, T., B. Iffath, R. Renjith Kumar, Natarajan Muninathan, Kuppusamy Baskaran, T. Srinivasan, and Shani T. John. 2022. "Insights on the Dynamics and Toxicity of Nanoparticles in Environmental Matrices." *Bioinorganic Chemistry and Applications* Article 348149, https://doi.org/10.1155/2022/4348149.

11. Holden, Patricia A., Jorge L. Gardea-Torresdey, Fred Klaessig, Ronald F. Turco, Monika Mortimer, Kerstin Hund-Rinke, Elaine A. Cohen Hubal, et al. 2016. "Considerations of Environmentally Relevant Test Conditions for Improved Evaluation of Ecological Hazards of Engineered Nanomaterials." *Environmental Science and Technology* 50 (12): 6124–45. https://doi.org/10.1021/acs.est.6b00608.
12. Tran, Thien Khanh, Minh Ky Nguyen, Chitsan Lin, Tuan Dung Hoang, Thanh Cong Nguyen, Aasif Mohmad Lone, Akhil Pradiprao Khedulkar, et al. 2024. "Review on Fate, Transport, Toxicity and Health Risk of Nanoparticles in Natural Ecosystems: Emerging Challenges in the Modern Age and Solutions toward a Sustainable Environment." *Science of the Total Environment* 912: 169331. https://doi.org/10.1016/j.scitotenv.2023.169331.
13. "Global Risks Report 2024 | World Economic Forum." n.d. https://www.weforum.org/publications/global-risks-report-2024/. (access January 2026).
14. Vale, Gonçalo, Kahina Mehennaoui, Sebastien Cambier, Giovanni Libralato, Stéphane Jomini, and Rute F. Domingos. 2016. "Manufactured Nanoparticles in the Aquatic Environment-Biochemical Responses on Freshwater Organisms: A Critical Overview." *Aquatic Toxicology* 170: 162–74. https://doi.org/10.1016/j.aquatox.2015.11.019.
15. Heinlaan, Margit, Angela Ivask, Irina Blinova, Henri Charles Dubourguier, and Anne Kahru. 2008. "Toxicity of Nanosized and Bulk ZnO, CuO and TiO2 to Bacteria Vibrio Fischeri and Crustaceans Daphnia Magna and Thamnocephalus Platyurus." *Chemosphere* 71 (7): 1308–16. https://doi.org/10.1016/j.chemosphere.2007.11.047.
16. Ma, Rui, John Stegemeier, Clément Levard, James G. Dale, Clinton W. Noack, Tiffany Yang, Gordon E. Brown, and Gregory V. Lowry. 2014. "Sulfidation of Copper Oxide Nanoparticles and Properties of Resulting Copper Sulfide." *Environmental Science: Nano* 1 (4): 347–57. https://doi.org/10.1039/c4en00018h.
17. Domingos, Rute F., Cristiana Franco, and José P. Pinheiro. 2015. "The Role of Charged Polymer Coatings of Nanoparticles on the Speciation and Fate of Metal Ions in the Environment." *Environmental Science and Pollution Research* 22 (4): 2900–2906. https://doi.org/10.1007/s11356-014-3546-8.
18. Thalmann, Basilius, Andreas Voegelin, Brian Sinnet, Eberhard Morgenroth, and Ralf Kaegi. 2014. "Sulfidation Kinetics of Silver Nanoparticles Reacted with Metal Sulfides." *Environmental Science and Technology* 48 (9): 4885–92. https://doi.org/10.1021/es5003378.
19. El-Kalliny, Amer S., Mahmoud S. Abdel-Wahed, Adel A. El-Zahhar, Ibrahim A. Hamza, and Tarek A. Gad-Allah. 2023. "Nanomaterials: A Review of Emerging Contaminants with Potential Health or Environmental Impact." *Discover Nano* 18: 68. https://doi.org/10.1186/s11671-023-03787-8.
20. Pourzahedi, Leila, Marina Vance, and Matthew J. Eckelman. 2017. "Life cycle assessment and release studies for 15 nanosilver-enabled consumer products: investigating hotspots and patterns of contribution." *Environmental Science and Technology* 51: 7148–58. https://doi.org/10.1021/acs.est.6b05923.
21. Mitrano, Denise M, Elisa Rimmele, Adrian Wichser, Rolf Erni, Murray Height, and Bernd Nowack. 2014. "Presence of Nanoparticles in Wash Water from Conventional Silver and Nano-silver Textiles." *ACS Nano* 8: 7208–19. https://doi.org/10.1021/nn502228w.
22. Gottschalk, Fadri, Tianyin Sun, and Bernd Nowack. 2013a. "Environmental Concentrations of Engineered Nanomaterials: Review of Modeling and Analytical Studies." *Environmental Pollution* 181: 287–300. https://doi.org/10.1016/j.envpol.2013.06.003.
23. Neal, Colin, Helen Jarvie, Philip Rowland, Alan Lawler, Darren Sleep, and Paul Scholefield. 2011. "Titanium in UK Rural, Agricultural and Urban/Industrial Rivers: Geogenic and Anthropogenic Colloidal/Sub-Colloidal Sources and the Significance of within-River Retention." *Science of the Total Environment* 409 (10): 1843–53. https://doi.org/10.1016/j.scitotenv.2010.12.021.

24. Sani-Kast, Nicole, Martin Scheringer, Danielle Slomberg, Jérôme Labille, Antonia Praetorius, Patrick Ollivier, and Konrad Hungerbühler. 2015. "Addressing the Complexity of Water Chemistry in Environmental Fate Modeling for Engineered Nanoparticles." *Science of the Total Environment* 535: 150–59. https://doi.org/10.1016/j.scitotenv.2014.12.025.
25. Tourinho, Paula S., Cornelis A.M. van Gestel, Stephen Lofts, Claus Svendsen, Amadeu M.V.M. Soares, and Susana Loureiro. 2012. "Metal-Based Nanoparticles in Soil: Fate, Behavior, and Effects on Soil Invertebrates." *Environmental Toxicology and Chemistry* 31 (8): 1679–92. https://doi.org/10.1002/etc.1880.
26. Sun, Tian Yin, Nikolaus A. Bornhöft, Konrad Hungerbühler, and Bernd Nowack. 2016. "Dynamic Probabilistic Modeling of Environmental Emissions of Engineered Nanomaterials." *Environmental Science and Technology* 50 (9): 4701–11. https://doi.org/10.1021/acs.est.5b05828.
27. Giese, Bernd, Fred Klaessig, Barry Park, Ralf Kaegi, Michael Steinfeldt, Henning Wigger, Arnim Von Gleich, and Fadri Gottschalk. 2018. "Risks, Release and Concentrations of Engineered Nanomaterial in the Environment." *Scientific Reports* 8 (1): 1–18. https://doi.org/10.1038/s41598-018-19275-4.
28. Gottschalk, Fadri, TianYin Sun, and Bernd Nowack. 2013b. "Environmental Concentrations of Engineered Nanomaterials: Review of Modeling and Analytical Studies." *Environmental Pollution* 181: 287–300. https://doi.org/10.1016/j.envpol.2013.06.003.
29. Metcalfe, Chris D, Tamanna Sultana, Jonathan Martin, Karla Newman, and Paul Helm. 2018. "Silver near Municipal Wastewater Discharges into Western Lake Ontario, Canada." *Environmental Monitoring and Assessment* 190: 555. https://doi.org/10.1007/s10661-018-6922-x.
30. Cong, Yi, Gary T. Banta, Henriette Selck, Deborah Berhanu, Eugenia Valsami-Jones, and Valery E. Forbes. 2014. "Toxicity and Bioaccumulation of Sediment-Associated Silver Nanoparticles in the Estuarine Polychaete, Nereis (Hediste) Diversicolor." *Aquatic Toxicology* 156: 106–15. https://doi.org/10.1016/j.aquatox.2014.08.001.
31. Luo, Zhuanxi, Zhenhong Wang, Qingzhao Li, Qikun Pan, Changzhou Yan, and Feng Liu. 2011. "Spatial Distribution, Electron Microscopy Analysis of Titanium and Its Correlation to Heavy Metals: Occurrence and Sources of Titanium Nanomaterials in Surface Sediments from Xiamen Bay, China." *Journal of Environmental Monitoring* 13 (4): 1046–52. https://doi.org/10.1039/c0em00199f.
32. Kiser, M. A., P. Westerhoff, T. Benn, Y. Wang, J. Pérez-Rivera, and K. Hristovski. 2009. "Titanium Nanomaterial Removal and Release from Wastewater Treatment Plants." *Environmental Science and Technology* 43 (17): 6757–63. https://doi.org/10.1021/es901102n.
33. Johnson, Andrew C., Michael J. Bowes, Alison Crossley, Helen P. Jarvie, Kerstin Jurkschat, Monika D. Jürgens, Alan J. Lawlor, et al. 2011. "An Assessment of the Fate, Behaviour and Environmental Risk Associated with Sunscreen TiO2 Nanoparticles in UK Field Scenarios." *Science of the Total Environment* 409 (13): 2503–10. https://doi.org/10.1016/j.scitotenv.2011.03.040.
34. Khosravi, Kambiz, M. Ehsanul Hoque, Brian Dimock, Holger Hintelmann, and Chris D. Metcalfe. 2012. "A Novel Approach for Determining Total Titanium from Titanium Dioxide Nanoparticles Suspended in Water and Biosolids by Digestion with Ammonium Persulfate." *Analytica Chimica Acta* 713: 86–91. https://doi.org/10.1016/j.aca.2011.11.048.
35. Luo, P., I. Morrison, A. Dudkiewicz, K. Tiede, E. Boyes, P. O'Toole, S. Park, and A. B. Boxall. 2013. "Visualization and Characterization of Engineered Nanoparticles in Complex Environmental and Food Matrices Using Atmospheric Scanning Electron Microscopy." *Journal of Microscopy* 250 (1): 32–41. https://doi.org/10.1111/jmi.12014.

36. Adam, Véronique, Alejandro Caballero-Guzman, and Bernd Nowack. 2018. "Considering the Forms of Released Engineered Nanomaterials in Probabilistic Material Flow Analysis." *Environmental Pollution* 243: 17–27. https://doi.org/10.1016/j.envpol.2018.07.108.
37. Zhao, Jian, Xinjie Wang, Son A. Hoang, Nanthi S. Bolan, M. B. Kirkham, Jingnan Liu, Xinghui Xia, and Yang Li. 2021. "Silver Nanoparticles in Aquatic Sediments: Occurrence, Chemical Transformations, Toxicity, and Analytical Methods." *Journal of Hazardous Materials* 418 (June): 126368. https://doi.org/10.1016/j.jhazmat.2021.126368.
38. Meier, Christoph, Andreas Voegelin, Ana Pradas Del Real, Geraldine Sarret, Christoph R. Mueller, and Ralf Kaegi. 2016. "Transformation of Silver Nanoparticles in Sewage Sludge during Incineration." *Environmental Science and Technology* 50 (7): 3503–10. https://doi.org/10.1021/acs.est.5b04804.
39. "World Health Organization (WHO). 2016. "Ambient Air Pollution: A Global Assessment of Exposure and Burden of Disease." https://www.who.int/publications/i/item/9789241511353 (Access January 2026).
40. Li, Chi, Aaron van Donkelaar, Melanie S. Hammer, Erin E. McDuffie, Richard T. Burnett, Joseph V. Spadaro, Deepangsu Chatterjee, et al. 2023. "Reversal of Trends in Global Fine Particulate Matter Air Pollution." *Nature Communications* 14 (1): 1–12. https://doi.org/10.1038/s41467-023-41086-z.
41. Nowack, Bernd, and Nicole C. Mueller. 2008. "Exposure Modeling of Engineered Nanoparticles in the Environment." *EMPA Activities* 41 (2008–2009): 63. http://hdl.handle.net/20.500.11850/76051.
42. Gottschalk, Fadri, Tobias Sonderer, Roland W. Scholz, and Bernd Nowack. 2009. "Modeled Environmental Concentrations of Engineered Nanomaterials ($TiO_2$, ZnO, Ag, CNT, Fullerenes) for Different Regions." *Environmental Science and Technology* 43 (24): 9216–22. https://doi.org/10.1021/es9015553.
43. Borzelleca, Joseph F. 2000. "Paracelsus: Herald of Modern Toxicology." *Toxicological Sciences* 53 (1): 2–4. https://doi.org/10.1093/toxsci/53.1.2.
44. Yang, Wen, Lin Wang, Evan M. Mettenbrink, Paul L. Deangelis, and Stefan Wilhelm. 2021. "Nanoparticle Toxicology." *Annual Review of Pharmacology and Toxicology* 61: 269–89. https://doi.org/10.1146/annurev-pharmtox-032320-110338.
45. Sabella, Stefania, Randy P. Carney, Virgilio Brunetti, Maria Ada Malvindi, Noura Al-Juffali, Giuseppe Vecchio, Sam M. Janes, et al. 2014. "A General Mechanism for Intracellular Toxicity of Metal-Containing Nanoparticles." *Nanoscale* 6 (12): 7052–61. https://doi.org/10.1039/c4nr01234h.
46. Lead, Jamie R., Shareen H. Doak, and Martin J.D. Clift. 2021. *Nanotoxicology in Humans and the Environment.* Springer Nature Switzerland. https://doi.org/10.1007/978-3-030-79808-6.
47. Magdolenova, Zuzana, Andrew Collins, Ashutosh Kumar, Alok Dhawan, Vicki Stone, and Maria Dusinska. 2014. "Mechanisms of Genotoxicity. A Review of in Vitro and in Vivo Studies with Engineered Nanoparticles." *Nanotoxicology* 8 (3): 233–78. https://doi.org/10.3109/17435390.2013.773464.
48. Pietroiusti, Antonio, Helene Stockmann-Juvala, Francesca Lucaroni, and Kai Savolainen. 2018. "Nanomaterial Exposure, Toxicity, and Impact on Human Health." *Wiley Interdisciplinary Reviews: Nanomedicine and Nanobiotechnology* 10 (5): 1–21. https://doi.org/10.1002/wnan.1513.
49. Sharma, Vyom, Ashutosh Kumar, and Alok Dhawan. 2012. "Nanomaterials: Exposure, Effects and Toxicity Assessment." *Proceedings of the National Academy of Sciences, India Section B: Biological Sciences* 82 (S1): 3–11. https://doi.org/10.1007/s40011-012-0072-7.

50. Zhu, Shuang, Linji Gong, Yijian Li, Haiwei Xu, Zhanjun Gu, and Yuliang Zhao. 2019. "Safety Assessment of Nanomaterials to Eyes: An Important but Neglected Issue." *Advanced Science* 6 (16): 1802289. https://doi.org/10.1002/advs.201802289.
51. Xie, Wenjing, Ziyi Xiong, Huimin Wang, Xiaoyi Liu, Hongyan Cui, Qiongyi Huang, and Ying Tang. 2023. "The Nanosafety Assessment of ENMs under a Dermal Exposure Scenario: From Key Molecular Events to in Silico Modeling Tools." *Environmental Science: Nano* 11 (3): 708–38. https://doi.org/10.1039/d3en00585b.
52. Yamini, V, Venkatkumar Shanmugam, M Rameshpathy, Ganesh Venkatraman, Gnanasambandan Ramanathan, Hakim Al, Ahmed Hashmi, Kathirvel Brindhadevi, and V Devi Rajeswari. 2023. "Environmental Effects and Interaction of Nanoparticles on Beneficial Soil and Aquatic Microorganisms." *Environmental Research* 236 (P1): 116776. https://doi.org/10.1016/j.envres.2023.116776.
53. Su, Yiming, Vanessa Ashworth, Caroline Kim, Adeyemi S Adeleye, Philippe Rolshausen, Caroline Roper, and David Jassby. 2019. "Delivery, Uptake, Fate, and Transport of Engineered Nanoparticles in Plants: A Critical Review and Data Analysis." *Environmental Science Nano* 6: 2311–31. https://doi.org/10.1039/c9en00461k.
54. Fernández, V, T Sotiropoulos, and P Brown. 2013. *Foliar Fertilization: Scientific Principles and Field Practices.* First Edit. International Fertilizer Industry Association.
55. Lopez, F B, and G F Barclay. 2017. *Plant Anatomy and Physiology. Pharmacognosy.* Elsevier Inc. https://doi.org/10.1016/B978-0-12-802104-0.00004-4.
56. Lead, Jamie R, Graeme E Batley, Pedro J J Alvarez, Richard D Handy, and Michael J Mclaughlin. 2018. "Nanomaterials in the Environment: Behavior, Fate, Bioavailability, and Effects—An Updated Review." *Environmental Toxicology and Chemistry* 37 (8): 2029–63. https://doi.org/10.1002/etc.4147.
57. Yang, Jie, Fuping Jiang, Yukui Rui, Mengmeng Rui, Muhammad Adeel, Weidong Cao, and Baoshan Xing. 2018. "Alteration of Crop Yield and Quality of Wheat upon Exposure to Silver Nanoparticles in a Life Cycle Study." *Journal of Agricultural and Food Chemistry* 66: 2589–97. https://doi.org/10.1021/acs.jafc.7b04904.
58. Vinković, Tomislav, Ivna Štolfa-čamagajevac, Monika Tkalec, Walter Goessler, and Darija Domazet-jurašin. 2018. "Does Plant Growing Condition Affects Biodistribution and Biological Effects of Silver Nanoparticles?" *Spanish Journal of Agricultural Research* 16 (4): e0803.
59. El-temsah, Yehia Sayed, and Erik J Joner. 2010. "Impact of Fe and Ag Nanoparticles on Seed Germination and Differences in Bioavailability During Exposure in Aqueous Suspension and Soil." *Environmental Toxicology* 27: 42–49. https://doi.org/10.1002/tox.20610.
60. Ghosh, Manosij, Maumita Bandyopadhyay, and Anita Mukherjee. 2010. "Genotoxicity of Titanium Dioxide ($TiO_2$) Nanoparticles at Two Trophic Levels: Plant and Human Lymphocytes." *Chemosphere* 81 (10): 1253–62. https://doi.org/10.1016/j.chemosphere.2010.09.022.
61. Sueli, Aline, De Lima Rodrigues, Paula Cristine, Silva Gomes, and Fabiano Guimar. 2023. "Plant-ZnO Nanoparticles Interaction : An Approach to Improve Guinea Grass (Panicum Maximum) Productivity and Evaluation of the Impacts of Its Ingestion by Freshwater Teleost Fish." *Journal of Hazardous Materials* 451: 131173. https://doi.org/10.1016/j.jhazmat.2023.131173.
62. Priester, John H, Shelly Cole, Katherine Espinosa, Yuan Ge, Ying Wang, Roger M Nisbet, Joshua P Schimel, A Susana Goggi, Jorge L Gardea-torresdey, and Patricia A Holden. 2017. "Damage Assessment for Soybean Cultivated in Soil with Either $CeO_2$ or ZnO Manufactured Nanomaterials." *Science of the Total Environment* 579: 1756–68. https://doi.org/10.1016/j.scitotenv.2016.11.149.
63. Arsenov, Danijela, Jelena Beljin, Danica Jovic, Snezana Maletic, Milan Borisev, and Ivana Borisev. 2023. "Nanomaterials as Endorsed Environmental Remediation Tools

for the next Generation: Eco-Safety and Sustainability." *Journal of Geochemical Exploration* 253: 107283. https://doi.org/10.1016/j.gexplo.2023.107283.

64. Rather, Muzamil A., Shuvam Bhuyan, Ratan Chowdhury, Rahul Sarma, Subham Roy, and Panchi Rani Neog. 2023. "Nanoremediation Strategies to Address Environmental Problems." *Science of the Total Environment* 886: 163998. https://doi.org/10.1016/j.scitotenv.2023.163998.
65. Al-Thabaiti, S.A., E.S. Aazam, Z. Khan, and O. Bashir. 2016. "Aggregation of Congo Red with Surfactants and Ag-Nanoparticles in an Aqueous Solution." *Spectrochimica Acta: Part A: Molecular and Biomolecular Spectroscopy* 156: 28–35. https://doi.org/10.1016/j.saa.2015.11.015.
66. Khani, R., B. Roostaei, G. Bagherzade, and M. Moudi. 2018. "Green Synthesis of Copper Nanoparticles by Fruit Extract of *Ziziphus Spina-Christi* (L.) Willd.: Application for Adsorption of Triphenylmethane Dye and Antibacterial Assay." *Journal of Molecular Liquids* 255: 541–49. https://doi.org/10.1016/J.MOLLIQ.2018.02.010.
67. Li, L.H., J. Xiao, P. Liu, and G.W. Yang. 2015. "Super Adsorption Capability from Amorphousization of Metal Oxide Nanoparticles for Dye Removal." *Scientific Reports* 5 (1): 9028. https://doi.org/10.1038/srep09028.
68. Karim, S., N. Ahmad, D. Hussain, Y.S. Mok, and G.U. Siddiqui. 2022. "Active Removal of Anionic Azo Dyes (MO, CR, EBT) from Aqueous Solution by Potential Adsorptive Capacity of Zinc Oxide Quantum Dots." *Journal of Chemical Technology and Biotechnology* 97 (8): 2087–97. https://doi.org/10.1002/jctb.7079.
69. Reshmy, R, Deepa Thomas, Eapen Philip, Sherely A. Paul, Aravind Madhavan, Raveendran Sindhu, Parameswaran Binod, et al. 2021. "Potential of Nanocellulose for Wastewater Treatment." *Chemosphere* 281 (October): 130738. https://doi.org/10.1016/j.chemosphere.2021.130738.
70. Bagheri, Hasan, Hanieh Fakhri, Reza Ghahremani, Meghdad Karimi, Tayyebeh Madrakian, and Abbas Afkhami. 2020. "Nanomaterial-Based Adsorbents for Wastewater Treatment." *Smart Nanocontainers: Micro and Nano Technologies*, Elsevier, 467–85. https://doi.org/10.1016/B978-0-12-816770-0.00028-9.
71. Huang, Dan, Jizi Wu, Lu Wang, Xingmei Liu, Jun Meng, Xianjin Tang, Caixian Tang, and Jianming Xu. 2019. "Novel Insight into Adsorption and Co-Adsorption of Heavy Metal Ions and an Organic Pollutant by Magnetic Graphene Nanomaterials in Water." *Chemical Engineering Journal* 358 (February): 1399–1409. https://doi.org/10.1016/J.CEJ.2018.10.138.
72. Singh, Harpreet, Neha Bhardwaj, Shailendra Kumar Arya, and Madhu Khatri. 2020. "Environmental Impacts of Oil Spills and Their Remediation by Magnetic Nanomaterials." *Environmental Nanotechnology, Monitoring & Management* 14 (December): 100305. https://doi.org/10.1016/J.ENMM.2020.100305.
73. R. Usha, Rani, Sharma Pratibba, Sharma Rajendra Kumar, Chahal Kavita, Selvan R. Shanmuga, and Pal Shankar Singh. 2024. "The Role of Nanotechnology in Environmental Remediation Opportunities." *African Journal of Biological Sciences* 6 (10): 4359–80. https://doi.org/10.33472/AFJBS.6.10.2024.4359-4380.
74. Mishra, Praveena, and Rajeev Jain. 2016. "Electrochemical Deposition of MWCNT-MnO2/PPy Nano-Composite Application for Microbial Fuel Cells." *International Journal of Hydrogen Energy* 41 (47): 22394–405. https://doi.org/10.1016/J.IJHYDENE.2016.09.020.
75. Bogdanowicz, Robert. 2022. "Functionalized Nanodiamonds as a Perspective Green Carbo-Catalyst for Removal of Emerging Organic Pollutants." *Current Opinion in Solid State and Materials Science* 26 (3): 100991. https://doi.org/10.1016/J.COSSMS.2022.100991.

76. Khan, Idrees, Adnan Ali Khan, Ibrahim Khan, Muhammad Usman, Muhammad Sadiq, Faiz Ali, and K. Saeed. 2020. "Investigation of the Photocatalytic Potential Enhancement of Silica Monolith Decorated Tin Oxide Nanoparticles through Experimental and Theoretical Studies." *New Journal of Chemistry* 44 (31): 13330–43. https://doi.org/10.1039/D0NJ00996B.
77. Oliveira, Larissa V.F., Simona Bennici, Ludovic Josien, Lionel Limousy, Marcos A. Bizeto, and Fernanda F. Camilo. 2020. "Free-Standing Cellulose Film Containing Manganese Dioxide Nanoparticles and Its Use in Discoloration of Indigo Carmine Dye." *Carbohydrate Polymers* 230 (February): 115621. https://doi.org/10.1016/J.CARBPOL.2019.115621.
78. Dai, Benlin, Wei Zhao, Haibao Huang, Shijie Li, Gang Yang, Hongwei Wu, Cheng Sun, and Dennis Y.C. Leung. 2022. "Constructing an Ohmic Junction of Copper@ Cuprous Oxide Nanocomposite with Plasmonic Enhancement for Photocatalysis." *Journal of Colloid and Interface Science* 616 (June): 163–76. https://doi.org/10.1016/J.JCIS.2022.02.056.
79. Kamaei, Morteza, Hamid Rashedi, Seyed Mohammad Mehdi Dastgheib, and Saeideh Tasharrofi. 2018. "Comparing Photocatalytic Degradation of Gaseous Ethylbenzene Using N-Doped and Pure $TiO_2$ Nano-Catalysts Coated on Glass Beads under Both UV and Visible Light Irradiation." *Catalysts* 8 (10): 466. https://doi.org/10.3390/CATAL8100466.
80. Mohammadnezhad, Farrokh, Mostafa Feyzi, and Sirus Zinadini. 2019. "A Novel Ce-MOF/PES Mixed Matrix Membrane; Synthesis, Characterization and Antifouling Evaluation." *Journal of Industrial and Engineering Chemistry* 71 (March): 99–111. https://doi.org/10.1016/J.JIEC.2018.09.032.
81. Shu, Lun, Lin Hua Xie, Yingshuang Meng, Tongxin Liu, Cui Zhao, and Jian Rong Li. 2020. "A Thin and High Loading Two-Dimensional MOF Nanosheet Based Mixed-Matrix Membrane for High Permeance Nanofiltration." *Journal of Membrane Science* 603 (May): 118049. https://doi.org/10.1016/J.MEMSCI.2020.118049.
82. Prado-Audelo, M. L. Del, I. García Kerdan, L. Escutia-Guadarrama, J. M. Reyna-González, J. J. Magaña, and G. Leyva-Gómez. 2021. "Nanoremediation: Nanomaterials and Nanotechnologies for Environmental Cleanup." *Frontiers in Environmental Science* 9 (December): 793765. https://doi.org/10.3389/fenvs.2021.793765.
83. Zhang, Chaoran, Lu Yao, Zhi Yang, Eric Siu Wai Kong, Xiaofeng Zhu, and Yafei Zhang. 2019. "Graphene Oxide-Modified Polyacrylonitrile Nanofibrous Membranes for Efficient Air Filtration." *ACS Applied Nano Materials* 2 (6): 3916–24. https://doi.org/10.1021/ACSANM.9B00806.
84. Willner, Marjorie R., and Peter J. Vikesland. 2018. "Nanomaterial Enabled Sensors for Environmental Contaminants." *Journal of Nanobiotechnology* 16 (1): 95. https://doi.org/10.1186/S12951-018-0419-1.
85. Khanmohammadi, Akbar, Arash Jalili Ghazizadeh, Pegah Hashemi, Abbas Afkhami, Fabiana Arduini, and Hasan Bagheri. 2020. "An Overview to Electrochemical Biosensors and Sensors for the Detection of Environmental Contaminants." *Journal of the Iranian Chemical Society* 17 (10): 2429–47. https://doi.org/10.1007/S13738-020-01940-Z.
86. Barreto, Francisco Contini, Martin Kassio Leme Silva, and Ivana Cesarino. 2022. "An Electrochemical Sensor Based on Reduced Graphene Oxide and Copper Nanoparticles for Monitoring Estriol Levels in Water Samples after Bioremediation." *Chemosensors* 10395. https://doi.org/10.3390/CHEMOSENSORS10100395.
87. Zahran, Moustafa, Ziad Khalifa, Magdy A.H. Zahran, and Magdi Abdel Azzem. 2021. "Recent Advances in Silver Nanoparticle-Based Electrochemical Sensors for Determining Organic Pollutants in Water: A Review." *Materials Advances* 2 (22): 7350–65. https://doi.org/10.1039/D1MA00769F.

# 9 Abandoned Mines as Sources of Metal Pollution

## *Eternity Costs and Everlasting Threats to the Aquatic Environment*

*Franz Jirsa, Jakob Windisch, and Andreas Gradwohl*

## 9.1 MINING AS SOURCE OF METAL POLLUTION IN SURFACE WATERS: A NOVEL PROBLEM?

The history of humankind on Earth is deeply linked to anthropogenic exploitation of the planet's resources. Metals in particular are linked with the development of civilization. The first metal to be extracted was lead and its mining is well documented to have started more than 5500 years before present (ybp) [1, 2]. Early metal production is mainly linked to lead and silver. Other metals came later and were so important that their names were used to characterize respective periods of time, most importantly the Copper, Bronze, and Iron ages. During these periods, humans refined the skills to extract metals from ores and form objects of utility from these metals, enhancing their quality of life [3]. Although metal production was rather local then, the amount of global Cu and Pb produced between 2500 and 175 ybp amounted to approximately $10^3$–$10^4$ metric tons per year [1, 4]. Due to the metal's volatility, hemispheric lead and copper pollution linked to mining is documented from 2500 ybp onward in Greenland ice [1, 4]. Starting in app. 1850, the demand for metals in general grew rapidly and outnumbered the amount produced in the heydays of Roman Antiquity [5]. Rising metal demand has not come to an end and still is ongoing: global yearly metal extraction has risen from 2.7 Gt in 1970 to almost 9.4 Gt in 2022 [6]. There is no doubt that the production and use of metals have brought enormous advantages to humanity, but metal mining has come with a cost: contamination of the environment. Pollution and health risks for humans associated with mining are not novel knowledge, but were known already in preindustrial times; e.g., mining operations were banned because of health concerns in some ancient cities in Greece already 4500 ybp [7]. The global scale of pollution was not recognized for

 DOI: 10.1201/9781003632092-10

many centuries and the call for conserving pristine nature as a necessity for human welfare was not recognized until the late 19th century. First attempts to save or protect "nature" in its pristine form started in the late 19th century in Central Europe [8] as well as in the United States [9]. Since the 1970s, scientists have tried to measure the anthropogenic impact on the environment and develop bioindicators that allow to observe the significance of measures taken to meliorate the human impact on the environment [10]. Only a few years later, scientists started to calculate inventories of metals in the environment and mass balances of their anthropogenic distribution, as well as their development over time [11]. Mining of metals is still a major contributor to the pollution of the environment, and even more so it has become a major global driver for environmental change [6].

The effects of mining are manifold, but pollution can be categorized in three different groups, following their order of effects: primary, secondary, and tertiary [12]. Primary pollution is found in open pit and underground mine water, in waste rocks, tailings, and slag material. Secondary pollution is found in groundwater, sediments, and soils, whereas tertiary contamination pertains to river sediments reworked from floodplain and groundwater from contaminated reservoir sediments [12]. These effects occur during the active phase of mining, but some of them continue after the mines are closed down or abandoned. One particular negative effect can occur after mining has stopped and the ground water, which is usually pumped off during the active phase, starts to enter and fill the open shafts. In cases where this water reacts with oxygen and the minerals in the shafts to form acid, then overflows and enters the environment, the term "acid mine drainage (AMD)" is used to describe this phenomenon [13].

## 9.2 MINING INFLUENCED WATER AND ACID MINE DRAINAGE FROM CHEMIST'S PERSPECTIVE

Acid mine drainage (AMD) is a naturally occurring phenomenon following a complex chemical process. This process is solely dependent on the chemicals involved – meaning that given the right circumstances, it may occur in mines around the globe [14]. Abandoned mines are often accompanied by mining dumps and piles in their vicinity, frequently contributing to the phenomenon. Chemically speaking, the main components leading to AMD are sulphide-bearing ores, oxygen, and water. The combination of these ingredients results in the specific AMD characteristics: low pH water with significantly elevated concentrations of dissolved iron and sulphate [15]. Water that is chemically affected through processing of minerals or mining is often referred to as mining influenced water (MIW) [16]. Amongst sulphide minerals, pyrite (iron (II) disulphide, $FeS_2$) plays the dominant role in generating MIW and subsequently AMD. The chemistry behind AMD including the oxidation of pyrite takes place in various stages, dependent on the pH of the MIW and the geochemistry of the surrounding rock [14] (Equation 1). Firstly, in the presence of oxygen, water reacts with pyrite, whereby the sulphur is oxidized forming ferrous sulphate and sulfuric acid (a). At lower pH values, the oxidation continues through the ferrous iron ($Fe^{2+}$) of the ferrous sulphate (b). The ferric iron produced acts as a more potent oxidizing agent towards pyrite (c), triggering an even higher generation of acidity [17]. This accelerated acidification leads to a further significant pH drop of MIW. The oxidation of ferrous to ferric iron (b) can also be strongly augmented by bacterial activity of, for example, *Leptospirillum* spp. or *Acidithiobacillus* spp. [18] Through those several reaction steps, and depending on the host rock composition, a number of meta-stable (e.g. schwertmannite between $Fe_8O_8(OH)_6SO_4$ and $Fe_{16}O_{16}(OH)_{10}(SO_4)_3$ and ferrihydrite) as well as more stable secondary products (e.g. jarosite and hemantite) are formed [19].

$$2FeS_2 + 7O_2 + 2H_2O \rightarrow 2Fe^{2+} + 4SO_4^{-2} + 4H^+ \, (a)$$

$$4Fe^{2+} + O_2 + 4H^+ \leftrightarrow 4Fe^{3+} + 2H_2O \, (b)$$

$$FeS_2 + 14Fe^{3+} + 8H_2O \rightarrow 15Fe^{2+} + 2SO_4^{2-} + 16H^+ \, (c)$$

**Equation 1:** (a) Primary oxidation of pyrite by oxygen and water leading to a release of $Fe^{2+}$, sulphate and protons. (b) oxidation of $Fe^{2+}$ to $Fe^{3+}$ in the presence of oxygen and potentially microbially catalyzed. (c) secondary oxidation of pyrite by $Fe^{3+}$, producing further acid (below pH=3 up to 10- to 100-fold more productive than primary oxidation [19, 20].

The process of mining creates a large surface area, which accelerates oxidation reactions. Through given circumstances, potential neutralization processes of the MIW by minerals such as dolomite ($CaMg(CO_3)_2$) and calcite ($CaCO_3$) are overpowered. As a consequence of the low pH conditions of the MIW and depending on the composition of the host rock, a significant leaching of trace elements and metals into MIW takes place. Thereby the dissolution of significant amounts of trace elements and metals into the MIW is inevitable [14]. In abandoned mines, pumping systems are no longer in place, which means that groundwater seeping in is no longer pumped out. This leads to the gradual rise in the level of the MIW, slowly filling mines until they eventually overflow, significantly contaminating local aquatic environments as well as groundwater resources [21]. At this point the MIW entering surface water is characterized by high electrical conductivity, particularly rich in ferrous and ferric iron, trace elements, as well as sulphate. The dilution of MIW as it enters rivers, streams, and other waterbodies leads to pH and conductivity changes, which further significantly influences the dissolution and (co-)precipitation behavior of metals and trace elements [22]. The dissolved components of the MIW are oxidized as well as hydrolyzed, and precipitation may occur as schwertmannite, ferrihydite, goethite, or jarosite – again depending on electrical conductivity and pH dynamics of surface water and MIW [23]. These yellow-orange iron hydroxides/-oxohydrid precipitates, also known as "yellow boy", leave a distinctive stain in riverbeds, giving the phenomenon its characteristic colour [24]. In addition, various other trace elements such as nickel can co-precipitate with iron [25], introducing a potential range of trace elements and metals from the MIW into the sediment. Through time and complex adsorption-desorption interactions between sediment and the potentially changing surrounding water conditions, a gradual mobilization and long-distance transport of trace elements and metals within river systems may occur, potentially causing contamination far from the original point source [26, 27].

## 9.3 THE CROCODILE RIVER SYSTEM IN SOUTH AFRICA: A CASE STUDY

### 9.3.1 Study Area, Materials, and Methods

South Africa is a country that has a remarkably rich mining history, especially regarding the mining of gold, and it is well aware of the issues associated with AMD [27, 28]. Especially one region, the Witwatersrand region close to Johannesburg, represents gold mining like no other place on Earth – nearly half the gold ever mined in human history originates from the Witwatersrand basin in South Africa [29]. Since the discovery of gold in 1886, mining activities began and continued for more than a century, greatly shaping the area's economy and turning former rural regions into densely populated urban areas. The abandonment of unprofitable mines in the 1990s left behind a network of interconnected voids [30].

In 1887, the company "West Rand Consolidated" started exploiting gold from the West Rand area 30 km west of Johannesburg, followed by the establishment of the town Krugersdorp. In 1890, the "Randfontein Estates Gold Mine" was established and operated for about 100 years until the 1990s. After that, the mining operations were abandoned, and the interlinked mine shaft voids of the Western Basin began to fill with inflowing natural water. In August 2002, rising water from one of the mines of the former "Randfontein Estates Gold Mine", later known as "Harmony Gold Mine" and then operated by "Rand Uranium", reached the surface near Krugersdorp and caused the overflow of mining influenced water (MIW) into the Tweelopiespruit, a small tributary feeding the Bloubankspruit and ultimately the Crocodile River (West) [27, 30, 31]. The first AMD surfaced at the "Black Reef Incline" mine shaft and several nearby boreholes in a north-easterly direction along the drainage of the Tweelopiespruit. The pH dropped to 3 and the Tweelopiespruit itself was rendered sterile [32–34]. Once the MIW was exposed to air and mixed with stream water, ferrihydrite and various other metals precipitated extensively, coating the riverbed with orange-red deposits typical for AMD-impacted systems [31, 32]. Early remediation measures included pumping of the MIW into holding ponds some distance away [32]. Later, a neutralization treatment of the MIW by "Rand Uranium", now called Sibanye Gold Water Project, was installed and is still operating today. This so-called "chemical process" uses limestone to increase the pH of the MIW to a slightly basic pH, which leads to the formation of ferric hydroxide precipitate after aeration. The precipitate, which contains various elements other than Fe, is collected in settling chambers to remove the toxic elements from the water. The treated MIW is then released into the Tweelopiespruit at several different discharge points [32, 35, 36].

Early scientific attention to the site was extensive but heterogeneous. Several studies examined water and sediment after the spill and found the metal and metalloids content to be extremely high and toxic for the aquatic environment [31–33]. Dzwairo and Mujuru [37] analyzed data records from the Department of Water and Sanitation (South Africa) from 2003–2008, documenting persistently high sulfate, iron, calcium, magnesium, and variable pH values of the Tweelopiespruit, implying severe impact of AMD up to the year 2008 and urging continuous research on this area. Durand et al. [31] highlighted potential risks of AMD-contaminated water to the Cradle of Humankind UNESCO World Heritage Site. Consequent hydrological studies revealed that the infiltration into the underlying karst is minor and very little stream water seems to infiltrate the dolomitic aquifer [36, 38]. The indication was that most of the MIW remains in surface flow, passing into the Bloubankspruit and further into the Crocodile River (West), posing no immediate threat to the Cradle of Humankind UNESCO World Heritage Site, but certainly not excluding threats to the surrounding natural environment. Van Deventer and Cho [49] reported high contents of toxic metals in the sediment of the Tweelopiespruit and furthermore in the rhizomes, stems, and leaves of reed *Phragmites australis* from the surrounding wetland. Complementing this, Singh et al. [46] identified heavily contaminated sediment in the Tweelopiespruit-Bloubankspruit river system, with significant adverse impacts on sediment bioassays. In order to complete the picture, a 13-month case study between January 2019 and February 2020 on this river system from our working group investigated the general condition as well as the fluctuations within the system during a calendar year. This included extensive water, sediment, and biota analysis of metals and metalloids in bulk and trace concentrations [27, 39]. This case study focused on the upper Crocodile River (West) system and its tributaries, the Tweelopiespruit and Bloubankspruit, located west of Krugersdorp in the western Witwatersrand Basin. The stream network represents a continuous flow path starting at the Tweelopiespruit and passing through the Bloubankspruit into the Crocodile River (West), shown in Figure 9.1.

Sampling sites 1–7 were chosen along the river system, especially considering confluences. Site 1 is located at Tweelopiespruit after the outflow of the treated MIW enters the system from the abandoned gold mine. Sites 2 and 3 are situated along Bloubankspruit before and after the inflow of Tweelopiespruit, an area influenced by agriculture. Sites 4, 5, and 6 are located within the area of a hotel resort. At site 4

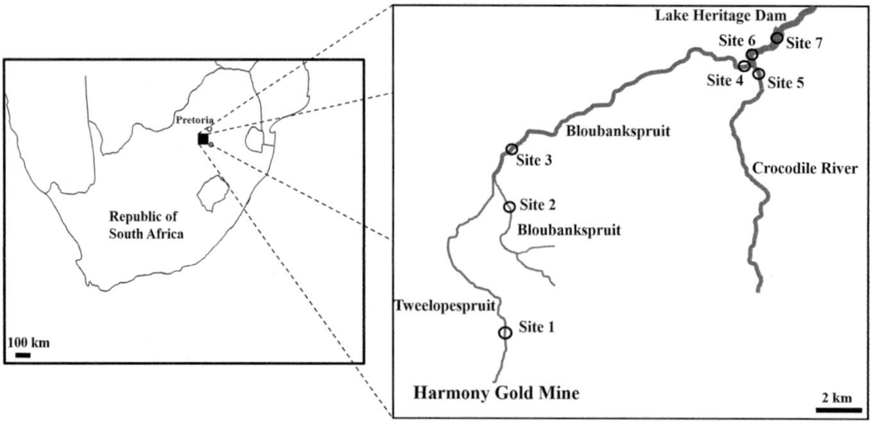


**FIGURE 9.1** The Republic of South Africa with a black square indicating the sampling area. Black rings specify the sampling sites 1–7. (Reprinted from Windisch et al. [27].)

the flow rate of Bloubankspruit is decreased as the river is slightly dammed. Site 5 is located along the Crocodile River (West) before the inflow of Bloubankspruit. Site 6 is located along the Crocodile River (West) shortly after the confluence of the rivers. Site 7 is situated where the Crocodile River (West) is dammed by the Lake Heritage Dam (Franz Richter wall). Water and surface sediments were sampled monthly from January 2019 to February 2020, thereby encompassing both wet and dry seasons. At each site, in-situ parameters such as pH, electrical conductivity (EC), temperature, total dissolved solids (TDS), and dissolved oxygen were recorded. Water (Na, Mg, K, Ca, Mn, Fe, Co, Ni, Zn, As, Sr, Cd) and sediment samples (Ti, V, Ni, Cu, Zn, As, Rb, Sr, Ag, Cd, Pt, Au, Hg, Pb, U) were analyzed for major cations and trace elements using total X-ray fluorescence spectrometry (TXRF), flame- and graphite furnace atomic absorbance spectroscopy (F-AAS, GF-AAS), and inductively coupled plasma mass spectrometry (ICP-MS) techniques following standard procedures [27]. Whenever possible, the analyzed concentrations were compared to the South African water-quality guidelines [40], providing a target water quality range (TWQR) and a chronic (CEV) and acute effect value (AEC), as well as the Canadian and Australian/New Zealand sediment guidelines, providing a threshold (TEL) and probable effect level (PEL) [41, 42] because parameters for sediment evaluation were not available from South African authorities.

A complementary sampling of fish and plants was performed in January 2020, targeting macrophytes (cattail *Typha* sp., reed *Phragmites australis*, water hyacinth *Pontederia crassipes*) at sites 1, 3, 5, and 7. Macrophytes were sampled where available, and parts of stems and rhizomes/roots were used for analyses. Tissue samples (muscle and liver) of 4 fish species (*Oreochromis mossambicus*, *Labeobarbus polylepis*, *Labeobarbus marequensis*, and *Clarias gariepinus*) were obtained from fish caught at site 7. Samples were frozen in situ, thereafter freeze-dried, digested (microwave-assisted) in half-concentrated nitric acid, and the elements Fe, Ni, Cu, Zn, As, Se, Ag, Cd, Hg, and Pb were measured using TXRF, GF-AAS, or cold vapor atomic absorption spectroscopy (CV-AAS). Details of sample preparation and quality control (respective recovery rates and LODs) are published elsewhere [27, 39]. The results of these analyses enabled the quantification of metal uptake and accumulation in the respective biota. Most relevant results in regard of AMD and its consequences to the local environment are discussed in the following.

### 9.3.2 Physico-chemical Parameters

The physicochemical water parameters from all sampling sites, including the means as well as the minimal and maximal values, are presented in Table 9.1. The mean EC and TDS content were significantly higher at site 1. This is explainable with the treated MIW flowing into this site from the treatment plant, especially with the addition of limestone ($CaCO_3$). Following the course of the river system, major dilution effects were observed during the whole observation period, resulting in significantly lower mean values for EC [27]. During the winter months, a significant drop in pH ($<7$) was observed, raising the question of whether the treatment plant was fully functional during this period. Analogous to the EC values, the mean concentrations of $Ca^{2+}$ as well as $Mg^{2+}$ decreased significantly from site 1 to site 7, showing clear

**TABLE 9.1**
**Mean and Min-Max Values of Physicochemical Parameters and Dissolved Bulk Elements (Na, Mg, K, Ca) from Sampling Sites 1–7 ($n$ = 13)**

| | Site 1 | Site 2 | Site 3 | Site 4 | Site 5 | Site 6 | Site 7 |
|---|---|---|---|---|---|---|---|
| **Temp. (°C)** | 20.1 | 17.3 | 17.4 | 17.8 | 19.0 | 18.2 | 20.9 |
| | 8.20–24.4 | 10.1–22.8 | 11.1–23.3 | 11.6–22.5 | 10.9–24.5 | 10.7–23.2 | 11.2–25.4 |
| **pH** | 8.20 | 7.91 | 7.76 | 8.23 | 8.36 | 8.27 | 8.47 |
| | 6.81–9.08 | 7.57–8.31 | 7.40–8.05 | 7.55–8.62 | 7.69–8.67 | 7.73–8.46 | 7.54–9.07 |
| **EC (μS/cm)** | 3562 | 932 | 1415 | 1006 | 433 | 788 | 696 |
| | 2981–3940 | 630–1188 | 702–1860 | 902–1140 | 326–493 | 650–881 | 403–790 |
| **TDS (mg/L)** | 2284 | 590 | 901 | 631 | 269 | 494 | 440 |
| | 1938–2580 | 310–772 | 340–1330 | 450–740 | 190–320 | 320–573 | 190–523 |
| **Salinity** | 1899 | 461 | 716 | 492 | 206 | 415 | 342 |
| | 1560–2100 | 310–590 | 640–1000 | 450–530 | 160–240 | 320–846 | 190–400 |
| **Na (mg/L)** | 131 | 82.1 | 80.9 | 46.3 | 44.4 | 45.7 | 33.2 |
| | 96.7–147 | 55.1–105 | 63.2–94.4 | 39.5–51.2 | 30.4–57.3 | 40.7–53.3 | 20.9–55.1 |
| **Mg (mg/L)** | 86.4 | 8.97 | 45.7 | 51.1 | 11.9 | 36.1 | 33.2 |
| | 34.2–174 | 5.23–11.1 | 37.6–57.4 | 36.6–58.4 | 4.90–13.4 | 26.7–41.6 | 17.5–48.2 |
| **K (mg/L)** | 11.0 | 14.9 | 11.9 | 5.78 | 9.90 | 7.17 | 7.43 |
| | 5.35–14.3 | 10.1–21.3 | 8.18–16.3 | 4.02–11.4 | 8.01–11.1 | 5.78–9.29 | 4.91–11.3 |
| **Ca (mg/L)** | 746 | 38.5 | 149 | 107 | 16.8 | 74.4 | 61.8 |
| | 443–972 | 23.6–50.9 | 85.2–239 | 88.1–130 | 12.7–22.4 | 59.4–90.9 | 27.8–79.8 |

*Source:* Redrawn after Windisch et al. [27].

dilution effects at the confluence of the Bloubankspruit with the Crocodile River (West). Similar dilution effects were also detected for $Na^+$, but not for $K^+$ at any of the sampling sites [27].

### 9.3.3 Water Quality

While several bulk and trace elements were recorded within the river system (Mn, Fe, Ni, Sr, Zn, Co, U, As, and Cd), the elements Mn, Fe, Ni, Zn, and U demonstrated the most pronounced and environmentally relevant patterns. The results for the dissolved elements from all sampling sites, including the means as well as the minimal and maximal values, are presented in Table 9.2. Similarly to bulk elements, the concentrations were significantly higher at site 1. Dissolved Pt, Au, and Pb could not be detected in any of the samples, and Cd and As did not occur in concentrations of concern [27].

The elements Mn, Fe, Ni, Zn, and U were consistently elevated above regional background concentrations and South African TWQR guidelines. Manganese exhibited the highest concentrations and temporal variation, reaching up to 178 mg $L^{-1}$ during the South African winter months, yet did not exceed 1 mg $L^{-1}$ in 9 out of 13 samples. The concentration declined downstream but remained above chronic-effect levels in most months. The maximum concentrations of dissolved Mn in sites 1–4 exceeded the acute effect levels, and the mean concentrations of these sites also

**TABLE 9.2**
**Mean and Min-Max Values of Dissolved Trace Elements from Sampling Sites 1–7 ($n$ = 13)**

| | Site 1 | Site 2 | Site 3 | Site 4 | Site 5 | Site 6 | Site 7 |
|---|---|---|---|---|---|---|---|
| Mn (mg/L) | 21.6<br>0.071–178 | 0.592<br>0.081–2.51 | 0.698<br>0.273–1.51 | 0.561<br>0.251–1.57 | 0.072<br>0.039–0.132 | 0.207<br>0.051–0.426 | 0.064<br>b.d.–0.194 |
| Fe (mg/L) | 0.988<br>0.059–3.71 | 0.264<br>0.107–0.759 | 0.132<br>0.095–0.226 | 0.213<br>0.086–0.742 | 0.163<br>0.081–0.291 | 0.136<br>0.086–0.187 | 0.109<br>0.079–0.158 |
| Ni (mg/L) | 0.391<br>0.011–2.84 | 0.043<br>0.012–0.089 | 0.023<br>0.011–0.048 | 0.022<br>0.015–0.032 | 0.017<br>0.012–0.052 | 0.028<br>0.012–0.124 | 0.016<br>0.009–0.037 |
| Zn (mg/L) | 0.106<br>0.024–0.441 | 0.035<br>0.019–0.062 | 0.040<br>0.027–0.091 | 0.034<br>0.016–0.063 | 0.034<br>0.012–0.058 | 0.027<br>0.011–0.037 | 0.028<br>0.006–0.048 |
| U (µg/L) | 18.1<br>1.94–62.5 | 0.492<br>0.26–0.92 | 0.922<br>0.31–2.77 | 0.804<br>0.37–1.30 | 0.274<br>0.14–1.29 | 0.624<br>0.34–1.15 | 0.595<br>0.39–0.88 |

*Source:* Redrawn after Windisch et al. [27].

surpassed chronic-effect levels, indicating potential toxic effects for aquatic organisms, especially at site 1. Iron peaked with 3.71 mg $L^{-1}$ at site 1 in June 2019, while the mean concentrations at the sites 3-7 were significantly lower. Nickel was found in concentrations up to 2.84 mg $L^{-1}$ and showed the least temporal variation at sites 2-7; it remained detectable at all sites, showing that a significant fraction remained in solution despite the alkaline pH. Zinc ranged from 0.04–0.44 mg $L^{-1}$ at site 1 and declined to a mean of 0.028 mg $L^{-1}$ downstream. Uranium, a key contaminant associated with the ores of the Witwatersrand region, reached up to 62 µg $L^{-1}$ at site 1, an order of magnitude above natural levels. It was still present at a mean concentration of 0.5 µg $L^{-1}$ at site 7 (Lake Heritage Dam) [27]. All other observed element concentrations were subject to high temporal variations, with the highest concentrations recorded for Mn, Fe, Ni, and Zn from June to August. At site 1, the temporal variation negatively correlated to the pH was especially pronounced (Figure 9.2). This temporary increase as well as the decrease in pH indicated that the MIW treatment was somehow flawed over a certain period of time [27].

### 9.3.4 Sediment Quality

While several bulk and trace elements were detected in the sediment samples (Ti, V, Cr, Ni, Cu, Zn, As, Rb, Sr, Ag, Cd, Au, Hg, Pb, U), the elements Ni, Zn, As, Pb, and U demonstrated the most pronounced and environmentally relevant patterns (Table 9.3 and Figure 9.3). Site 1 showed a significantly higher mean content for Zn, Ni, As, and U. Fe and Mn in sediment were within the range described in the literature for AMD and mining-impacted areas [43].

Nickel showed one of the most pronounced patterns. Sediments at site 1, directly after the treated MIW discharge, contained the highest concentrations, with mean values exceeding 100 mg $kg^{-1}$. These values are consistent with rapid co-precipitation of Ni and Fe under the pH conditions of treated MIW [25]. A

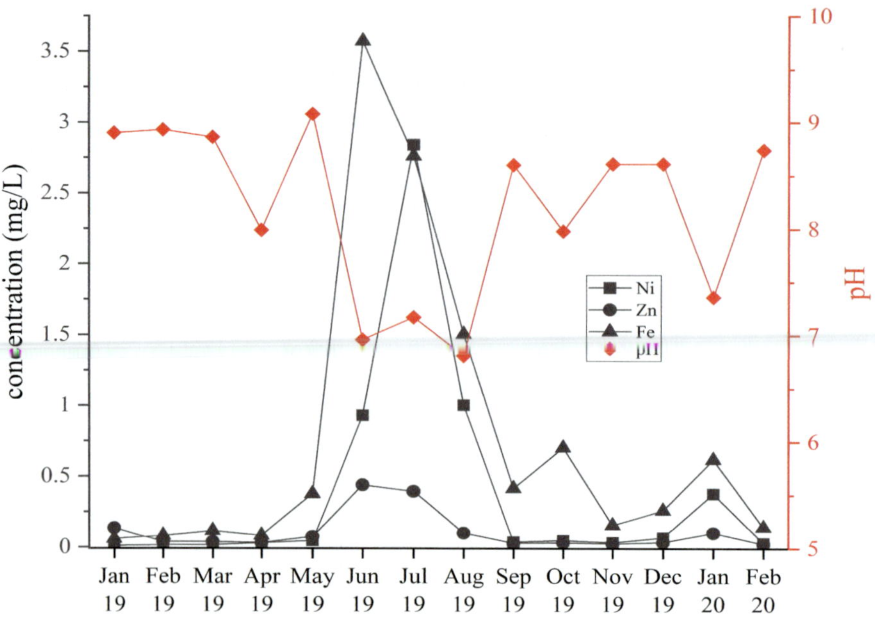


**FIGURE 9.2** Temporal variation of dissolved trace elements at sampling Site 1. (Reprinted from Windisch et al. [27]).

**TABLE 9.3**
**Content of Trace Elements (mg kg$^{-1}$ dw) in Surface Sediment Samples from All Sampling Sites, ($n$ = 13)**

| | Site 1 | Site 2 | Site 3 | Site 4 | Site 5 | Site 6 | Site 7 |
|---|---|---|---|---|---|---|---|
| **Ni** | 107 ± 43.0 | 32.3 ± 16.2 | 25.7 ± 9.09 | 62.1 ± 28.3 | 15.5 ± 7.8 | 20.6 ± 10.2 | 36.8 ± 26.8 |
| **Zn** | 171 ± 100 | 63.0 ± 45.0 | 50.7 ± 28.4 | 56.8 ± 30.9 | 26.4 ± 10.4 | 28.4 ± 13.4 | 57.7 ± 16.2 |
| **As** | 76 ± 39.7 | 6.38 ± 2.13 | 6.75 ± 3.45 | 5.14 ± 3.50 | 2.76 ± 2.07 | 1.99 ± 1.29 | 15.4 ± 7.7 |
| **Pb** | 17.6 ± 18.49 | 5.60 ± 2.08 | 5.50 ± 1.15 | 13.7 ± 9.0 | 4.85 ± 2.71 | 4.65 ± 2.13 | 6.43 ± 4.65 |
| **U** | 12.7 ± 7.9 | <2.00 | <2.00 | 3.29 ± 2.12 | <2.00 | <2.00 | <2.00 |

*Source:* Redrawn after Windisch et al. [27].

second, smaller maximum appeared at site 4, where locally reduced flow velocities enhanced sedimentation processes. Downstream attenuation was evident, with site 6 and 7 displaying the lowest Ni concentrations. Zinc followed a similar but somewhat less pronounced trend. Although none of the sediment samples exceeded the PEL, the TEL was occasionally surpassed at sites 1 and 4. The elevated concentrations at site 1 indicate rapid precipitation or sorption processes immediately after the discharge. Arsenic content also peaked sharply at site 1, where concentrations were significantly higher than at all other locations and above the PEL. This reflects the strong affinity of As for Fe and its tendency to accumulate in zones of intense Fe precipitation [44, 45]. Downstream, As declined

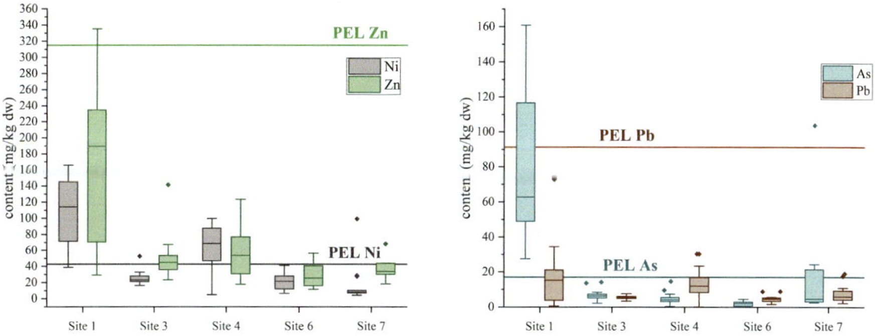


**FIGURE 9.3** Box plots including mean ± SD in mg $kg^{-1}$ dw, outliers as well as the Possible effect level (PEL) for Ni and Zn (left) and for Pb and As (right) in sediments from sites along the course of the mining influenced water (excluding sites 2 and 5, because the metal contents there had no apparent impact on sediment levels in the following system). (From [27]).

markedly, and only site 4 showed minor secondary enrichment in As content. Earlier studies had reported lower As values near site 1 [46], suggesting temporal variability or an increased co-precipitation during the time of our study. Overall, the data confirmed that also the As accumulation is tightly linked to a high Fe content of the sediment and that the sediment at site 1 is still heavily polluted with As. Additionally, the mean As content at sites 2, 3, and 7 were above the TEL [27]. Lead exhibited the same trend, with the highest mean concentrations at site 1 and slightly elevated values at site 4, where the slower flow promoted the sedimentation of this element. Although Pb levels occasionally exceeded TEL values at these two sites, they never approached the PEL. Downstream sediments showed significantly lower concentrations [27].

Uranium displayed a distinctively different pattern. While dissolved U remained relatively constant across sites 2–7, U in sediment was strongly concentrated at site 1 and, to a lesser degree, site 4. U in sediment at site 1 averaged around 13 mg $kg^{-1}$, consistent with rapid sorption of dissolved U species onto iron oxides. The secondary peak at site 4 reflects the conditions favorable for sedimentation, including locally reduced flow and high organic carbon content of the sediment [47]. At all other sites, U remained below detection limits. In general, the sediment contamination in this case study was most pronounced at sites 1 and 4. These findings demonstrate that site 1 remains heavily polluted as a direct consequence of the metal (co)-precipitation associated with AMD and with, at least temporarily, insufficient treatment. Our results further suggest the continuous mobilization of toxic elements from site 1, leading to increased and potentially harmful concentrations of metals in sediments at the sites further downstream.

### 9.3.5 Trace Elements in Macrophytes

The trace element concentrations in macrophytes strengthened the observations gained from the water and sediment samples. The mean content (± SD) of all species at the respective sites is given in Table 9.4. In general, the metal contents were higher in roots than in stems. At site 1, only *P. australis* was abundant, while at sites 3, 5, and 7 the

**TABLE 9.4**
**Mean (±SD) Metal and Metalloid Content in Roots and Stems of *P. australis* (site 1), *Typha* sp. at Sites 3, 5, and 7, as well as *P. crassipes* at Sites 5 and 7.**

| Element | Tissue | *P. australis* Site 1 (*n* = 13) | *Typha* sp. Site 3 (*n* = 9) | *Typha* sp. Site 5 (*n* = 4) | *Typha* sp. Site 7 (*n* = 4) | *P. crassipes* Site 5 (*n* = 3) | *P. crassipes* Site 7 (*n* = 13) |
|---|---|---|---|---|---|---|---|
| Fe | Root | $11064^{*} \pm 7444$ | $3262^{a*} \pm 2929$ | $1053^{a*} \pm 1196$ | $1601^{b*} \pm 483$ | $1225^{a*} \pm 969$ | $14690^{b*} \pm 8550$ |
| | Stem | $112 \pm 95$ | $1138^{a} \pm 1594$ | $502^{a} \pm 381$ | $61.2^{b} \pm 15.9$ | $161^{a} \pm 46.6$ | $265^{a} \pm 151$ |
| Ni | Root | $28.9^{*} \pm 29.5$ | $32.0^{a} \pm 27.1$ | $28.9^{a*} \pm 8.4$ | $160^{b*} \pm 21$ | $47.4^{a} \pm 75.4$ | $99.8^{a*} \pm 41.7$ |
| | Stem | $4.36 \pm 4.41$ | $10.1^{a} \pm 12.2$ | $2.44^{b} \pm 0.41$ | $5.60^{a} \pm 1.48$ | $75.2^{a} \pm 60.9$ | $17.5^{a} \pm 34.6$ |
| Cu | Root | $14.7^{*} \pm 10.6$ | $11.8^{a} \pm 8.5$ | $18.8^{a*} \pm 10.7$ | $34.4^{a*} \pm 10.2$ | $22.7^{a*} \pm 27.5$ | $37.1^{a*} \pm 24.4$ |
| | Stem | $4.65 \pm 3.87$ | $8.86^{a} \pm 4.18$ | $2.65^{b} \pm 0.85$ | $20.7^{a} \pm 20.7$ | $18.3^{a} \pm 11.4$ | $5.13^{a} \pm 4.03$ |
| Zn | Root | $68.6 \pm 48.1$ | $292^{a} \pm 293$ | $154^{a*} \pm 74.0$ | $51.7^{a*} \pm 22.4$ | $164^{a} \pm 217$ | $87.2^{a*} \pm 33.4$ |
| | Stem | $48.8 \pm 33.0$ | $116^{a} \pm 120$ | $27.4^{a} \pm 18.8$ | $11.7^{a} \pm 1.3$ | $263^{a} \pm 190$ | $21.1^{b} \pm 8.00$ |

*Note:* All values are given in mg $kg^{-1}$ dw. Significant differences between mean values of root and stem samples for *P. australis* are indicated by an asterisk "*" at the root mean value ($p < 0.05$). For *Thypha* sp. and *P. crassipes*, different superscript letters indicate significant differences between the sites for the respective tissue and species ($p < 0.05$).

*Source:* Redrawn after Windisch et al. [39].

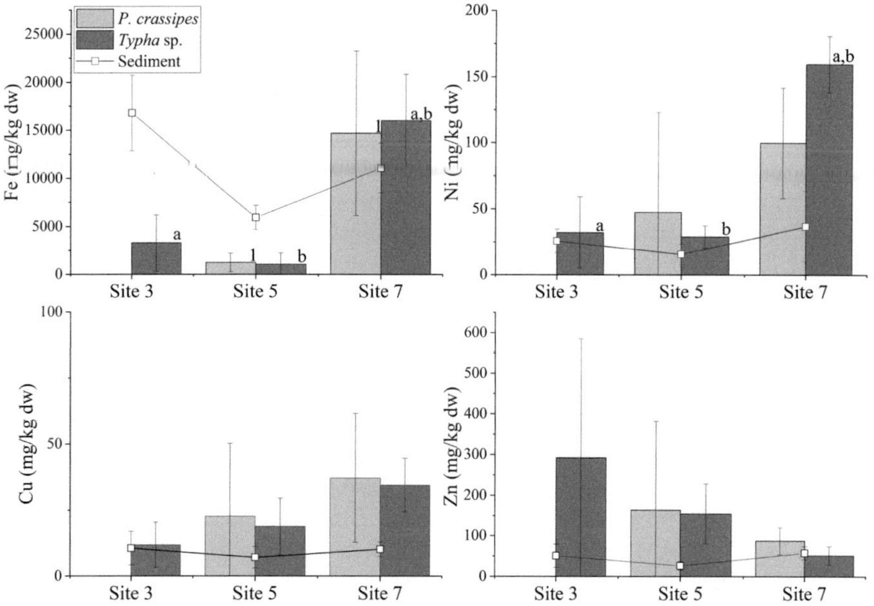


**FIGURE 9.4** Mean and SD of the metal content in the roots of *P. crassipes* and *Typha* sp., as well as the metal content of the respective sediments for Ni, Cu, Zn, and Fe at sampling sites 3, 5, and 7. (Reprinted after Windisch et al. [39].)

macrophytes *P. crassipes* and *Typha* sp. were dominant. A comparative overview of Fe, Ni, Cu, and Zn contents in the roots of the latter two species is given in Figure 9.4. Au and U were below the detection limits in all plant samples (0.035 mg $kg^{-1}$ and 2 mg $kg^{-1}$ for Au and U, respectively) [39].

Most abundant in all samples was Fe. At site 1, where only *P. australis* could be analyzed, a high accumulation of Fe in roots was apparent (11 064 ± 7445 mg $kg^{-1}$). Compared with values from unpolluted sites and with values from the same site in 2011/12 [48, 49], our study confirmed the excessive and ongoing accumulation of Fe at site 1. These findings imply the ongoing impact of the MIW on the local biota even years after the implementation of the water treatment. Further downstream at the AMD-influenced sites, Fe contents in the roots of *P. crassipes* and *Typha* sp. directly correlated with the concentrations in the sediment and were highest at site 7. Reflecting the Fe values, the mean Ni content in plant roots was highest at site 7 and lower at sites 3 and 5 for *Typha* sp. and *P. crassipes*. This was ascribed in both cases to the reduced flow velocity at site 7. The mean Ni root content of *P. australis* at site 1 was lower than that of *P. crassipes* and *Typha* sp. at site 7 but not compared to sites 3 and 5 [39]. Interestingly, compared to the results of Van Deventer and Cho [49] the values in our study point to a significant decrease in Ni content at this site over time. Nonetheless, individual root samples exceeded the mean metal content significantly, emphasizing an inhomogeneous distribution of Ni in *P. australis* roots at site 1. The Cu content in the roots *P. crassipes* and *Typha* sp. followed the same trend as for Fe and Ni, yet less pronounced. Cu contents in the roots of *Typha* sp. and *P. crassipes* showed a positive correlation with the Cu content in the sediments. The highest Cu content was found for *Typha* sp. and *P. crassipes* at site 7, with mean concentrations > 30 mg $kg^{-1}$. The Cu content at other sites and of *P. australis* was < 30 mg $kg^{-1}$. Comparing the Cu content of roots and stems revealed a rather pronounced translocation of Cu. This translocation is in line with literature [50] demonstrating the high permeability of the root-stem barrier for this element. Concentration ranges in the underground parts of all macrophytes investigated are in line with the literature [50–52]. Significant differences between root/stem were found only for *P. crassipes* and *Typha* sp. at sites 5 and 7.

Regarding the Zn content, no significant differences were detected between root and stems in all macrophytes investigated. Comparisons to previously published data on the Zn content show that especially the levels in the roots are in the same range [50, 53, 54]. These comparisons also strongly indicate an unchanged Zn burden at site 1 compared to earlier investigations [49]. Our results furthermore point to a strong location-dependent, inhomogeneous distribution of Zn in roots and stems [39]. In summary, the roots of *Typha* sp. and *P. crassipes* showed elevated levels of Fe, Ni, and Cu at AMD-influenced sites. Furthermore, Fe, Cu, and Zn levels were in general comparable to those reported a decade earlier, demonstrating a persistent metal burden on the vegetation despite the treatment efforts.

### 9.3.6 Trace Elements in Fish

The fish species *O. mossambicus* (4 specimens), *L. polylepis* (13 specimens), *L. marequensis* (28 specimens), and *C. gariepinus* (30 specimens) were exclusively sampled at site 7 using gill nets. Hereby, the total length of the two closely related species *L. polylepis* and *L. marequensis* were very comparable, implying a similar age range [39]. The Fe, Ni, Cu, and Zn contents were particularly significant in relation to AMD and are presented in Table 9.5, distinguishing between the contents in liver and muscle tissue.

For all four fish species, the mean Fe content was higher in the liver tissue than in the muscle tissue, yet *C. gariepinus* and *O. mossambicus* accumulated significantly higher amounts of Fe in the liver. These values are comparable to other metal-contaminated river systems in South Africa [55, 56]. The high Fe concentrations in the sediment at sampling site 7 as a consequence of AMD provides a suitable explanation for the increased values in the liver tissues and reflects the results obtained for the macrophytes. The mean Ni content in liver was higher than in muscle tissue for all four fish species. Thereby, the *O. mossambicus* liver values were significantly higher, also compared to values reported for the closely related *Oreochromis niloticus* from Lake Victoria (Kenya) [57]. In contrast, the mean Ni content in the liver of *C. gariepinus* was significantly lower than that reported for this species in heavily polluted systems [56]. Compared to cyprinids from a rather unpolluted South African waterbody [58], the Ni levels in liver and muscle of *L. marequensis* and *L. polylepis* were in a very similar range. In contrast to the macrophytes, the Ni content in fish muscle cannot be seen as an indicator for Ni contamination in the system. The copper content in liver was significantly higher compared to muscle, which is

**TABLE 9.5**
**Trace Element Contents in Muscle and Liver Tissue in mg kg$^{-1}$ dw**

| | Tissue | *O. mossambicus* ($n = 4$) | *L. polylepis* ($n = 13$) | *L. marequensis* ($n = 28$) | *C. gariepinus* ($n = 30$) |
|---|---|---|---|---|---|
| Fe | Liver | $1008^{a*} \pm 293$ | $398^{b*} \pm 175$ | $321^{b*} \pm 138$ | $3444^{a*} \pm 3445$ |
| | Muscle | $20.5^{a} \pm 6.79$ | $20.6^{a} \pm 13.2$ | $26.9^{a} \pm 32.0$ | $55.2 \pm 27.2$ |
| Ni | Liver | $2.98^{*} \pm 1.24$ | $0.277^{a,b} \pm 0.323$ | $0.447^{a*} \pm 0.326$ | $0.284^{b*} \pm 0.335$ |
| | Muscle | $0.323^{a} \pm 0.214$ | $0.214^{a} \pm 0.361$ | $0.071^{a} \pm 0.104$ | $0.117^{a} \pm 0.146$ |
| Cu | Liver | $184^{a*} \pm 124$ | $72.7^{a*} \pm 27.2$ | $41.1^{b*} \pm 16.8$ | $35.7^{b*} \pm 20.0$ |
| | Muscle | $1.15^{a,b} \pm 0.24$ | $3.74^{a} \pm 10.34$ | $4.16^{a,c} \pm 8.79$ | $1.37^{b,c} \pm 0.355$ |
| Zn | Liver | $65.3^{a} \pm 42.4$ | $158^{b*} \pm 26$ | $116^{a,b*} \pm 39$ | $105^{a*} \pm 31$ |
| | Muscle | $59.6^{a} \pm 45.1$ | $20.1^{a} \pm 5.4$ | $34.4^{a} \pm 41.8$ | $38.9^{a} \pm 37.4$ |

*Note:* Tissues of different species followed by the same letter do not differ significantly for the respective element ($p > 0.05$); significant differences between liver and muscle at the same site are indicated by an asterisk "*" at the liver mean value.

*Source:* Redrawn after Windisch et al. [39].

consistent with the literature [59]. However, in contrast to the investigated plants, the fish did not reflect the rather mild Cu contamination at site 7. The only exception hereby was the Cu content in the liver of *O. mossambicus*, whose range of 184 ± 124 mg kg$^{-1}$ was in the same order of magnitude as values for the same species from Cu-polluted areas in India [60]; the reasons for these high Cu levels remain unexplained but might be species-specific. Finally, and consistent with the literature [59], Zn levels were generally higher in liver tissue than in muscle tissue, ranging from 158 ± 26 (*L. polylepis*) to 65.3 ± 42.3 (*O. mossambicus*) with significant differences between these two species. Values for *L. marequensis* and *C. gariepinus* were in between and did not differ significantly from each other. In general, the Zn content in liver tissue was within the same range or lower than values reported in the literature [61–64]. We therefore conclude that site 7 is not polluted with Zn, a finding in line with the results obtained for the macrophytes. In summary, macrophytes can serve as better bioindicators for metals and metalloids associated with AMD. Our findings point to no excessive accumulation of Ni, Cu, and Zn in fish as a consequence of the chronic low-level pollution. Nevertheless, bioaccumulation of *e.g.* Ni and Cu was recorded in the liver tissue of *O. mossambicus* and of Fe in tissues of *C. gariepinus*, emphasizing the potential use of these species as bioindicators for metals associated with AMD and MIW.

## 9.4 CONCLUSION

The monitoring of the Crocodile River (West) system for more than year, starting at the Tweelopiespruit and continuing to the Lake Heritage dam, clearly demonstrated that legacy of the mining history in the Western Basin of the Witwatersrand remains a burden for the aquatic environment until today. Almost two decades after the initial spill of MIW into the Tweelopiespruit and despite remediation measures, the river system continues to exhibit elevated concentrations of elements associated with AMD in water, sediment, and biota. From our findings, we conclude that the pH neutralization of the MIW mostly succeeds in preventing a strong acidification of the water of the Tweelopiespruit itself. Yet, it appears that in some situations, the capacity of the treatment plant simply is exceeded and the pH in the Tweelopiespruit can drop below pH 7, as *e.g.* observed in August 2019. This can mobilize already precipitated and presumably immobilized toxic elements from the sediment of the Tweelopiespruit further downstream. In any case, the water chemistry remains influenced by the treated MIW. Furthermore, the investigation of biota confirmed that the bioavailability of potentially toxic trace elements did not disappear. The macrophytes analyzed in this case study accumulated especially Fe, Ni, and Cu, predominantly in roots at the sites impacted by AMD and MIW. Fish, in contrast, generally appeared less suitable as bioindicators for trace metal pollution here. Nonetheless, especially *O. mossambicus* displayed elevated mean contents of Ni and Cu in the liver tissue, emphasizing the presence of a chronic low-level pollution with metals associated with AMD. In a broader context, the case study exemplifies the title-giving "eternity costs" associated with abandoned mines. Even after implementing MIW treatment, when pH values stabilize and after the visible pollution with toxic elements disappears, the dissolved metal concentrations fluctuate and call for indefinite maintenance and oversight.

## 9.5 FUTURE PERSPECTIVES: WHERE DO WE STAND REGARDING METAL POLLUTION BY ABANDONED MINES?

There is no doubt that (metal) pollution from mines is a current problem and will remain one in the future. In particular AMD and its influence on surface waters will continue to challenge mining companies and policy makers for the next generations to come [6]. The public, and even more so policymakers, have been more or less concerned about the long-lasting costs caused by mining companies after mines have been closed down. International institutions have reacted and installed guidelines for preparing mine closure plans [65, 66]. These guidelines include the mining companies' obligation to plan ahead during the active mining period; in some countries laws or guidelines have been installed to provide legal security. Amongst those are Canada [67], Australia [68], and the European Union [69]. For South Africa in particular, the water research commission has recently published a proposal titled "developing mine closure risk rating and a post-closure opportunity framework for South Africa" [70], an update of the mine water management policy from 2022 [71]. Although these guidelines or laws are in power, their consequent execution is still not guaranteed. On the one hand, new rules do not apply for "old" or "historical" mines, which were closed before the respective legislation was installed; in this case mostly governmental institutions or communities must take on the burden of restoration/remediation of the impacted areas. On the other hand, loopholes were found: mines can be put into "Care and maintenance (C&M)" status, which refers to mines that have closed temporarily. This is a status during which no special care must be taken, and it has no expiry date. Accordingly, numerous cases are reported in which mining companies have put a mine into this status and left it there for several years to decades, avoiding the costly official closure [72].

Putting the problem into a global perspective is difficult because recent global numbers are scarce. One example could serve for others: back in 1993 the US Forest Service stated that "AMD has been described by some as the largest environmental problem facing the U.S. mining industry", with an estimated 20,000–50,000 mines in western USA alone generating acid and impacting up to 16,000 km of streams [73]. In 2020, the Governmental Accountability Office of the US published data stating that of 140,000 recorded abandoned mines, 22,500 may pose environmental hazards, causing yearly costs of approximately 250 million US dollars for addressing these [74]. The most recent publication stated "there is no comprehensive national inventory of abandoned mine features and the physical and environmental hazards they present, so the total risks and liabilities from abandoned mine features in the United States is unknown" [75].

Overall, our case study and the given examples suggest that the entire scale, costs, and environmental risks of closed mines, long after they operated, remain diffuse and continuously problematic, and are rarely accounted for in their full extent. Addressing this problematic gap legally via strict regulatory enforcement of appropriately adapted laws as well as ensuring long-term monitoring and surveillance well beyond the time a mine is operating can therefore be considered of utmost importance.

## ACKNOWLEDGMENTS

We thank all collaborators who joined us in the investigations on heavy metals in the environment. In particular, we want to thank our colleagues Bernhard K. Keppler, Annemariè Avenant-Oldewage, Gabriele Wallner, Beric M Gilbert, Quinton Dos Santos, and (former) students Jabulani Ndaba, Matthias Yeh, and Maximilian Eiser, who contributed with their work. Michael Stachowitsch is gratefully acknowledged for reviewing the English of this text. Parts of this work were funded by the OeAD, Austria's Agency for Education and Internationalization, with Grant no. WTZ ZA-072019.

## REFERENCES

1. Hong, Sungmin, Jean-Pierre Candelone, Clair C. Patterson, and Claude F. Boutron. 1994. "Greenland Ice Evidence of Hemispheric Lead Pollution Two Millennia Ago by Greek and Roman Civilizations." *Science* 265 (5180):1841–1843.
2. Nriagu, Jerome O. 1998. "Tales Told in Lead." *Science* 281 (5383):1622–1623. doi: doi:10.1126/science.281.5383.1622.
3. Barzun, J., T.C. Champion, P.N. Stearns, G.R.R. Treasure, M. Frassetto, E. Peters, D. Herlihy, R.J. Mayne, J.H.M. Salmon, N.G. Parker, M.S. Sørensen, H. Aubin, D. Weinstein, and J.E. Herrin. 2025. History of Europe. In *Encyclopedia Britannica*.
4. Hong, Sungmin, Jean-Pierre Candelone, Clair C. Patterson, and Claude F. Boutron. 1996. "History of Ancient Copper Smelting Pollution During Roman and Medieval Times Recorded in Greenland Ice." *Science* 272 (5259):246–249. doi: 10.1126/science.272.5259.246.

5. Neukirchen, Florian. 2016. "Industrielle Revolution und Hightech." In *Von der Kupfersteinzeit zu den Seltenen Erden: Eine kurze Geschichte der Metalle*, edited by Florian Neukirchen, 113–152. Berlin, Heidelberg: Springer Berlin Heidelberg.
6. Giljum, Stefan, Victor Maus, Laura Sonter, Sebastian Luckeneder, Tim Werner, Stephan Lutter, Julia Gershenzon, Megan J. Cole, Juliana Siqueira-Gay, and Anthony Bebbington. 2025. "Metal mining is a global driver of environmental change." *Nature Reviews Earth & Environment* 6 (7):441–455. doi: 10.1038/s43017-025-00683-w.
7. Lessler, Milton A. 1988. "Lead and Lead Poisoning from Antiquity to Modern Times." *Ohio Journal of Science* 88 (3):78–84.
8. Röscheisen, Helmut. 2005. "Der Deutsche Naturschutzring - Geschichte, Interessensvielfalt, Organisationsstruktur und Perspektiven." *Forschungsjournal Soziale Bewegungen* 18 (3):7–18. doi: doi:10.1515/fjsb-2005-0303.
9. Worster, Donald. 2005. "John Muir and the Modern Passion for Nature." *Environmental History* 10 (1):8–19.
10. Winiwarter, Verena, and Hans-Rudolf Bork. 2015. *Geschichte unserer Umwelt.* Freiburg: wbg.
11. Nriagu, Jerome O, and Jozef M Pacyna. 1988. "Quantitative assessment of worldwide contamination of air, water and soils by trace metals." *Nature* 333 (6169):134–139.
12. Moore, Johnnie N., and Samuel N. Luoma. 1990. "Hazardous wastes from large-scale metal extraction. A case study." *Environmental Science & Technology* 24 (9):1278–1285. doi: 10.1021/es00079a001.
13. USEPA. 1994. *Acid Mine Drainage Prediction*, edited by U.S. Environmental Protection Agency. Washington, DC.
14. Park, I., C. B. Tabelin, S. Jeon, X. Li, K. Seno, M. Ito, and N. Hiroyoshi. 2019. "A review of recent strategies for acid mine drainage prevention and mine tailings recycling." *Chemosphere* 219:588–606. doi: 10.1016/j.chemosphere.2018.11.053.
15. McCarthy, Terence. 2011. "The impact of acid mine drainage in South Africa." *South African Journal of Science* 107 (5/6):7 pages.
16. Wildeman, Thomas, and Ron Schmiermund. 2004. "Mining influenced waters: Their chemistry and methods of treatment." *Journal of American Society of Mining and Reclamation* 2004. doi: 10.21000/JASMR0402001.
17. Dold, Bernhard. 2014. "Evolution of acid mine drainage formation in sulphidic mine tailings." *Minerals* 4 (3):621–641.
18. Bryner, L. C., R. B. Walker, and Palmer R. 1968. "Minerals Beneficiation: Some Factors Influencing the Biological and Non-Biological Oxidation of Sulfide Minerals." The American Institute of Mining, Metallurgical, and Petroleum Engineers.
19. Dold, Bernhard. 2010. "Basic Concepts in Environmental Geochemistry of Sulfidic Mine-Waste Management." In *Waste Management*, edited by Sunil Kumar. London: IntechOpen.
20. Ritchie, A.I.M. 1994. "Sulfide Oxidation Mechanisms: Controls and Rates of Oxygen Transport." In *Short Course Handbook on Environmental Geochemistry of Sulfide Mine-Waste*, edited by David W. Blowes, J. L. Jambor and Charles N. Alpers, 201–244. Nepean, Ont., Canada: Mineralogical Association of Canada.
21. Simate, Geoffrey S., and Sehliselo Ndlovu. 2014. "Acid mine drainage: Challenges and opportunities." *Journal of Environmental Chemical Engineering* 2 (3):1785–1803. doi: https://doi.org/10.1016/j.jece.2014.07.021.
22. Miranda, Lorena S., Buddhi Wijesiri, Godwin A. Ayoko, Prasanna Egodawatta, and Ashantha Goonetilleke. 2021. "Water-sediment interactions and mobility of heavy metals in aquatic environments." *Water Research* 202:117386. doi: https://doi.org/10.1016/j.watres.2021.117386.

23. Bigham, J. M., U. Schwertmann, S. J. Traina, R. L. Winland, and M. Wolf. 1996. "Schwertmannite and the chemical modeling of iron in acid sulfate waters." *Geochimica et Cosmochimica Acta* 60 (12):2111–2121. https://doi.org/10.1016/0016-7037(96)00091-9.
24. Bigham, J. M., and D. Kirk Nordstrom. 2000. "Iron and aluminum hydroxysulfates from acid sulfate waters." *Reviews in Mineralogy and Geochemistry* 40 (1):351–403. doi: 10.2138/rmg.2000.40.7.
25. Cravotta III, Charles A., and Mary Kay Trahan. 1999. "Limestone drains to increase pH and remove dissolved metals from acidic mine drainage." *Applied Geochemistry* 14 (5):581–606. https://doi.org/10.1016/S0883-2927(98)00066-3.
26. Mayes, W. M., H. A. Potter, and A. P. Jarvis. 2010. "Inventory of aquatic contaminant flux arising from historical metal mining in England and Wales." *Sci Total Environ* 408 (17):3576–83. doi: 10.1016/j.scitotenv.2010.04.021.
27. Windisch, Jakob, Andreas Gradwohl, Beric Michael Gilbert, Quinton Marco Dos Santos, Gabriele Wallner, Annemarie Avenant-Oldewage, and Franz Jirsa. 2022. "Toxic Elements in Sediment and Water of the Crocodile River (West) System, South Africa, Following Acid Mine Drainage." *Applied Sciences* 12 (20):10531.
28. Sorensen, Paul. 2011. "Mining in South Africa: a mature industry?" *International Journal of Environmental Studies* 68 (5):625–649. doi: 10.1080/00207233.2011.591242.
29. Kirk, Jason, Joaquin Ruiz, John Chesley, Spencer Titley, and Spence Titley. 2003. "The Origin of Gold in South Africa: Ancient rivers filled with gold, a spectacular upwelling of magma and a colossal meteor impact combined to make the Witwatersrand basin a very special place." *American Scientist* 91 (6):534–541.
30. Stoch, Eliezer Joshua, and Frank Winde. 2010. "Threats and opportunities for post-closure development in dolomitic gold mining areas of the West Rand and Far West Rand (South Africa) - a hydraulic view part 3: planning and uncertainty – lessons from history." *Water SA* 36 (1):83–88. doi: 10.4314/wsa.v36i1.50910.
31. Durand, Jf, J. Meeuvis, and Marike Fourie. 2010. "The threat of mine effluent to the UNESCO status of the Cradle of Humankind World Heritage Site." *The Journal for Transdisciplinary Research in Southern Africa* 6. doi: 10.4102/td.v6i1.125.
32. Cobbing, Jude. 2008. "Institutional linkages and acid mine drainage: the case of the Western basin in South Africa." *International Journal of Water Resources Development* 4:451–462. doi: 10.1080/07900620802127374.
33. Coetzee, Henk, Uwe Horstmann, Gabriel Ntsume, and Leon Croukamp. 2003. "The potential environmental impact of the decant of water from Witwatersrand." 8th International Congress on Mine Water & the Environment, Johannesburg.
34. Hobbs, P. J., & Cobbing, J. E. (2007). Hydrogeological assessment of acid mine drainage impacts in the West Rand Basin, Gauteng Province.
35. Akcil, Ata, and Soner Koldas. 2006. "Acid Mine Drainage (AMD): causes, treatment and case studies." *Journal of Cleaner Production* 14 (12):1139–1145. doi: https://doi.org/10.1016/j.jclepro.2004.09.006.
36. Mugova, Elke, and Christian Wolkersdorfer. 2022. "Identifying potential groundwater contamination by mining influenced water (MIW) using flow measurements in a sub-catchment of the "Cradle of Humankind" Unesco World Heritage Site, South Africa." *Environmental Earth Sciences* 81 (3):104. doi: 10.1007/s12665-022-10224-z.
37. Dzwairo, Bloodless, and Munyaradzi Mujuru. 2017. "Assessment of Impacts of Acid Mine Drainage on Surface Water Quality of Tweelopiespruit Micro-Catchment, Limpopo Basin." In *Water Quality*, edited by Hlanganani Tutu. IntechOpen.
38. Wolkersdorfer, C., & von Hünefeld-Mugova, E. (2018). "Salt dilution flow measurement in the Tweelopiespruit rivulet, Witwatersrand, South Africa." In *Mine Water – Risk to Opportunity* (Vol II), edited by Wolkersdorfer, Ch.; Sartz, L.; Weber, A.; Burgess, J. & Tremblay, G., pp. 1024–1027. Pretoria, South Africa: Tshwane University of Technology.

39. Windisch, Jakob, Andreas Gradwohl, Beric Michael Gilbert, Quinton Marco Dos Santos, Annemariè Avenant-Oldewage, and Franz Jirsa. 2025. "Metals and metalloids in macrophytes and fish from an acid mine drainage-impacted river system in South Africa: aspects of bioindication and phytoremediation." *Environmental Science: Advances* 4 (9):1501–1513. doi: 10.1039/D5VA00136F.
40. DWAF. 1996. *South African Water Quality Guidelines, Aquatic Ecosystems*, edited by CSIR (Council for Scientific and Industrial Research) Environmental Services Pretoria Department of Water Affairs and Forestry, Republic of South Africa, Pretoria.
41. ARMCANZ. 2000. Australian and New Zealand Guidelines for Fresh and Marine Water Quality. edited by Australian and New Zealand Environment and Conservation Council and the Agriculture and Resource Management Council of Australia and New Zealand. Canberra.
42. CCME. 1999. *Canadian Sediment Quality Guidelines for the Protection of Aquatic Life*, edited by Canadian Council of Ministers of the Environment. Winnipeg, MB, Canada.
43. Filipek, L. H., D. K. Nordstrom, and W. H. Ficklin. 1987. "Interaction of acid mine drainage with waters and sediments of West Squaw Creek in the West Shasta Mining District, California." *Environ Sci Technol* 21 (4):388–96. doi: 10.1021/es00158a009.
44. Cruz-Hernández, Pablo, Rafael Peréz-López, and José Miguel Nieto. 2017. "Role of arsenic during the aging of acid mine drainage precipitates." *Procedia Earth and Planetary Science* 17:233–236. https://doi.org/10.1016/j.proeps.2016.12.079.
45. Paikaray, Susanta. 2015. "Arsenic Geochemistry of Acid Mine Drainage." *Mine Water and the Environment* 34 (2):181–196. doi: 10.1007/s10230-014-0286-4.
46. Singh, P., A. Nel, and J. F. Durand. 2017. "The use of bioassays to assess the toxicity of sediment in an acid mine drainage impacted river in Gauteng (South Africa)." *Water SA* 43 (4):673–683. doi: 10.4314/wsa.v43i4.15.
47. Raji, I. B., E. Hoffmann, A. Ngie, and F. Winde. 2021. "Assessing Uranium Pollution Levels in the Rietspruit River, Far West Rand Goldfield, South Africa." *International Journal of Environmental Research and Public Health* 18 (16). doi: 10.3390/ijerph18168466.
48. Bonanno, Giuseppe, and Jan Vymazal. 2017. "Compartmentalization of potentially hazardous elements in macrophytes: Insights into capacity and efficiency of accumulation." *Journal of Geochemical Exploration* 181:22–30.
49. Van Deventer, Heidi, and Moses A Cho. 2014. "Assessing leaf spectral properties of Phragmites australis impacted by acid mine drainage." *South African Journal of Science* 110 (7–8):1–12.
50. Klink, Agnieszka, Aurelia Macioł, Magdalena Wisłocka, and Józef Krawczyk. 2013. "Metal accumulation and distribution in the organs of *Typha latifolia* L.(cattail) and their potential use in bioindication." *Limnologica* 43 (3):164–168.
51. Duman, F., M. Cicek, and G. Sezen. 2007. "Seasonal changes of metal accumulation and distribution in common club rush (*Schoenoplectus lacustris*) and common reed (*Phragmites australis*)." *Ecotoxicology* 16(6), 457–463.
52. Parzych, Agnieszka Edyta and Zbigniew Sobisz. 2024. "Accumulation of Macronutrients and Trace Elements in Leaves and Stems of *Reynoutria japonica* and *R. sachalinensis*" *Journal of Ecological Engineering* 25 (11):306–316. doi: 10.12911/22998993/193408.
53. Klink, A. (2017). A comparison of trace metal bioaccumulation and distribution in Typha latifolia and Phragmites australis: implication for phytoremediation. *Environmental Science and Pollution Research* 24(4), 3843–3852.
54. Kamari, A., Yusof, N., Abdullah, H., Haraguchi, A., & Abas, M. F. 2017. "Assessment of heavy metals in water, sediment, *Anabas testudineus* and *Eichhornia crassipes* in a former mining pond in Perak, Malaysia." *Chemistry and Ecology* 33 (7):637–651. https://doi.org/10.1080/02757540.2017.1351553.

55. Avenant-Oldewage, A., and HM Marx. 2000. "Bioaccumulation of chromium, copper and iron in the organs and tissues of *Clarias gariepinus* in the Olifants River, Kruger National Park." *Water SA* 26 (4):569–582.
56. Coetzee, L, HH Du Preez, and JHJ Van Vuren. 2002. "Metal concentrations in *Clarias gariepinus* and *Labeo umbratus* from the Olifants and Klein Olifants River, Mpumalanga, South Africa: Zinc, copper, manganese, lead, chromium, nickel, aluminium and iron." *Water SA* 28 (4):433–448.
57. Outa, James Omondi, Chrispin O. Kowenje, Annemariè Avenant-Oldewage, and Franz Jirsa. 2020. "Trace Elements in Crustaceans, Mollusks and Fish in the Kenyan Part of Lake Victoria: Bioaccumulation, Bioindication and Health Risk Analysis." *Archives of Environmental Contamination and Toxicology* 78:589–603. doi: 10.1007/s00244-020-00715-0.
58. Gilbert, Beric M, Ebrahim Hussain, Franz Jirsa, and Annemariè Avenant-Oldewage. 2017. "Evaluation of trace element and metal accumulation and edibility risk associated with consumption of *Labeo umbratus* from the Vaal Dam, South Africa." *International Journal of Environmental Research and Public Health* 14 (7):678.
59. Hofer, R., and R. Lackner. 1995. *Fischtoxikologie: Theorie und Praxis*. Fischer.
60. Chatterjee, S, Buddhadeb Chattopadhyay, and SK Mukhopadhyay. 2006. "Trace metal distribution in tissues of cichlids (*Oreochromis niloticus* and *O. mossambicus*) collected from wastewater-fed fishponds in East Calcutta Wetlands, a Ramsar site." *Acta Ichthyologica et Piscatoria* 36:119–125.
61. Crafford, D., and A. Avenant-Oldewage. 2011. "Uptake of selected metals in tissues and organs of *Clarias gariepinus* (sharptooth catfish) from the Vaal River system: chromium, copper, iron, manganese and zinc." *Water SA* 37 (2):181–200.
62. Gilbert, BM, and A Avenant-Oldewage. 2014. "Arsenic, chromium, copper, iron, manganese, lead, selenium and zinc in the tissues of the largemouth yellowfish, Labeobarbus kimberleyensis (Gilchrist and Thompson, 1913), from the Vaal Dam, South Africa, and associated consumption risks." *Water Sa* 40 (4):739–748.
63. Jabeen, Farhat, and Abdul Shakoor Chaudhry. 2013. "Metal uptake and histological changes in gills and liver of *Oreochromis mossambicus* inhabiting Indus River." *Pakistan Journal of Zoology* 45 (1):9–18.
64. Plessl, C., B. M. Gilbert, M. F. Sigmund, S. Theiner, A. Avenant-Oldewage, B. K. Keppler, and F. Jirsa. 2019. "Mercury, silver, selenium and other trace elements in three cyprinid fish species from the Vaal Dam, South Africa, including implications for fish consumers." *Science of the Total Environment* 659:1158–1167.
65. ICMM. 2025. Integrated Mine Closure. In *Good Practice Guide*. London: ICMM.
66. IISD. 2023. *Relinquishment of Closed Mine Sites: Policy steps for governments*. edited by The International Institute for Sustainable Development. Winipeg, Manitoba: International Institute for Sustainable Development.
67. Government of Canada.. 2009. Environmental Code of Practice for Metal Mines. edited by Environment Canada.
68. Government of Australia. 2025. *Guideline for Preparing Mine Closure Plans*, edited by Petroleum and Exploration Department of Mines. East Perth: Government of Western Australia.
69. EU. 2021. *Guidelines for Mine Closure Activities and Calculation and Periodic Adjustment of Financial Guarantees*. European Commission: Publications Office.
70. Cole, M., T. Chimbganda, M. Esau, A. Abrams, and J. Broadhurts. 2024. *Developing Mine Closure Risk Ratings and Post-closure Opportunity Framework for South Africa*. Cape Town: University of Cape Town.
71. Government of South Africa. 2022. *Mine Water Management Policy*, edited by Water and Sanitation. Pretoria: Republic of South Africa.

72. Pepper, Mia, Michael Hughes, and Yvonne Haigh. 2021. "Loophole or lifeline? The policy challenges of mines in care and maintenance." *The Extractive Industries and Society* 8 (3):100879. doi: https://doi.org/10.1016/j.exis.2021.01.014.
73. USDA, Forest Service. 1993. Acid Mine Drainage From Mines on the National Forests, A Management Challenge.
74. GAO. 2020. Abandoned hardrock mines, Information on Number of Mines, Expenditures, and Factors That Limit Efforts to Address Hazards. United States Government Accountability Office. Report Nr GAO-20-238.
75. Mauk, Jeffrey, Nick Karl, Justin S. Pierson, and Carma San Juan. 2025. The abandoned mine inventory of the United States—A brief summary. In *Fact Sheet*. Reston VA.

# 10 Use of Natural Materials for Arsenic and Antimony Removal from Water

*Ariana M.A. Pintor, Mariko A. Carneiro, Clara C. Costa, Isabella T. Tomasi, Ricardo M. Ferreira, and Cidália M.S. Botelho*

## 10.1 INTRODUCTION

Arsenic (As) and antimony (Sb) are metalloids of group 15 of the periodic table. They share chemical similarities with other elements of the same group, such as phosphorus and nitrogen, namely the tendency to form anionic oxygen species in water. Both As and Sb are commonly found in aqueous systems in the +3 or +5 oxidation states, with the species varying according to Eh and pH as shown in Figure 10.1.

As and Sb are highly toxic to living organisms because they disrupt important biological processes governed by N and P ions, which have similar chemical properties. In both cases, the trivalent state is estimated to be ten times more toxic than the pentavalent form [3]. In addition, As and Sb are classified as priority pollutants and pose various health risks, including cancer. The International Agency for Research on Cancer (IARC) categorizes As as a known carcinogen and Sb as a probable carcinogen [4].

For these reasons, the levels of As and Sb in drinking water are heavily regulated, with maximum limits, recommended by the World Health Organization, of 10 and 20 $\mu g\ L^{-1}$, respectively [5]. However, drinking water sources often exceed these limits due to both natural factors and human activities. High concentrations of As are found in groundwater in many regions worldwide due to the mineral characteristics of soils. This issue affects over 230 million people globally [6]. Conversely, Sb contamination is primarily associated with anthropogenic activities, as it is leached into water from soils near mining and smelting operations, shooting ranges, and other sources. In addition, Sb can be mobilized through weathering of Sb-bearing minerals and improper disposal of industrial waste, leading to its persistence in surface and groundwater systems [7].

In this context, it is very important to provide adequate treatment for drinking water to remove As and Sb in order to comply with maximum levels and reduce risks to human health. Many methods have been used for this purpose, including ion exchange, membrane separation, chemical oxidation, bioremediation, coagulation/flocculation, and adsorption. The latter techniques are often the most used due to their simplicity, low cost, and effectiveness. Removal of As has been studied more

DOI: 10.1201/9781003632092-11

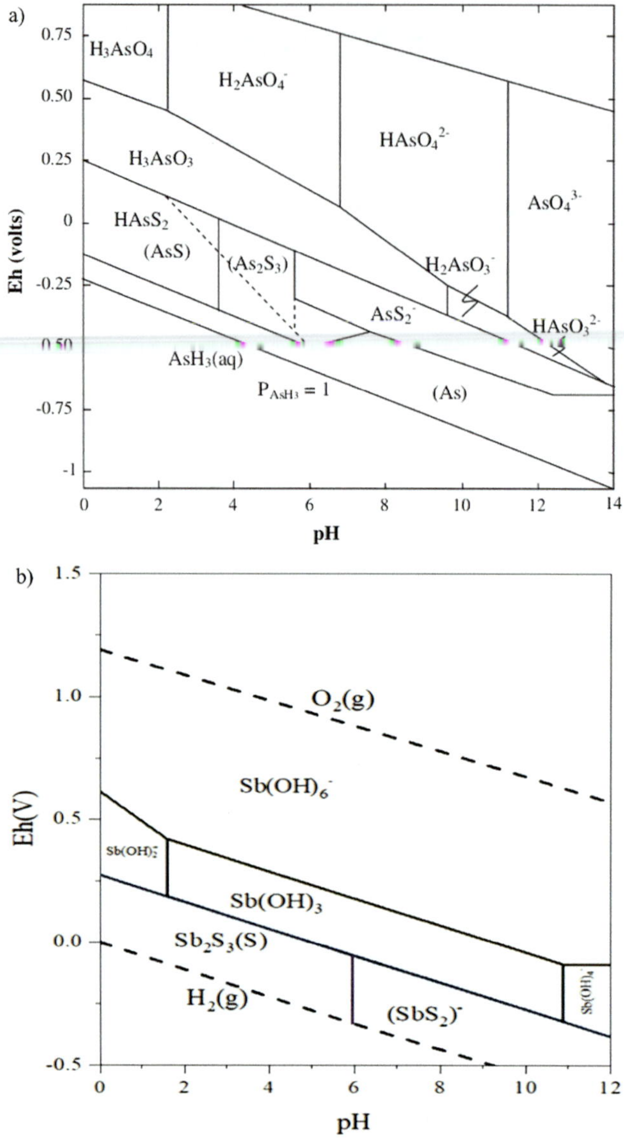


**FIGURE 10.1** Speciation according to Eh and pH for (a) As (reprinted from [1]) and (b) Sb (reprinted from [2]).

extensively because of the widespread nature of the As contamination issue in drinking water. However, most of the literature suggests that many technologies developed for removing As are also effective in removing Sb [4].

Previous studies clearly demonstrate that iron and aluminum are elements with a special affinity for As, and, due to chemical similarity, for Sb [2]. Therefore, coagulation using Fe and Al salts (such as chlorides, sulfates, and Fe or Al-based inorganic polymers) is a common removal methodology. Similarly, adsorbents like granular ferric hydroxide and activated alumina can be very effective in capturing As and Sb. However, the main drawbacks of these methods are the high material consumption and the large volume of generated sludge/waste, which is also highly toxic because it contains the entrapped metalloids and often requires costly, targeted treatment [8].

Current research trends in water and wastewater treatment favor the use of natural and organic materials to replace costly and potentially toxic chemicals. New coagulants have been developed based on organic polymers or natural substances, such as tannins [9]. While natural coagulants have often been tested for reducing turbidity or removing cationic metals, such as iron, zinc, copper, or cadmium, their application in the treatment of anionic metalloids has rarely been investigated. Existing studies mainly focus on mucilage extracted from *Opuntia ficus-indica*, which has shown promising results, achieving approximately 64% removal of As from tailings pond water and demonstrating higher efficiency compared with metallic coagulants [10]. A couple of studies have also referenced the use of chitosan-based coagulants [11, 12]. However, tannin-based adsorbents also appear to be a promising alternative for Sb removal by adsorption [13, 14], providing valuable insights for future studies on the use of tannins in the development of coagulants for As and Sb removal.

In adsorption processes, natural and waste materials have been explored as novel adsorbents due to their lower cost and environmental benefits. However, these materials often show limited efficiency without proper modifications [15]. Producing activated carbons typically involves thermal treatment, which consumes a large amount of energy. A common approach has been to dope or coat natural materials with elements that have high adsorption affinity, such as Fe and Al. This method offers the economic and environmental advantages of a natural carrier material combined with the high efficiency of targeted chemical interactions [16].

In this study, we examine novel coagulants and adsorbents made from natural materials for the removal of As and Sb, aiming to improve the environmental friendliness of the water treatment process without compromising its effectiveness. Tannin-based coagulants, including both commercial products (TANFLOC) and those derived from *Castanea sativa* shell extracts, were evaluated as alternatives and compared to traditional iron and aluminum salt coagulants. For adsorption, we used cork granulates as carriers for iron oxide coatings for the uptake of both As and Sb and compared their performance to granular ferric hydroxide adsorbents such as Bayoxide E33.

## 10.2 MATERIALS AND METHODS

### 10.2.1 Materials

#### 10.2.1.1 Arsenic Solution

As stock solutions were prepared from commercially available standards: As(III) (1003 ± 3 mg $L^{-1}$) obtained from $As_2O_3$ in 4% $HNO_3$ (SCP Science, Baie-D'Urfe, QC, Canada) and As(V) (994 ± 10 mg $L^{-1}$) prepared by dissolving $HAsNa_2O_4{\cdot}7H_2O$ (analytical grade, Sigma-Aldrich, St. Louis, MO, USA) in 2% HCl. These solutions were subsequently diluted to obtain the desired working concentrations.

#### 10.2.1.2 Antimony Solution

Antimony stock solutions (1 g $L^{-1}$ Sb) were prepared using antimony trichloride ($SbCl_3$, ACS reagent, Acrós Organics, Queluz, Portugal) dissolved in 0.1% (v/v) HCl for Sb(III), and potassium hexahydroxoantimonate ($KSb(OH)_6$, Aldrich, St. Louis, MO, USA) dissolved in distilled water for Sb(V). The stock solutions were subsequently diluted with distilled water or NaCl solutions to obtain the desired working concentrations.

### 10.2.2 Coagulation/Flocculation

#### 10.2.2.1 Tannin-based Coagulant Synthesis

Chestnut shells (*Castanea sativa*) were supplied by Agromontenegro, a chestnut dehydration company based in Trás-os-Montes, Portugal. The extraction was performed under solid–liquid conditions at 90°C for 10 min, with a liquid-to-solid ratio of 50 mL·$g^{-1}$, as described in previous studies [17]. Because tannins naturally behave as anionic compounds, the extracts obtained from chestnut shells were aminomethylated through the Mannich reaction using formaldehyde and $NH_4Cl$, following a method adapted from [18–20] using a tannin:$NH_4Cl$:formaldehyde molar ratio of 1:2.5:7.5. For this step, 2.5 g of freeze-dried tannin extract was dissolved in 6 mL of distilled water and stirred at about 55°C until a uniform solution was achieved. Separately, a 37% (w/w) aqueous formaldehyde solution (reagent grade, José Manuel Gomes dos Santos, Odivelas, Portugal) was combined with $NH_4Cl$ (reagent grade, VWR Chemicals, Carnaxide, Portugal) and refluxed for 3 h to ensure the formation of the iminium ion required for cationization. After cooling the Mannich solution (MS) to 55°C, it was added to the tannin solution. The reaction mixture was then heated to 85°C and gently stirred for 5 h. The produced coagulant will hereafter be referred to as CC.

#### 10.2.2.2 Jar Testing

Coagulation–flocculation–sedimentation (CFS) experiments were performed in batch mode using jar tests to evaluate the removal of arsenic and antimony from synthetic groundwater. The synthetic water was prepared by dispersing soil (6 g $L^{-1}$ for the coagulation tests with the chestnut coagulant and 3 g $L^{-1}$ for the other coagulants) in tap water, followed by vigorous stirring for 15 min. The mixture was then allowed to stand for about 1 h, after which the supernatant was collected. Arsenic and antimony were subsequently introduced at concentrations of 1 mg·$L^{-1}$ each, and this solution was used for the coagulation experiments. A programmable jar test unit (VELP, JLT6, Usmate, Italy) was employed for all trials. The assays were performed in 800 mL beakers, each containing 300 mL of the test solution. Preliminary trials were first conducted to determine the lowest coagulant dose capable of initiating visible floc formation across a pH range of 4 to 9. The pH of the water samples was adjusted with hydrochloric acid (HCl) and sodium hydroxide (NaOH) solutions at concentrations of 0.1 mol·$L^{-1}$ and 1.0 mol·$L^{-1}$. Coagulant additions were made stepwise in 50 μL increments. After each addition, the suspension was subjected to rapid mixing at 150 rpm for 1 min, followed by slow mixing at 20 rpm for 3 min, after which the presence of flocs was checked. A subsequent series of tests was designed to evaluate how coagulant dosage influenced treatment efficiency. Beginning with the minimum effective dose, progressively larger quantities of coagulant were introduced into the jars containing water adjusted to the previously determined optimal pH. The mixing–settling procedure was performed in three steps: rapid agitation at 150 rpm for 3 min, gentle stirring at 20 rpm for 15 min, and a final quiescent settling phase of 15 min. The performance of the produced coagulant (CC) was evaluated in comparison with conventional chemical coagulants, $Al_2(SO_4)_3{\cdot}14H_2O$ and $FeCl_3{\cdot}6H_2O$, as well as with Tanfloc (TF), a tannin-based coagulant derived from the *Acacia mearnsii* tree.

### 10.2.3 Adsorption

The adsorption of As and Sb was investigated in both batch and continuous modes using iron-coated cork granulates (ICG) as adsorbent. Extensive research on the adsorption of As and Sb oxyanions has been reported in several studies [21–28]. This chapter summarizes the effect of pH on the adsorption of As and Sb onto ICG in batch and continuous modes. In addition, the adsorption of arsenic from a groundwater matrix is presented. In-depth studies on the adsorption of As and Sb onto ICG, such as isotherms, kinetics, breakthrough curves, and thermodynamic studies are reported in the previous works cited above.

#### 10.2.3.1 Iron-Coated Cork Granulates

The adsorbent, ICG, was prepared using raw cork granulates with a particle size of 0.5-1.0 mm. The procedure was developed and optimized by Pintor, Vieira et al. [28] and adapted to a larger scale by Carneiro, Pintor et al. [22]. In brief, cork granulates were placed in contact with a 0.05 M $FeCl_3$ solution of pH 7.0, at 20 g $L^{-1}$, in a rotating shaker at 20 rpm, for 24 h, at room temperature. Following the coating time, the granulates were washed with distilled water until the excess iron was removed and dried in an oven for 24 h. The iron content was on average $25 \pm 2$ mg $g^{-1}$.

#### 10.2.3.2 Adsorption in Batch Mode

The effects of pH on the adsorption of As(III), As(V), Sb(III), and Sb(V) adsorption onto ICG were investigated in aqueous solution. For each assay, 45 mL of As or Sb solution was placed in contact with the ICG at an S/L ratio of 2.5 g $L^{-1}$, in a rotating shaker (20 rpm), at a controlled temperature of $20.0 \pm 0.5$ °C. After the desired contact times, the samples were filtered and total As and Sb were determined by FAAS.

#### 10.2.3.3 Adsorption in Continuous Mode

The adsorption of metalloids onto ICG was investigated in continuous mode, as adsorption is typically applied under continuous-flow conditions at full-scale plants. The adsorption in continuous mode was carried out in a lab installation at the Faculty of Engineering of the University of Porto (Portugal). The lab-scale adsorption system was the same as described by Carneiro, Pintor et al. [23], and it is illustrated in Figure 10.2. The optimal operational parameters of the lab-scale system were identified by Carneiro, Pintor et al. [23] for arsenic adsorption, and the same parameters were applied for antimony assays. The operating parameters

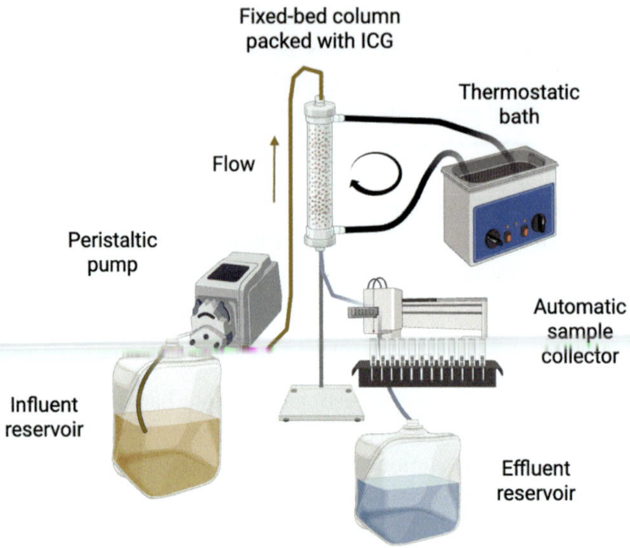


**FIGURE 10.2** Setup of the continuous adsorption system at laboratory scale. (Reprinted with permission from Carneiro, Silva et al. [24]. Copyright © Elsevier, 2024.

were as follows: flow rate: 5 mL $min^{-1}$, bed height: 12 cm, diameter: 2.5 cm, adsorbent mass: 6 g, temperature: 20°C, bed volume: 59 mL, empty bed contact time (EBCT): 12 min. The fixed-bed column operated in a downward flow due to the low density of the adsorbent.

Breakthrough time was defined as the time required to reach an effluent concentration of 10 µg $L^{-1}$ for As and 20 µg $L^{-1}$ for Sb, which are the maximum levels allowed in the European Union and the recommended guideline by the World Health Organization [5, 29].

### 10.2.4 Analytical Methods

The analyses of As and Sb were performed using ICP-OES (Inductively Coupled Plasma Optical Emission Spectroscopy) with a, to quantify metalloid concentrations in both raw and treated water samples. Measurements were carried out at wavelengths of 193.7 nm for As and 217.5 nm for Sb, within concentration ranges of 0.005–5.00 mg·$L^{-1}$ for As and 0.50–5.00 mg·$L^{-1}$ for Sb.

For batch adsorption studies, the metalloid concentration in solution was measured by flame atomic absorption spectrometry (FAAS) in the working range 3–50 mg $L^{-1}$ for As and 2–30 mg $L^{-1}$ for Sb.

## 10.3 RESULTS AND DISCUSSION

### 10.3.1 Arsenic and Antimony Removal Through Coagulation

The removal of As and Sb from a binary system through coagulation with different coagulants was investigated, and the results are shown in Figure 10.3. The optimal

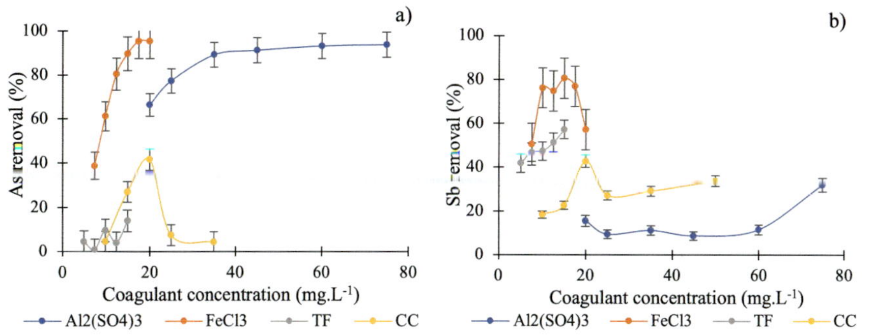


**FIGURE 10.3** Removal of (a) arsenic and (b) antimony in a binary system as a function of coagulant concentration. Uncertainties were determined from three independent measurements.

pH selected for $Al_2(SO_4)_3$ was 7, while an optimal pH of 5 was identified for the remaining three coagulants. For arsenic, the metallic coagulants performed similarly and effectively, achieving around 90% reduction at 15 mg·L$^{-1}$ for the $FeCl_3$ coagulant and 35 mg·L$^{-1}$ for the $Al_2(SO_4)_3$. However, for the removal of antimony, the results indicated substantially lower efficiency when the aluminum-based coagulant was used. In contrast, $FeCl_3$ maintained high efficiency, reaching approximately 80% reduction at 10 mg·L$^{-1}$, whereas the aluminum one achieved only 30% even at 75 mg·L$^{-1}$. These results are consistent with the literature, which reports that iron-based coagulants generally exhibit higher efficiency in the removal of both metalloids [4].

As shown in Figure 10.3, the Al-based coagulants demonstrated significantly higher efficiency in arsenic removal compared with antimony. This behavior can be attributed to differences in chemical speciation, adsorption affinity, and hydrolysis characteristics of the two elements. The removal of arsenic, particularly in its pentavalent form As(V), is promoted by the strong adsorption and coprecipitation [10, 30] of arsenate ions ($AsO_4^{3-}$) onto aluminum hydroxide flocs formed during the hydrolysis of $Al_2(SO_4)_3$ [31, 32]. Under the tested conditions, the coagulant achieved over 90% removal of As(V) at concentrations above 40 mg L$^{-1}$, whereas for Sb(V), although higher coagulant dosages enhanced removal, the maximum efficiency obtained was approximately 30%. When comparing these results with the removal rates obtained using the iron chloride coagulant, it can be confirmed that the latter is more efficient for both studied metalloids. Regarding the efficiency of $FeCl_3$, it was observed that concentrations above 15 mg L$^{-1}$ led to a decrease in Sb(V) removal. This reduction can be attributed to the lower floc formation observed at higher coagulant doses, as evidenced by the increase in initial turbidity from approximately 2 to nearly 4 NTU at 20 mg L$^{-1}$ of $FeCl_3$, indicating an excess of coagulant. When coagulants are applied in excess, charge re-estabilization of the suspended particles may occur, thereby reducing floc formation and overall removal efficiency [33]. Regarding the natural coagulants, Sb(V) removal was higher compared with As(V), in contrast to the behavior observed with the metallic ones. Among them, Tanfloc achieved nearly 60% removal of Sb(V) at a dosage of 15 mg L$^{-1}$, whereas As(V) removal was inconsistent, showing fluctuating values and reaching a maximum of approximately 14% at the same concentration. The chestnut shell-based coagulant showed a similar trend to Tanfloc. At a concentration of 20 mg $L^{-1}$, the CC coagulant exhibited the highest efficiency for both metalloids, achieving removal rates superior to 40%. However, further increases in CC dosage resulted in a decrease in removal efficiency, which can once again be attributed to reduced floc formation. At the optimal concentration, the CC coagulant reduced the initial turbidity by approximately 30%, whereas at higher dosages, the initial turbidity nearly doubled, indicating overdosage and subsequent floc destabilization.

### 10.3.2 Arsenic and Antimony Removal Through Adsorption onto ICG

#### 10.3.2.1 Adsorption in Batch Mode

The effect of pH for As and Sb oxyanion adsorption onto ICG is presented in Table 10.1.

The effect of pH on the As and Sb oxyanions adsorption is different. While the increase in pH caused an increase in adsorption capacity for the less oxidized species of As, it caused a reduction in the adsorption capacity for As(V) and Sb(V) and didn't affect the adsorption capacity of Sb(III). A different trend was observed in continuous adsorption, where the increase of pH led to the reduction of the adsorption capacity of both Sb species [23, 24].

Moreover, the adsorption capacities of antimony oxyanions are higher than the respective As species. This behavior is opposite to the one observed during coagulation of As and Sb, in which coagulation was more effective to remove As than Sb (Figure 10.4).

#### 10.3.2.2 Adsorption in Continuous Mode

The study of As and Sb adsorption onto ICG was also carried out in continuous mode. Figures 10.4 and 10.5 illustrate the effect of pH in arsenic and antimony adsorption from aqueous solution.

The trends were similar to those observed in batch mode, where the lower pH favored the adsorption of As(V) and both antimony species. Also, in general, the adsorption capacity of Sb was higher than As onto ICG.

**TABLE 10.1**
**Effect of pH for Adsorption of As(III), As(V), Sb(III), Sb(V) in Aqueous Solution**

| Species | pH range | Adsorption capacity (mg $L^{-1}$) | Reference |
|---|---|---|---|
| **As(III)** | 2–9 | ~1.0 to 3.7 mg $L^{-1}$ | [28] |
| **As(V)** | 2–10 | ~4.7 to 0.9 mg $L^{-1}$ | [28] |
| **Sb(III)** | 4–10 | ~3.9 to 4.0 mg $L^{-1}$ | [26] |
| **Sb(V)** | 2–10 | ~6.8 to 0.5 mg $L^{-1}$ | [26] |

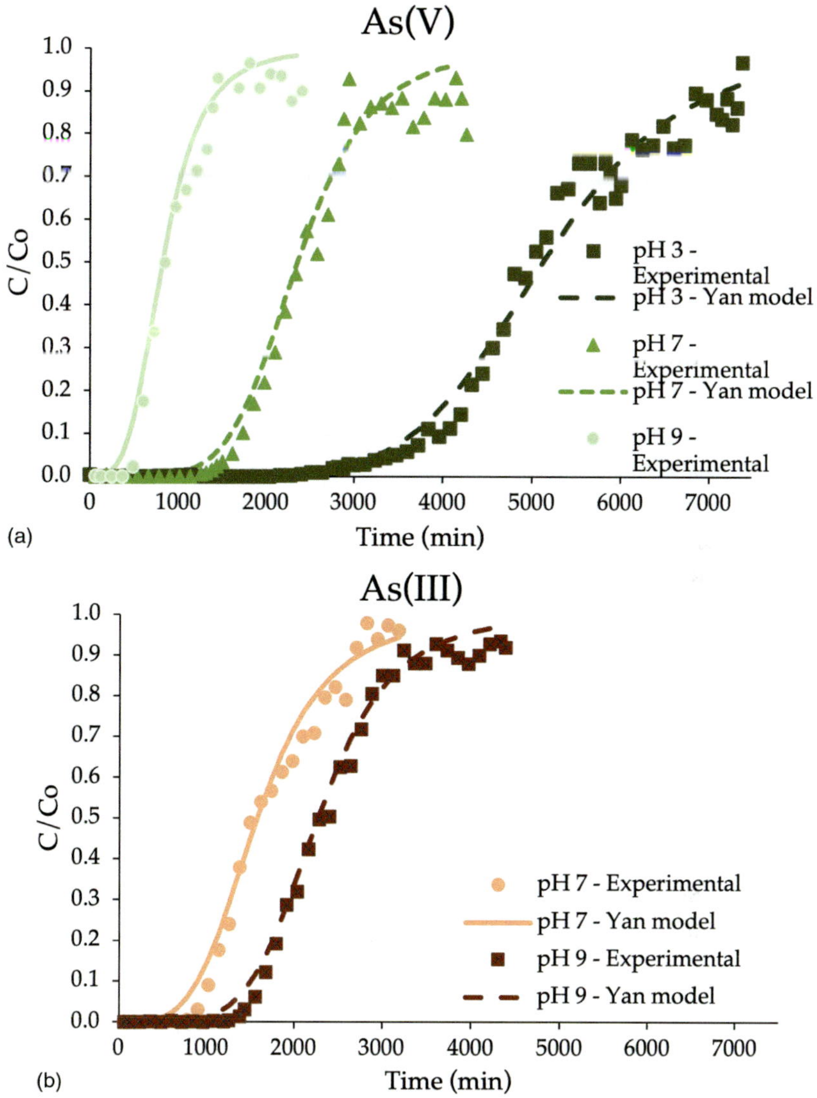


**FIGURE 10.4** Experimental and theoretical (Yan model) breakthrough curves of arsenic adsorption onto ICG for different pH values: (a) As(III), pH of 7 and 9; (b) As(V) pH of 3, 7 and 9. (Dots = experimental data, lines = Yan model predicted curves) (Inlet As(III)/As(V) concentration 1 mg/L; Bed height: 12 cm; Flow rate: 5 mL min$^{-1}$; Temperature: 20°C). (Reprinted with permission from Carneiro, Pintor et al. [23].)

The removal efficiency of As and Sb through adsorption onto ICG exceeded 99% before column breakthrough [23, 34]. This efficiency is higher than that reported in section 10.3.1 for coagulation. The combination of coagulation and adsorption to remove arsenic and antimony from water may be more cost-effective and can minimize the impacts and waste generation inherent to each method.

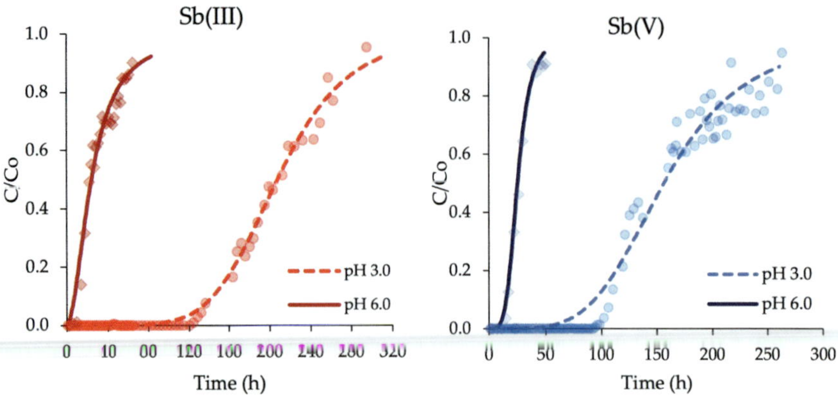


**FIGURE 10.5** pH effect on the Sb(III) and Sb(V) removal in the fixed-bed column. (Dots = experimental data, lines = Yan model predicted curves) (Inlet Sb(III)/Sb(V) concentration 1 mg/L; Bed height: 12 cm; Flow rate: 5 mL min$^{-1}$; Temperature: 20°C). (Reprinted with permission from Carneiro, Silva et al. [24].)

Further investigation of As adsorption onto ICG was carried out using groundwater matrices to validate the method in a real scenario. Table 10.2 shows the summary of the assays carried out in continuous mode using groundwater matrices. The fixed-bed column performance was evaluated by breakthrough curves (BTC), according to the following parameters: breakthrough time ($t_b$), adsorption capacity at breakthrough ($q_b$), bed volume treated at breakthrough ($BV_b$), breakthrough volume ($V_b$),

**TABLE 10.2**
**Experimental parameters of arsenic continuous adsorption from different groundwater matrices onto ICG (flow rate 5 mL min$^{-1}$ and 6 g of ICG, EBCT of 12 min, breakthrough point of 10 µg L$^{-1}$)**

| Feed | Treated As-Spiked Groundwater | Untreated As-Spiked Groundwater | Untreated As-Spiked Groundwater | Untreated Groundwater |
|---|---|---|---|---|
| **Groundwater** | The Netherlands | The Netherlands | Portugal | Portugal |
| **pH** | 7.6 ± 0.2 | 7.4 ± 0.1 | 7.09 ± 0.01 | 6.95 |
| **$C_0$, average (mg L$^{-1}$)** | 0.3 ± 0.1 | 0.18 ± 0.04 | 0.1 | 0.011 ± 0.002 |
| **$t_b$ (h)** | 31 ± 1 | 36 ± 1 | 26 ± 2 | 321 |
| **$BV_b$** | 202 ± 27 | 189 ± 6 | 140 ± 13 | 1706 ± 78 |
| **$V_b$ (L)** | 9.3 ± 0.3 | 10.9 ± 0.4 | 7.9 ± 0.5 | 96.3 ± 0.1 |
| **$q_b$ (mg g$^{-1}$)** | 0.45 ± 0.07 | 0.33 ± 0.08 | 0.12 ± 0.01 | 0.09 ± 0.04 |
| **$y_b$ (%)** | 96.9 ± 0.6 | 100.0 | 99.2 ± 0.6 | 53 ± 14 |
| **Ref.** | [22] | [22] | [21] | [22] |

*Source:* From [22].

and metalloid removal at breakthrough ($y_b$). The initial metalloid concentration ($C_o$) varied in each assay.

The performance of new adsorbents in real water is not frequently discussed in the literature. Compared to other modified biosorbents like iron-chitosan spacer granulates (ICS) [35], xanthated chitosan flakes [36], and iron chitosan granules (ICB) [37], ICG performed better. Additionally, compared to biochars and nanobio chars, the adsorption capacity was higher [38].

In another study, the performance of ICG was compared to commercial adsorbent Bayoxide E33 [34]. ICG exhibited lower performance, treating 1.233 L $g^{-1}$ of arsenic-contaminated water, compared with 3.257 L $g^{-1}$ for Bayoxide E33. This outcome is expected, as the active sites of iron hydroxides account for approximately 3% of the mass of ICG, whereas they constitute approximately 99% of the mass of Bayoxide E33. However, ICG showed superior environmental performance. A Life Cycle Assessment (LCA) showed that the treatment of arsenic-contaminated water using ICG had a lower impact on 13 out of 18 categories compared to the use of the commercial adsorbent Bayoxide E33 [34].

## 10.4 CONCLUSIONS

This study evaluated the use of natural materials for the removal of As and Sb from aqueous systems using both coagulation and adsorption methods. The results demonstrate that removal efficiency strongly depends on the treatment method, coagulant or adsorbent type, pH conditions, and the chemical speciation of the metalloids.

Novel tannin coagulants were tested for both As and Sb removal, but with limited efficiency when compared to metallic coagulants. Among the latter, iron-based coagulants presented higher efficiencies when compared with aluminum-based coagulants in the removal of both As and Sb in binary mixtures. $FeCl_3$ was highly effective with removal percentages of around 90% for As and 80% for Sb, requiring a comparatively lower dosage. $Al_2(SO_4)_3$ was effective in the removal of As, but not significantly effective in the removal of Sb, even at higher dosages. Natural coagulants tended to behave in a different manner, showing high selectivity toward Sb rather than As. The tannin coagulants from chestnut shell were promising at optimum dosage levels, removing both metalloids, but showed lower efficacy at higher dosages owing to the breakdown of flocs.

Adsorption studies presented a summary of the research done using ICG, a novel, cork-based material with affinity for both As and Sb. The results demonstrated that removal efficiency is strongly influenced by pH and metalloid speciation. In general, Sb was adsorbed in larger amounts than As, and an increasing pH enhanced As(III) adsorption but reduced As(V), Sb(III), and Sb(V) removal. These trends were confirmed under continuous fixed-bed conditions, with the material showing a stable adsorption performance, with lower pH favoring the removal of As(V) and Sb.

Overall, iron-based coagulation and adsorption onto ICG are effective for As and Sb removal, with $FeCl_3$ suitable for rapid treatment of binary systems and ICG representing a promising bio-based adsorbent for continuous groundwater treatment.

Future work should explore the application and scale-up of the adsorption technology using cork-based adsorbents. Experiments using natural groundwater matrices have already shown that ICG has good potential for As decontamination, competing with existing adsorbents in terms of adsorption capacity and environmental impact. An on-site pilot column study should be carried out to verify the viability of this technology for drinking water treatment at a larger scale.

## REFERENCES

1. Wang, S., & Mulligan, C. N. (2006). "Occurrence of arsenic contamination in Canada: Sources, behavior and distribution." Science of the Total Environment, **366**(2): 701–721. doi: 10.1016/j.scitotenv.2005.09.005
2. Xiang, S., Z. Fu, H. Lu, Y. Sun, Y. Shen and F. Wu (2025). "Antimony and arsenic interactions with iron oxides and aluminum oxides in surface environment: A review focused on processes and mechanisms." Science of The Total Environment **979**: 179423.
3. Guo, X., Wu, Z., He, M., Meng, X., Jin, X., Qiu, N., & Zhang, J. (2014). Adsorption of antimony onto iron oxyhydroxides: Adsorption behavior and surface structure. Journal of Hazardous Materials **276**: 339–345.
4. Ungureanu, G., S. Santos, R. Boaventura and C. Botelho (2015). "Arsenic and antimony in water and wastewater: Overview of removal techniques with special reference to latest advances in adsorption." Journal of Environmental Management **151**: 326–342.
5. WHO (2022). *Guidelines for drinking-water quality: incorporating the first and second addenda*. Geneva, World Health Organization.
6. Shaji, E., M. Santosh, K. V. Sarath, P. Prakash, V. Deepchand and B. V. Divya (2021). "Arsenic contamination of groundwater: A global synopsis with focus on the Indian Peninsula." Geoscience Frontiers **12**(3): 101079.
7. Filella, M., N. Belzile and Y.-W. Chen (2002). "Antimony in the environment: a review focused on natural waters: I. Occurrence." Earth-Science Reviews **57**(1): 125–176.
8. Ghosh, S., C. A. Igwegbe, A. Malloum, M. A. E. Elmakki, H. Onyeaka, A. H. Fahmy, M. O. Aquatar, S. Ahmadi, B. M. Alameri, S. Ghosh, N. A. Khan, L. Singh, N. M. Mubarak and M. H. Dehghani (2025). "Sustainable technologies for removal of arsenic from water and wastewater: A comprehensive review." Journal of Molecular Liquids **427**: 127412.
9. Tomasi, I. T., C. A. Machado, R. A. R. Boaventura, C. M. S. Botelho and S. C. R. Santos (2022). "Tannin-based coagulants: Current development and prospects on synthesis and uses." Science of The Total Environment **822**: 153454.
10. Wan, J., T. Chakraborty, C. C. Xu and M. B. Ray (2019). "Treatment train for tailings pond water using *Opuntia ficus-indica* as coagulant." Separation and Purification Technology **211**: 448–455.
11. Gan, Y., C. Ding, B. Xu, Z. Liu, S. Zhang, Y. Cui, B. Wu, W. Huang and X. Song (2023). "Antimony (Sb) pollution control by coagulation and membrane filtration in water/wastewater treatment: A comprehensive review." Journal of Hazardous Materials **442**: 130072.
12. Venegas-García, D. J., B. G. K. Steiger and L. D. Wilson (2023). "A pyridinium-modified chitosan-based adsorbent for arsenic removal via a coagulation-like methodology." RSC Sustainability **1**(5): 1259–1269.
13. Bacelo, H., S. C. R. Santos, A. Ribeiro, R. A. R. Boaventura and C. M. S. Botelho (2022). "Antimony removal from water by pine bark tannin resin: Batch and fixed-bed adsorption." Journal of Environmental Management **302**: 114100.
14. Bacelo, H., B. R. C. Vieira, S. C. R. Santos, R. A. R. Boaventura and C. M. S. Botelho (2018). "Recovery and valorization of tannins from a forest waste as an adsorbent for antimony uptake." Journal of Cleaner Production **198**: 1324–1335.

15. Zoroufchi Benis, K., A. Motalebi Damuchali, K. N. McPhedran and J. Soltan (2020). "Treatment of aqueous arsenic: A review of biosorbent preparation methods." Journal of Environmental Management **273**: 111126.
16. Hao, L., M. Liu, N. Wang and G. Li (2018). "A critical review on arsenic removal from water using iron-based adsorbents." RSC Advances **8**(69): 39545–39560.
17. Tomasi, I. T., R. A. R. Boaventura and C. M. S. Botelho (2024). "Solid-Liquid Extraction of Polyphenols from Chestnut Shells: A Sustainable Approach for Coagulant Synthesis." Sustainable Chemistry and Pharmacy **42**: 101806.
18. Lamb, L. H. and O. G. Decusati (2002). Manufacturing Process for Quaternary Ammonium Tannate, a Vegetable Coagulant/Flocculant Agent. Brazil, Tanac S.A. **6478986**.
19. Quamme, J. E. and A. H. Kemp (1985). Stable Tannin Based Polymer Compound. USA, Dearborn Chemical Company.
20. Tomasi, I. T., S. C. R. Santos, A. Ribeiro, V. Homem, R. A. R. Boaventura and C. M. S. Botelho (2025). "Coagulants from chestnut shell tannins: Synthesis, characterization and performance on water treatment." Journal of Water Process Engineering **69**: 106818.
21. Carneiro, M. A., J. F. R. Coelho, A. M. A. Pintor, R. A. R. Boaventura and C. M. S. Botelho (2022). "Multi-cycle regeneration of arsenic-loaded iron-coated cork granulates for water treatment." Journal of Water Process Engineering **50**: 103291.
22. Carneiro, M. A., A. M. A. Pintor, L. L. F. Agostinho, R. A. R. Boaventura and C. M. S. Botelho (2025). "Advances in the treatment of arsenic-contaminated groundwater by the adsorption onto iron-coated cork granulates on fixed-bed column." Groundwater for Sustainable Development **30**: 101488.
23. Carneiro, M. A., A. M. A. Pintor, R. A. R. Boaventura and C. M. S. Botelho (2022). "Efficient removal of arsenic from aqueous solution by continuous adsorption onto iron-coated cork granulates." Journal of Hazardous Materials: 128657.
24. Carneiro, M. A., T. A. Silva, P. J. S. Teixeira, A. M. A. Pintor, R. A. R. Boaventura and C. M. S. Botelho (2024). "New insights on antimony removal and recovery from water and wastewater using iron-coated cork granulates: Desorption on batch and fixed-bed column systems." Chemical Engineering Science: 120207.
25. Pintor, A. M. A., C. C. Brandão, R. A. R. Boaventura and C. M. S. Botelho (2021). "Multicomponent adsorption of pentavalent As, Sb and P onto iron-coated cork granulates." Journal of Hazardous Materials **406**: 124339.
26. Pintor, A. M. A., B. R. C. Vieira, R. A. R. Boaventura and C. M. S. Botelho (2020). "Removal of antimony from water by iron-coated cork granulates." Separation and Purification Technology **233**: 116020.
27. Pintor, A. M. A., B. R. C. Vieira, C. C. Brandão, R. A. R. Boaventura and C. M. S. Botelho (2020). "Complexation mechanisms in arsenic and phosphorus adsorption onto iron-coated cork granulates." Journal of Environmental Chemical Engineering **8**(5): 104184.
28. Pintor, A. M. A., B. R. C. Vieira, S. C. R. Santos, R. A. R. Boaventura and C. M. S. Botelho (2018). "Arsenate and arsenite adsorption onto iron-coated cork granulates." Science of The Total Environment **642**: 1075–1089.
29. European Union, C. (2020). "Directive (EU) 2020/2184 of the European Parliament and of the Council of 16 December 2020 on the Quality of Water Intended for Human Consumption (Recast)." Off. JL **435**: 1–62.
30. Dias, A. C. and M. P. F. Fontes (2020). "Arsenic (V) removal from water using hydrotalcites as adsorbents: A critical review." Applied Clay Science **191**.
31. Baskan, M. B. and A. Pala (2009). "Determination of arsenic removal efficiency by ferric ions using response surface methodology." Journal of Hazardous Materials **166**(2–3): 796–801.

32. Yang, N., X. Qi, Y. Li, G. Li and X. Duan (2023). "Highly effective remediation of high arsenic-bearing wastewater using aluminum-containing waste residue." Journal of Environmental Management **325**: 116417.
33. Santos, J. D. d., M. T. Veit, P. T. Juchen, G. d. C. Gonçalves, S. M. Palácio and M. Fagundes-Klen (2018). "Use of different coagulants for cassava processing wastewater treatment." Journal of Environmental Chemical Engineering **6**(2): 1821–1827.
34. Carneiro, M. A., I. T. Tomasi, R. A. R. Boaventura, C. M. S. Botelho and A. M. A. Pintor (2025). "Environmental impact and cost analysis of arsenic removal from water using iron-based adsorbents: Comparison between natural and commercial materials." Journal of Environmental Chemical Engineering **13**(1): 115044.
35. Gupta, A. and N. Sankararamakrishnan (2010). "Column studies on the evaluation of novel spacer granules for the removal of arsenite and arsenate from contaminated water." Bioresource Technology **101**(7): 2173–2179.
36. Gupta, A., M. Yunus and N. Sankararamakrishnan (2012). "Equilibrium and dynamic studies of the removal of As(III) and As(V) from contaminated aqueous systems using a functionalized biopolymer." Journal of Chemical Technology & Biotechnology **87**(4): 546–552.
37. Gupta, A., V. S. Chauhan and N. Sankararamakrishnan (2009). "Preparation and evaluation of iron–chitosan composites for removal of As(III) and As(V) from arsenic-contaminated real-life groundwater." Water Research **43**(15): 3862–3870.
38. Kumari Sharma, P., R. Kumar Singh, R. Kumar, N. Kumar, A. Ghosh, P. Sharma and A. Kumar (2023). "Synthesis and exploration of physical properties of nanobiochar synthesized from rice straw for its applications in arsenic remediation from contaminated water environments." *Materials Today: Proceedings*, 113, 299–306. https://doi.org/10.1016/j.matpr.2023.10.030.

# 11 Dermal Bioaccessibility and Percutaneous Absorption of Metal(loid)s in Contaminated Soils

## *Mechanisms, Experimental Evidence, and Implications for Human Health Exposure Assessment*

*Gerald J. Zagury*

## 11.1 INTRODUCTION: RECONSIDERING THE DERMAL PATHWAY

Human exposure to metal(loid)s in contaminated soils has traditionally been evaluated primarily through ingestion and inhalation pathways which dominate most human health exposure and risk assessment frameworks. This emphasis is justified in some cases, particularly for children, for whom soil ingestion represents a major exposure route, and for adults in occupational settings where inhalation of contaminated dust can be significant. In contrast, dermal exposure has long been regarded as a minor or negligible pathway for inorganic contaminants and is often treated using conservative and sometimes inaccurate default assumptions or omitted from detailed exposure assessments. Nevertheless, accumulating experimental evidence indicates that such default assumptions inadequately represent the complex physicochemical processes governing soil–metal(loid)–sweat–skin interactions and may lead to inaccurate exposure estimates under some exposure scenario and environmental and physiological conditions [1–3].

Contrary to some authors who see the dermal absorption of chemicals from soil as a one-step process [4], where the chemical is released from the soil and simultaneously absorbed into the skin, we and other authors feel that the concept of dermal bioaccessibility is highly relevant to dermal bioavailabity of metal(loid)s [3, 5–10]. Dermal exposure to soil-bound metal(loid)s is governed by a sequence of interconnected steps. Soil particles must first adhere to the skin surface, after which metals may dissolve into the thin layer of skin surface film liquids (SSFLs) composed

DOI: 10.1201/9781003632092-12 

primarily of sweat and sebum [9]. Only the fraction that dissolves under these conditions—the dermally bioaccessible fraction—can potentially penetrate the stratum corneum and reach viable skin layers and eventually the systemic circulation. Importantly, dissolution does not equate to skin absorption, as percutaneous penetration depends on additional constraints such as metal speciation in SSFLs and skin barrier properties. Yet, without dissolution in SSFLs, dermal absorption is generally not possible. Therefore, an appropriate in vitro dermal bioavailability estimation should involve the coupling of bioaccessibility data with skin permeation data in order to characterize risks more accurately [1, 3, 6, 10].

Recent advances in in vitro methodologies have enabled investigation of metal(loid) dermal bioaccessibility and percutaneous penetration and absorption using synthetic sweat and sebum formulations, standardized soil fractions, and diffusion cell systems equipped with human skin and skin surrogates. Applied to both contaminated field soils and soil reference materials, these approaches demonstrate that sweat chemistry, sebum content, soil properties, and metal speciation strongly influence dermal exposure potential, and that total soil metal concentrations are often poor predictors of dermal exposure relevance [1–3, 6, 7, 11].

This chapter provides an integrative synthesis of current knowledge on dermal bioaccessibility and percutaneous absorption of metal(loid)s from contaminated soils, with particular emphasis on incorporating dermal bioaccessibility and skin permeation data into human health exposure assessment frameworks.

## 11.2 THE SKIN–SOIL INTERFACE: A REACTIVE ENVIRONMENTAL BOUNDARY

The skin is frequently described as a highly efficient physical barrier; however, from an exposure science perspective, it functions more accurately as a chemically reactive interface between the environment and the human body. The outermost layer of the skin, the stratum corneum, consists of corneocytes embedded in a structured lipid matrix that confers resistance to penetration by many hydrophilic substances [12]. Despite this, the skin is not homogeneous. Appendageal structures such as hair follicles, sweat ducts, and sebaceous glands interrupt the continuous barrier and may act as preferential transport pathways under specific conditions [13]. The absorption of foreign substances into the skin takes place by dissolution and molecular diffusion through a composite, heterogeneous membrane, comprising of a two-phase protein–lipid interaction [4].

Covering the skin surface is a thin but chemically active layer of SSFLs composed of eccrine sweat, lipid-rich sebum, desquamated cells, and trace intracellular components. Although microscopically thin, this layer governs the initial dissolution and speciation of contaminants upon contact with the skin. SSFL composition varies with body location, physical activity, ambient conditions, and individual physiology, resulting in substantial variability in pH, ionic strength, and organic ligand content [8, 11, 14].

When contaminated soil contacts skin, particularly under moist or sweaty conditions, fine particles readily adhere to the SSFL layer. Metal(loid)s may then desorb from soil surfaces, dissolve into the aqueous sweat phase, potentially undergo complexation with organic ligands, or experience redox transformations. As a result, SSFLs act not merely as passive solvents, but as dynamic microenvironments that selectively control which metal species become dermally bioaccessible.

## 11.3 METAL(LOID)S IN CONTAMINATED SOILS: SOURCES, SPECIATION, AND REACTIVITY

Among the metal(loid)s commonly found in contaminated sites, arsenic (As), chromium (Cr), nickel (Ni), lead (Pb), and zinc (Zn) warrant special consideration in studies of dermal exposure because of their harmful properties as skin sensitizers or their other toxic effects including carcinogenicity [11, 15]. Although their toxicological profiles vary widely, their dermal exposure relevance is governed primarily by chemical speciation and soil binding rather than total concentration [2]. In soils, metal(loid)s partition among mineral surfaces, organic matter, oxide phases, and crystalline structures. Soil pH, organic carbon content, cation exchange capacity, buffer capacity, and mineralogy strongly influence metal(loid) retention and release [16, 17].

Speciation exerts a particularly strong influence for certain elements. Chromium may exist as Cr(III), which forms insoluble hydroxides and exhibits minimal mobility, or as the more soluble and toxic Cr(VI), which has demonstrated capacity for dermal penetration [18, 19]. Arsenic speciation also affects solubility and toxicity, with As(V) generally predominating under oxic soil conditions [13]. Metal(loi)d speciation influences dermal bioaccessibility and therefore potential percutaneous absorption.

Soil particle size further modulates exposure relevance. Fine soil fractions (<425 μm) adhere more readily to skin and possess higher surface area to mass ratios [20]. Moreover, this particle size covers most of fine fractions for dermal exposure [21]. Consequently, fine fractions are typically used in dermal bioaccessibility assessments as environmentally realistic exposure surrogates [2, 7, 11].

## 11.4 DERMAL BIOACCESSIBILITY: CONCEPTUAL BASIS AND METHODOLOGICAL APPROACHES

Dermal bioaccessibility is defined as the fraction of a contaminant that dissolves from soil into SSFLs and is thus available for potential skin penetration. It is distinct from dermal bioavailability, which encompasses subsequent transport across the skin and systemic uptake. Dermal bioaccessibility represents a necessary precondition for absorption: without dissolution, dermal uptake cannot occur [5, 6, 10].

The use of in vitro dermal bioaccessibility tests aims to replace overly conservative assumptions with site-specific and exposure-relevant, chemically grounded metrics. Analogous to oral bioaccessibility testing for ingestion exposure assessment, dermal bioaccessibility accounts for incomplete dissolution of soil-bound metals under realistic soil-skin contact conditions [5].

Typical in vitro protocols involve incubating fine soil fractions with various synthetic sweat or sweat–sebum mixtures at controlled temperature and agitation, followed by quantification of dissolved metals. Early methodological work involved the development of SSFLs and demonstrated the feasibility and reproducibility of this approach [5, 8, 14], while more recent studies compared various SSFL formulations (with and without sebum) and evaluated soil- and metal-specific determinants of bioaccessibility [1, 2, 11].

## 11.5 INFLUENCE OF SWEAT COMPOSITION AND ON DERMAL BIOACCESSIBILITY

Human sweat is composed of electrolytes, ionic constituents, organic acids, carbohydrates, amino acids, nitrogenous substances, vitamins and other constituents in highly variable amounts [8]. The more acidic eccrine sweat is produced by healthy humans and has a thermoregulatory function, and sebum is secreted through sebaceous glands and forms a very thin layer (0.5 to 4 μm) depending on the area in the body [8].

Among the variables controlling dermal bioaccessibility, sweat composition—particularly pH—exerts a dominant influence. Numerous studies have reported increased metal dissolution under acidic sweat conditions, reflecting enhanced proton-driven desorption and solubility [6, 11]. Comparisons among synthetic sweat formulations such as EN-1811 (pH 6.5), NIHS-9610 (pH 4.7), various sebum formulations, and more chemically complex models reveal that dermal bioaccessibility depends on sweat, sebum, element and soil characteristics rather than a metal-specific behavior [2, 11].

These findings highlight the importance of selecting sweat formulations that reflect relevant exposure scenarios and caution against reliance on a single standardized extractant for dermal exposure assessment.

## 11.6 THE ROLE OF SEBUM IN MODULATING DERMAL BIOACCESSIBILITY AND PERCUTANEOUS ABSORPTION

Sebum has traditionally been viewed as a hydrophobic barrier limiting dermal transport of water-soluble substances. However, emerging evidence indicates that sebum should be considered as an integral component of SSFLs capable of influencing both metal dissolution and skin permeation [1, 9, 11, 22].

Sebum alone exhibits low extraction capacity for most metals, consistent with the ionic nature of dissolved metal species [11]. Nonetheless, when combined with sweat, sebum modifies the physicochemical environment at the skin surface. Ghislain and Zagury [1] showed that sebum's presence in synthetic sweat formulations significantly modified bioaccessibility percentages for As, Cr, and Cu in reference soils. However, sebum proportion in both sweat formulations did not influence the bioaccessibility of Pb and Zn. Some metal(loid)s, namely As and Cu, permeated synthetic skin (Strat-M® membrane) during permeation tests when sebum was added to sweat while no permeation was observed without sebum in sweat formulations. Depending on sweat formulation, the addition of sebum (1% v/v) increased or decreased Cr permeation coefficients ($K_p$). Ni transdermal permeation was not influenced by the presence of sebum, and no permeation was observed for Pb and Zn. These findings indicate that sebum influences not only dermal bioaccessibility but also membrane partitioning and transport behavior.

## 11.7 FROM BIOACCESSIBILITY TO TRANSPORT: PERCUTANEOUS PENETRATION AND ABSORPTION

Percutaneous penetration introduces an additional constraint beyond dermal bioaccessibility. While many soil-bound metal(loid)s might dissolve into SSFLs, only a subset of dissolved species can cross the stratum corneum. The most common method to evaluate dermal diffusion was introduced by Franz in 1975 [23]. This method has been adapted and modified by many researchers to study the dermal pathway for drugs, cosmetics, and chemicals due to its low cost, short time, and good reproducibility [24]. In vitro diffusion systems such as Franz cells enable controlled evaluation of this process using human skin, animal skin, or validated synthetic membranes [25].

In a pioneered work, Marin Villegas and Zagury [6] assessed the diffusion parameters of the bioaccessible fraction of metal(loid)s such as As, Cr, Cu, Ni, Pb, and Zn found in reference and field collected soils through artificial skin (Strat-M® membrane) using the static Franz cell methodology. They demonstrated that Cr and Cu could permeate Strat-M® membranes when delivered via sweat extracts of contaminated soils, while Pb and Zn exhibited no detectable permeation. Ghislain and Zagury [1] confirmed and expanded these findings, highlighting the importance of sweat composition and sebum presence for As and Ni percutaneous absorption.

## 11.8 METAL SPECIATION AS THE MECHANISTIC LINK BETWEEN DISSOLUTION AND TRANSPORT

Metal speciation provides the mechanistic bridge linking dermal bioaccessibility to percutaneous penetration and absorption. Numerous studies demonstrate that transdermal permeability depends on aqueous speciation rather than total dissolved concentration [10, 26]. Chromium exemplifies this principle: Cr(VI) is far more permeable than Cr(III) due to differences in solubility and charge [18].

Marin Villegas and Zagury [7] combined soil-bound metal(loid) dermal bioaccessibility experiments with geochemical modeling of metal(loid) speciation using VMINTEQ software. They reported that permeation patterns of bioaccessible metals through synthetic skin were consistent with predicted dominant soluble species, highlighting the need for speciation-aware exposure assessment. Arsenic was almost entirely found as As(V) in two tested sweat formulations. In sweat EN-1811 (pH 6.5), comparable concentrations of Cr(III) and Cr(VI) were found whereas in more acidic sweat NIHS-9610 (pH 4.7), Cr was primarily present as Cr(III). Therefore, when extracted with NIHS-9610, soil-bound Cr could hardly permeate the membrane. Copper was primarily found as $Cu^{2+}$ but different soluble species were observed depending on sweat formulation. Bioaccessible Cr extracted from nearly all soils permeated through the Strat-M® membrane when it was coated with sebum highlighting again the importance of metal speciation and potential interaction with organic ligands. The Cr permeation coefficient ($K_p$) ranged between 0.004 and 0.13 cm/h and the $K_p$ for Cu was higher (0.024–0.52 cm/h). Further studies on the speciation of metal(loid)s in bioaccessible extracts in the presence of sebum are strongly recommended.

## 11.9 INCORPORATING DERMAL BIOACCESSIBILITY INTO HUMAN HEALTH EXPOSURE ASSESSMENT

The USEPA [21] recommends two approaches to estimate the dermal absorbed dose (DAD in mg/kg bw/day, Eq. 11.1) depending on the contaminant matrix: (1) contact with soil and (2) contact with water. Both strategies consider total metal(loid) concentrations ((mg/kg) in the soil approach and ($mg/cm^3$) for the water approach).

$$DAD = DA_{event} \times EV \times ED \times EF \times SA/BW \times AT \tag{11.1}$$

Where $DA_{event}$ is the dermal absorbed dose per event ($mg/cm^2 \times$ event); EV is event frequency (event/day); ED is exposure duration (year); EF is exposure frequency (day/year); SA: is the exposed skin surface area ($cm^2$); BW is body weight (kg); and AT is averaging time (day).

Since the dermal pathway in case of contact with soil relies on the use of uncertain default dermal-soil absorption values ($ABS_d$) to modulate the total soil metal concentrations (Eq. 11.2), our research group modified the water approach to incorporate field measured dermal bioaccessibility of soil-bound metal(loid)s and site-specific percutaneous permeability coefficient [2, 3, 6]. In this customised soil approach, $DA_{event}$ is calculated using equation Eq. 11.3 rather than the conventional Eq. 11.2 suggested by USEPA [21].

$$DA_{event} = C_{total} \times CF \times AF \times ABS_d \tag{11.2}$$

Where $C_{total}$ is the total concentration of metal(loid) in soil (mg/kg); CF is a conversion factor; AF is the adherence factor ($mg/cm^2$); and $ABS_d$ is a generic dermal-soil absorption value (unitless, 0.03 for As and 0.001 for other elements).

$$DA_{event} = K_p \times C_{sweat} \times t_{event} \tag{11.3}$$

Where $C_{sweat}$ is the site-specific bioaccessible concentration of the metal(loid) in synthetic sweat solution ($mg/cm^3$); $K_p$ is the site-specific measured permeability coefficient of metal(loid) in sweat using a Franz cell (cm/h) instead of USEPA generic values (Cr(III): $1 \times 10^{-3}$; Cr (VI): $2 \times 10^{-3}$; Ni: $2 \times 10^{-4}$; Pb: $1 \times 10^{-4}$; Zn: $6 \times 10^{-4}$); and $t_{event}$ (h) is the event duration (2 h or 8 h/event).

Once $DA_{event}$ is calculated using $C_{sweat}$ obtained from dermal bioaccessibility tests and $K_p$ obtained from percutaneous permeation tests, DAD is determined (Eq. 11.1) using the specific parameters for the evaluated scenario following the methodology presented by USEPA [21].

Marin Villegas and Zagury [6], using the Strat-M® synthetic membrane as a skin surrogate after extraction of soil-bound metal(loid)s in synthetic sweat EN-1811 (pH 6.5), reported much higher $K_p$ values than the generic ones summarized by USEPA [21].

In another study, the influence of sebum proportion (1% v/v and 3% v/v) in two commonly used synthetic sweat formulations (EN-1811 and NIHS-9610) on soil dermal bioaccessibility and subsequent diffusion through synthetic skin was assessed [1]. In most cases, the experimentally obtained transdermal permeation coefficients

(Kp) for metal(loid)s (As, Cr, Cu, and Ni) were higher than the generic values for inorganic chemicals suggested by USEPA. Other studies reported higher metal permeability coefficients than the USEPA default values. Examples of these are Larese Filon et al. [27] who reported a $K_p$ value of 0.0124 for Cr in powder, and Fullerton et al. [28] who reported $K_p$ values for Ni of 0.0015 for epidermis and 0.23 for the dermis. This suggests that risk assessors should be cautious when using generic $K_p$ values since transdermal diffusion of metal(loid)s in polluted soils appears to be not only site-specific but also sweat formulation-dependent. Moreover, in a field study based on ten soil samples collected near CCA-treated utility poles, the hazard quotient was higher when dermal bioaccessibility of field-collected samples was considered (Eq. 11.3) compared to the conventional method using total metal(loid) concentrations in the soil and the assumed generic value for $ABS_d$ (Eq. 11.2) [3]. These recent studies demonstrate that substituting total concentrations with dermal bioaccessible fractions and substituting soil-absorption default values with percutaneous permeation coefficients obtained experimentally yields more realistic, soil- and metal-specific exposure estimates while remaining compatible with regulatory frameworks.

## 11.10 RESEARCH GAPS AND FUTURE DIRECTIONS

The results presented in this chapter show that the Strat-M® membrane is suited for preliminary dermal permeation studies of metal(loid)s from contaminated soils. Synthetic membranes are commonly available, produce less variability, and are significantly less expensive than human skin [6, 29]. However, even though the Strat-M® membrane was engineered to mimic the layered structure and lipid chemistry of human skin [29], this membrane only models the diffusion of chemicals and is not suitable to model other dermal pathways such as appendages (through sweat glands and hair follicles). Moreover, the Strat-M® will not exactly match the absolute permeability behavior of human skin. Therefore, further research to compare the diffusion results with real human skin (intact and damaged) is necessary to reduce uncertainty in the calculation of dermal exposure to metal(loid)s present in contaminated soils. Moreover, additional studies are needed to assess metal(loid) speciation (including interaction with organic sebum components) in various SSFLs after soil extraction to explain differences in bioaccessible metal(loid)s' ability to permeate synthetic skin.

Finally, even if metal(loid)s do not fully permeate the skin barrier to reach the systemic circulation, this does not exclude the occurrence of potential cytotoxic effects. Therefore, in vitro toxicological studies using human skin keratinocytes (HaCaT) exposed to soil extracts to assess cytotoxicity, intracellular reactive oxygen species (ROS) generation, and apoptosis are recommended.

## 11.11 CONCLUSION

Dermal exposure to metal(loid)s in contaminated soils is governed by complex interactions between soil physicochemical properties, SSFL composition, metal speciation, and skin barrier properties. The evidence synthesized in this chapter demonstrates that dermal bioaccessibility along with site-specific percutaneous permeation parameters are the critical exposure-relevant metric linking soil contamination to potential human exposure. Inclusion of sebum in SSFLs and mechanistic permeation data further refines understanding of dermal uptake. Incorporating dermal bioavailability into exposure assessment frameworks represents a scientifically robust and regulatory-relevant advance toward more accurate protection of human health.

## REFERENCES


1. Ghislain M, Zagury GJ. 2023. Influence of sebum proportion in synthetic sweat on dermal bioaccessibility and on permeation of metal(loid)s from contaminated soils. *Environ. Sci. Poll. Res.* 30:86762–86772.
2. Chaparro Leal LT, Guney M, Zagury GJ. 2018. In vitro dermal bioaccessibility of selected metals in contaminated soil and mine tailings and human health risk characterization. *Chemosphere* 197:42–49.
3. Marin Villegas CA, Zagury GJ. 2023a. Incorporating oral, inhalation and dermal bioaccessibility into human health risk characterization following exposure to Chromated Copper Arsenate (CCA)-contaminated soils. *Ecotoxicol. Environment. Safety* 249: 114446.
4. Beriro DJ, Cave MR, Wragg J, Thomas R, Wills G, Evans F. 2016. A review of the current state of the art of physiologically-based tests for measuring human dermal in vitro bioavailability of polycyclic aromatic hydrocarbons (PAH) in soil. *J. Hazard Mater* 305:240–259.
5. Hillwalker WE, Anderson KA. 2014. Bioaccessibility of metals in alloys: evaluation of three surrogate biofluids. *Environ. Pollut* 185:52–58.
6. Marin Villegas CA, Zagury GJ. 2021. Comparison of synthetic sweat and influence of sebum in the permeation of bioaccessible metal(loid)s from contaminated soils through a synthetic skin membrane. *Environ. Sci. Technol* 55:8215–8222.
7. Marin Villegas CA, Zagury GJ. 2023b. Metal(loid) speciation in dermal bioaccessibility extracts from contaminated soils and permeation through synthetic skin. *J. Hazard. Mater.* 455: 131523.
8. Stefaniak AB, Harvey CJ. 2006. Dissolution of materials in artificial skin surface film liquids. *Toxicol. In Vitro* 20:1265–1283.
9. Stefaniak AB, Harvey CJ, Wertz PW. 2010. Formulation and stability of a novel artificial sebum under conditions of storage and use. *Int. J. Cosmetic Sci* 32: 347–355.
10. Stefaniak AB, Duling MG, Geer L, Virji MA. 2014. Dissolution of the metal sensitizers Ni, Be, Cr in artificial sweat to improve estimates of dermal bioaccessibility. *Environ. Sci. Process Impacts* 16:341–351.
11. Marin Villegas CA, Guney M, Zagury GJ. 2019. Comparison of five artificial skin surface film liquids for assessing dermal bioaccessibility of metals in certified reference soils. *Sci. Total Environ.* 692:595–601.
12. Gray H. 2000. *Anatomy of the Human Body.* Philadelphia: Lea & Febiger, 1918; Twentieth ed. Bartleby.com
13. WHO (World Health Organization). 2006. Environmental Health Criteria 235, Dermal Absorption. http://www.inchem.org/documents/ehc/ehc/ehc235.pdf.
14. Harvey CJ, LeBouf RF, Stefaniak AB. 2010. Formulation and stability of a novel artificial human sweat under conditions of storage and use. *Toxicol. in Vitro* 24 (6): 1790–1796.
15. Guy, RH. 1999. *Metals and the Skin: Topical Effects and Systemic Absorption.* CRC Press.

16. Balasoiu CF, Zagury GJ, Deschênes L. 2001. Partitioning and speciation of chromium, copper and arsenic in CCA-contaminated soils: influence of soil composition, *Sci. Tot. Environ* 280:239–255.
17. Dobran S, and Zagury GJ. 2006. Arsenic speciation and mobilization in CCA-contaminated soils: influence of organic matter content, *Sci. Tot. Environ* 364:239–250.
18. Van Lierde V, Chéry CC, Moens L,Vanhaecke F. 2005. Capillary electrophoresis hyphenated to inductively coupled plasmasector field-mass spectrometry for the detection of chromium species after incubation of chromium in simulated sweat. *Electrophoresis*, 26: 1703–1711.
19. Van Lierde V, Chery CC, Roche N, Monstrey S, Moens L, Vanhaecke F. 2006. In vitro permeation of chromium species through porcine and human skin as determined by capillary electrophoresis–inductively coupled plasma–sector field mass spectrometry. *Anal. Bioanal. Chem.* 384: 378–384.
20. Horowitz SB, Finley BL. 1994. Setting health-protective soil concentrations for dermal contact allergens: a proposed methodology. *Regul. Toxicol. Pharmacol* 19:31–47.
21. USEPA, 2004. Risk Assessment Guidance for Superfund Volume I: Human Health Evaluation Manual (Part E, Supplemental Guidance for Dermal Risk Assessment). EPA/540/R/99/005. Office of Superfund Remediation and Technology Innovation, Washington, DC.
22. Pawar G, Abdallah MA-E, de Sáa EV, Harrad S. 2017. Dermal bioaccessibility of flame retardants from indoor dust and the influence of topically applied cosmetics. *J. Expo. Sci. Environ. Epidemiol* 27 (1): 100–105.
23. Franz TJ. 1975. Percutaneous absorption. On the relevance of in vitro data. *J. Invest. Dermatol* 64:190–195.
24. Sartorelli P, Andersen HR, Angerer J, Corish, J, Drexler, H, Göen, T, Griffin, P, Hotchkiss, SAM, Larese, F, Montomoli, L, Perkins, J, Schmelz, M, van de Sandt, J, Williams, F. 2000. Percutaneous penetration studies for risk assessment. *Environ. Toxicol. Pharmacol* 8:133–152.
25. Franken A, Eloff FC, Du Plessis J, Du Plessis JL. 2015. In vitro permeation of metals through human skin: A review and recommendations. *Chem. Res. Toxicol* 28 (12): 2237–2249.
26. Nikpour S, Hedberg YS. 2021. Using chemical speciation modelling to discuss variations in patch test reactions to different aluminium and chromium salts. *Contact Dermat.* 85:415–420.
27. Larese Filon F, D'Agostin F, Crosera M, Adami G, Bovenzi M, Maina G. 2008. In vitro percutaneous absorption of chromium powder and the effect of skin cleanser. *Toxicol. in Vitro* 22:1562–1567.
28. Fullerton A, Andersen JR, Hoelgaard A. 1988. Permeation of nickel through human skin in vitro: Effect of vehicles. *Br. J. Dermatol* 118:509–516.
29. Haq A, Goodyear B, Ameen D, Joshi V, Michniak-Kohn B. 2018. Strat-M® synthetic membrane: Permeability comparison to human cadaver skin. *Int. J. Pharmaceutics* 547:432–437.

# 12 Is Phytoremediation of Heavy Metal-Contaminated Soils Using Industrial Crops a Sustainable Solution?

*Leandro A. Gomes, Bruno Barbosa, Barbara Rachele Ciaramella, Jorge Costa, and Ana Luisa Fernando*

## 12.1 INTRODUCTION

Globally, millions of hectares of land are affected by soil contamination resulting from intensive agriculture practices, waste disposal, mining activities, and industrial operations. Heavy metal-contaminated soils, including post-industrial or mining sites, often contain elevated concentrations of potentially hazardous elements, such as cadmium (Cd), lead (Pb), zinc (Zn), copper (Cu), chromium (Cr), nickel (Ni), and arsenic (As), among others, that can leach to groundwater and surface water and get absorbed by crops, and in case of food crops, these can enter in the food chain causing human and ecological health risks [1, 2]. These lands represent, at the same time, an underutilized resource and an environmental problem.

The remediation of heavy metal-contaminated soils involves various physical, chemical, and biological techniques. These methods, applied either in situ (on-site) or ex situ (off-site), aim to restore soil quality, reduce environmental risks, and manage liability for future land use. As a solar-driven and environmentally compatible technology, phytoremediation (Figure 12.1) uses plants and their associated microbes to remediate soil, water, and air [3]. It is particularly attractive for the remediation of heavy metal-contaminated soils due to its lower cost and greater economic feasibility compared to physicochemical methods.

Phytoremediation is a versatile method used against different pollutants, at different levels of single or multiple contaminants to promote soil or wastewater decontamination. The recent development in this field can be attributed to improvements in the biotechnology field, with new methods to promote the stabilization or volatilization of the contaminants [4].

This technique can be divided into different groups, depending on the biochemical mechanisms that the plant uses to remediate. Figure 12.1 shows the different


DOI: 10.1201/9781003632092-13

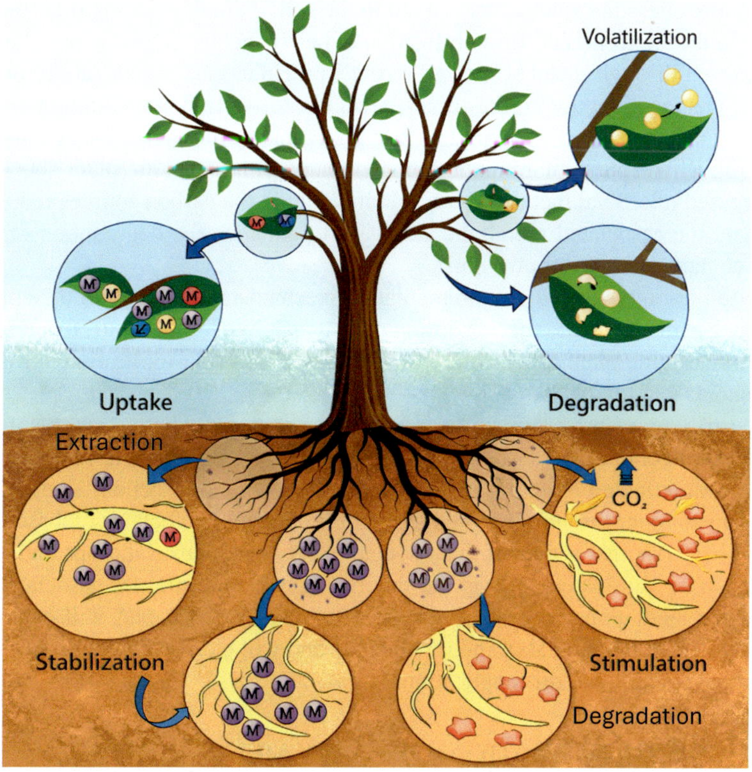


**FIGURE 12.1** Mechanisms of phytoremediation.

paths in phytoremediation and how they work. They are phytoextraction, phytostabilization, phytodegradation, phytostimulation, and phytovolatilization.

Plants can promote phytoextraction in different ways, such as phytoaccumulation, phytosequestration, or photoabsorption. In the process of phytoaccumulation, the plants' root system encourages the uptake of the contaminants that can remain accumulated in the root system or be translocated to the aerial parts of the plants. Phytoextraction refers to the uptake and translocation of heavy metals from soil into harvestable plant tissues. It can be designed for different contaminated soils due to the particularities of each rapid crop growth, the potential for metal accumulation, extensive root system, high tolerance to contaminants, broad geographic distribution, adaptation of plant to prevailing climatic conditions, and repulsion to herbivores to avoid food chain contamination [5, 6].

Phytovolatilization is the process of plant uptake of organic compounds through their roots, followed by the translocation of pollutants and subsequent release of those volatile organic compounds (VOCs) from the stem/trunk and leaf surfaces. Indirectly, phytovolatilization occurs due to the increased flow of volatile contaminants from the soil or subsurface water due to the activity of the plant root [7]. Phytostabilization/phytoimmobilization is the transformation of toxic compounds into nontoxic/less poisonous forms fixed in the soil, thereby reducing their bioavailability in the environment. Immobilization of heavy metals in soils using plants can be achieved through metal sorption by roots, precipitation, complexation, or metal valence change/reduction in the rhizosphere. Rhizosphere bioremediation or rhizodegradation is enhanced biodegradation of pollutants by root-associated bacteria and fungi [8]. Rhizosphere bioremediation is efficient within specific plant species due to increased microbial metabolic rates. Plants supply the carbon sources required to stimulate the associated metabolic pathways to enhance the degradation of heavy metal contaminants within the soil.

Phytofiltration is also a phytoremediation mechanism and refers to the approach of using a plant's biomass for contaminants removal, like toxic metals, from polluted water by utilizing plant roots (rhizofiltration), seedlings (blastofiltration), or excised plant shoots (caulofiltration) [7].

Industrial crops are promising candidates for phytoremediation, as they tolerate heavy metal-contaminated soils [9, 10]. Moreover, considering the further utilization of contaminated biomass, industrial crops become an excellent alternative for phytoremediation, through the production of bioenergy, biofuels or biobased products [11–13]. Another aspect of the association between industrial crops and its production in contaminated soils is the environmental opportunities associated with the phytoremediation process [14, 15]. The following sub-chapter will explore the phytoremediation potential of industrial crops and will clarify the opportunities and challenges related with this application.

## 12.2 PHYTOREMEDIATION OF HEAVY METAL CONTAMINATED SOILS WITH INDUSTRIAL CROPS

Energy/industrial crops present several characteristics that make them suitable candidates for cultivation in contaminated lands. These includes fast growth rate, extensive root system, high biomass yields, low need for fertilization, and a vast range of crops with different adaptability strategies for different conditions. Many of the energy crops are perennial, which reduces the need for soil disturbance in this way limiting contaminant leaching and erosion risks [16]. Research shows that several energy crops can tolerate elevated concentration of contaminants while maintaining physiological function and biomass productivity. Another advantage lies in the ability of these crops to translocate the contaminants from its underground to its aboveground fraction, promoting the extraction of the contaminants from the soil, in this way, mitigating soil contamination [17]. However, the amount concentrated in the plants is relatively low compared to hyperaccumulating plants [18]. The significance of the phytoextraction is, nevertheless, significant when the amount extracted per ha is evaluated, mainly due to the high yields obtained in the contaminated soils due to the high tolerance of these crops to contamination [19]. Energy crops consist in plant species primarily cultivated for biomass production to feed energy conversion systems. These crops include perennial grasses (e.g., *Miscanthus* × *giganteus*, *Arundo donax*, *Panicum virgatum*), annual crops (e.g., sorghum), oilseed crops (e.g., Brassica species), fiber crops (e.g., hemp, kenaf), and woody species (e.g., willow, poplar) [9, 18] (Figure 12.2). Usually characterized as resistant crops, they are known

**FIGURE 12.2** Energy crops. Top row, from left to right: giant reed, hemp, carinata. Bottom row, from left to right: kenaf, switchgrass, miscanthus.

by their high biomass productivity, adaptability to diverse environmental conditions, and compatibility to mechanized agricultural practice [20].

When evaluating the phytoextraction process of these crops, results show that there are differences mostly related to the level of metals concentration in the soil, their speciation, and eventual treatments applied to enhance the mobility of the metals in the soil and their accumulation in the different fractions of the crops [18, 19]. However, for most metals, it is observed a higher concentration in the biomass with increasing levels of metals in the soil [e.g. 19]. However the distribution between aboveground organs and belowground organs differs depending on the studied element. Regarding zinc concentration in plant organs, usually no differences are observed between above and belowground fractions. This indicates that zinc is easily mobilized within the plant, being transported from the belowground organs to the aboveground organs. Kabata-Pendias [21] indicates that this mobility can be associated with the bounding of Zn to light organic compounds in xylem fluids. Regarding less mobilized elements in the soil, e.g. Cr and Pb, concentrations of these elements in the belowground fraction are usually higher than in the aerial fraction. This reflects the limited translocation of these metals from belowground organs to the aerial fraction. Moreover, the concentration of these elements in the belowground organs, is related to the propensity of $Cr^{3+}$ and Pb pyrophosphate to bind to cell walls [21].

Results obtained for energy crops showed that heavy metal–contaminated soils reduces aboveground biomass production [18, 19]. However, differences can be observed due to the amount of the contamination and due to the degree of mobilization of the contaminants. Also, the different crops also show different capacities to phytoremediate the soils. For example, in the study of Barbosa et al. [19], *Miscanthus* x *giganteus* showed significantly higher yields than *M. sinensis* and *M. floridulus*, for the same level of contamination. And a side by side comparison of giant reed in zinc contaminated soils to Miscanthus indicates that *M.* x *giganteus* was significantly more productive than giant reed, and giant reed significantly more productive than *M. floridulus* and *M. sinensis*. Several authors reported tolerance of energy crops to heavy metal contamination [14, 22]. However, the tolerance to heavy metal exposure is also governed by the element concentration in soil. Guo and Miao [22] observed only phytotoxic effects on giant reed when Pb concentration in soil was higher than 1000 mg $kg^{-1}$. Below this concentration no effects were observed on the shoot dry biomass production. According to the different studies related with phytoremediation of heavy metal–contaminated soils by energy/industrial crops, the majority of the results show that exposure of plants to significant amount of heavy metals tested resulted in the reduction of biomass productivity. According to Kabata-Pendias [21], heavy metals in excess in soils contribute to damages on the photosynthesis apparatus and belowground organs, leading to retarded growth of the entire plant. Anatomical changes are mechanisms that help to quench direct ionic metals toxicity, such as development of secondary sheath bundles and tissue scarification [22]. However, information on the interactions between the belowground organs of these crops and its growing medium is still lacking, which might provide more insights on the tolerance mechanisms associated with these plants on contaminated soils.

Their utilization through biomass production for energy production offers a pathway to restore their ecosystem functions, reduce environmental risks, and generate economic value without competing with food production [23]. In fact, the need to mitigate climate change by reducing greenhouse (GHG) gases emissions and fossil fuels dependency, has driven a significant transformation in the global energy production systems. International frameworks such as the Paris Agreement emphasize the need for deep decarbonization and the expansion of renewable energies production [24]. At COP 28, the parties agreed to double energy efficiency and triple the renewable energy share by 2030, to phase down fossil fuels [25]. At regional level, European climate law associated with Fit for 55 aims to reduce emissions by 55% until 2030, projecting reaching carbon neutrality by 2050, using strategies such as REPowerEU that set bindings renewable energy targets and promotes the increase of renewable energy sources, including bioenergy [26]. Biomass based energy processes contribute to the reduction of GHG emissions by replacing fossil based energy, promoting carbon cycling, and enable negative emissions when coupled with carbon capture systems [27]. However, the sustainability of the bioenergy systems depends on the feedstock selection, the chosen conversion process, and the land-use dynamics. The selection of the biomass can significantly affect the economical, environmental, and social impact of the energy system [28]. Despite their potential, the large scale cultivation of energy crops raises concerns over land-use competition with energy crops, commonly cited as "food vs fuel" debate. Its production in productive land can cause a cascade effect that put in risk food security by reducing the available productive land size, causing the increase of food prices and induce indirect land-use change (ILUC), whereby natural ecosystems must be replaced by agricultural land to compensate the reduction in the productive area [29]. A viable alternative that can both incentivise energy crops while avoiding competition with food production is the utilization of marginal lands, including contaminated lands [14].

## 12.3 ENVIRONMENTAL SUSTAINABILITY OF THE PHYTOREMEDIATION OF HEAVY METAL CONTAMINATED SOILS WITH INDUSTRIAL CROPS

Even when the phytoextraction is not significant, the phytoremediation process benefits from the environmental advantages of introducing a crop in a contaminated land. Environmentally, there are opportunities brought by the cultivation of industrial crops in heavy metal–contaminated soils. Besides the amount that can be harvested from the soil via accumulation of heavy metals in the aerial part of the crops, several ecosystem services benefit from these cropping systems. In fact, the introduction of these crops in the contaminated soils help to control soil erosion, contribute to carbon sequestration and minimization of nutrient leaching as well as to the restoration of soil properties (fertility, structure, organic matter). Compared with annual systems, perennial crops present additional advantages, namely regarding erodibility and biodiversity due to the extensive root and rhizome system. However, the intensive cultivation of these grasses may lead to soil nutrient and water depletion, as well as to degradation and pollution. These impacts gain importance, especially in water-scarce regions, like the Mediterranean, where the competition for land and water among agriculture, industry and urbanization are leading to environmental desertification [11]. In addition, water and land resources must be included in an integrated management strategy in order to ensure food and water security, as well as biological and landscape diversity [28].

The valorization of the contaminated biomass is central to the sustainability of phytoremediation systems. Thermochemical and biochemical conversion pathways differ in their handling of contaminants and associated risks.

Thermochemical process such as combustion, pyrolysis, and gasification offers high energy recovery efficiencies and significant reduction of the outcome residues. The relevance of these characteristics remains in the concentration of the contamination, allowing a higher efficiency treatment. During combustion, heavy metals can be found in different parts of the residues: bottom ash, fly ash, and flue gas, depending on their volatility and oxidation state. To mitigate this effect reducing the release of these contaminants, advanced techniques can be used to minimize emissions allowing even the recovery of these contaminants from the solid residues [30].

Biochemical pathways, including anaerobic digestion and fermentation, convert the biomass into biogas or biofuels, concentrating the contaminants in the digested residues, facilitating their recovery or stabilization. Pre-treatment processes, such as ionic liquid extraction have further improved the remove of the contaminants prior to the conversion process, becoming also an option [31]. The integration of bioenergy production and phytoremediation can be seen following the circular economy principles, which emphasize resource efficiency, waste valorization, among others. By converting marginal lands into productive assets, phytomanagement systems contribute to land restoration, renewable energy generation, and, as an option, metal recovery.

Life cycle assessment (LCA) studies suggest that bioenergy processes that use contaminated lands as biomass source can achieve favorable environmental performance, however, specificities from the contaminated sites need to be addressed to obtain the best performance of the combination of both techniques. Following, results from environmental sustainability studies on the thermochemical and biochemical valorization of the biomass harvested from heavy metal–contaminated soils will be presented and discussed.

### 12.3.1 Environmental Sustainability of the Thermochemical Valorization of the Biomass Harvested from Contaminated Soils

The biomass produced via phytoremediation has increasingly been considered a viable feedstock for energy production and material recovery [30]. The crops can grow in soils under several contaminants' exposition, such as Cd, Pb, Zn, Cu, Ni, Cr, and As, accumulating them in their tissues. When this accumulation occurs in the belowground part, the contaminants remain in the soil, usually in a non-bioavailable form, when the contaminants are translocated to the aboveground fraction of the plant, it occurs the phytoextraction, the contaminants are removed from the soil in the harvest moment [32].

The thermochemical valorization of this biomass offers a path to at the same time reduce the volume occupied by the contaminants, while produce energy, and manage the contaminants under controlled conditions. The environmental impacts of these processes depend however on the chosen conversion process, operating condition, and fate of metals across product stems.

Combustion is a well established thermochemical and its usually considered due to its simplicity, which reduces biomass volume and concentrating the heavy metals in the ash fraction. A study elaborate by Xue et al. used combustion path over sorghum cultivated in Cd, Pb, and Cu contaminated soils. The biomass was reduced to 6% of the initial feedstock, while the heavy metals presented two particular behaviours: Cd and Pb, being more volatile metals were preferred transfer to fly ash, reaching approximately concentrations of 63 $mg.kg^{-1}$ and 800 $mg.kg^{-1}$, on the other hand, Cu, being less volatile remained mostly in the bottom ash (85%) [33]. The combustion of contaminated biomass was also done by Sui et al. which observed among different biomasses cultivated under Zn, Pb, and Cd contaminated mining soils, showing that the combustion behaviour was comparable to the traditional biomass. The study also pointed for the possibility of Zn retention in the ash due to its Ca rich components, while points out that Cd and Pb presented a more volatile behaviour [34]. After observing the same behaviour by the hyperaccumulator *Sedum plumbizincicola* exposed to the same contaminants (Zn, Pb, and Cd), Zha et al. pointed that heavy metal emissions during combustion can be mitigated by controlling temperature and heating rate during early devolatilization, while promoting in situ stabilization trough interactions with Ca and Si rich ash components that convert the metals in less mobile phases [35].

Despite some concerns Espada and collaborators demonstrate that the utilization of contaminated biomass in cogeneration systems can improve the environmental viability when compared with landfill disposal. The utilization of phytoremediation biomass in energy combustion processes helps to reduce fossil fuels utilization, leading to a reduction of greenhouse gases emissions, acidification potential, and cumulative energy demand [36]. In addition to conventional heavy metals, phytoremediation biomass can be used also for the recovery of noble metals (NMs) and rare earth elements (REEs). Dinh et al. demonstrate the possibility to recovery NMs due to its accumulation in the fly ashes, with the help of cleaning techniques, while REEs elements are easier to recovery due to their accumulation in the bottom ash, opening the possibility for the phytomining [37]. Ma et al. showed that toxic elements such as Cd, As, Pb, and Hg, have a high chance of atmospheric dispersion due to its concentration in the fly ash. The ecological risk assessment indicates that fly ashes represent a moderated to high environmental risks mainly due to the toxicity and mobility of Cd and Hg, while slag presents lower risks due to the better metal stabilization [38].

Pyrolysis is a path also suitable to valorize contaminated biomass. Using temperatures in the range of 300°C to 900°C, lower than in combustion processes, most of the heavy metals in the contaminated biomass, reaching up to 98% of metals such Ni, Zn, Cu, Co, and Cr, remain in the char when the pyrolysis is conducted around 600°C, as observed by Zheng et al. [39]. Park and collaborators investigated both batch based and continuous pyrolysis of contaminated biomass at temperatures in the range of 300°C and 900°C, comparing one stage and two stage configurations. Their results show that most heavy metals, including Zn, Pb, Cu, and Ni are retained in the biochar fraction [40]. A study focused on Pb contaminated biomass was developed by Amato et al. using poplar, and slow pyrolysis at 465, 600, and 800°C, and showed that the retention of lead in biochar decreased with the increase of temperature, due to enhance volatilization. The experiment also observed that pyrolysis reduced the soluble and exchangeable Pb fractions compared to the raw biomass, confirming its stabilization effect. The overall result indicated that both the chemical form and the contamination mechanism must be considered when applying pyrolysis to Pb contaminated biomass [41]. Zinc was another heavy metal studied with a higher focus. An experiment conducted by Ozkan et al. applied sequential phytoremediation and pyrolysis using sunflower, corn, and rape cultivated on Zn contaminated soil and a fixed bed pyrolysis at 500°C. The contaminant was predominantly fixed in the solid phase, with a lower amount in the liquid phase and no significant amount in the gas phase. The Zn observed in the solid fraction was effectively immobilized, with low leaching values, while the bio-oil, once tested for energy production, did not compromise environmental safety [42]. Han et al. demonstrated that the pyrolysis of *Broussonetia papyrifera* contaminated with Cd and Cu allows the recovery of the metals at the same time of the bioenergy generation and pointed that despite the presence of these components slightly increased the energy needs to start the reaction, they do not alter the solid fraction yield [43].

Dabela et al., draw attention to the fact that pyrolysis converts biomass into a carbon-rich, aromatized char, with higher temperatures enhancing structural stability and promoting the immobilization of heavy metals within the solid matrix. From the investigation of heavy metal contaminated soils with woody biomass, Dabela observed a reduction in the contaminants leaching up to 93%, attributed to the biochar formation and the associated encapsulation and chemical stabilization of the metals. It was suggested that the co-pyrolysis of contaminated biomass can achieve similar results, resulting in a higher immobilization of heavy metals [44].

Voglar et al. investigated the pyrolysis of industrial hemp (*Cannabis sativa* L.) cultivated in contaminated lands demonstrating that Cd, Pb, and Zn can increase when reducing agents like CO are present in the reaction, resulting in a concentrated gaseous fraction that can be separated using the proper techniques, supporting the mitigation of the heavy metal-contaminated soils [45].

Process modifications aimed to increase metal stabilization have been proposed as an effective strategy to reduce the environmental risks associated to high temperature pyrolysis. One of the approaches is the addition of stabilizing agents prior to thermal treatment has received particular attention. For example, He et al., pointed to the use of ferric salts showed to significantly influence metal behaviour during pyrolysis by the formation of thermally stable metal-iron compounds and mineral phases, reducing the tendency for metals to volatilize and migrate into gaseous or condensate products, reducing the risk of environmental contamination [46].

Gasification has been studied as a suitable path for the valorization of biomass cultivated on contaminated soils, especially when it comes to phytoextraction. When comparing to pyrolysis, gasification operates in higher temperatures and under a sub stoichiometric oxygen, conditions that considerably affect metals present in the biomass. From an environmental point of view, contaminated biomass gasification contributes to generate a combustible gas from a renewable resource cultivated on a marginal land. On the other hand, it may promote the mobilization of potentially toxic substances, thereby necessitating the adoption of gas cleaning and residue management strategies.

Literature indicates that typical gasification temperatures favours the volatilization of heavy metals, particular Cd and Zn, while metals like Cu and Ni presents significant lower volatility. On reducing environment, Cd, and Zn, usually forms elemental species or semi-volatiles composites with high vapor pressure, being easily transferable to the gaseous phase [47]. Experimental studies on fluidized bed gasifiers using steam demonstrate that most of the heavy metals remains in the solid fraction, the ash in the gasifier bed and cyclone, while a small fraction is transferred to the syngas either in gaseous phase or in small particles. It was observed that Cd is the most volatile contaminant, followed by Pb and Zn, while Cu and Ni, present limited volatilization. This behaviour shows that for certain metals, there is the need for the implementation of additional cleaning gas steps (such as cyclones, filters, or condensators) to avoid environmental issues. In fact, the type of the gasifier also affects the behaviour of the metals. Fluidized reactors have a higher interaction between solid and gaseous phase, resulting in a higher transport of low volatile metals. In a fixed bed gasifier, there is a lower interaction between solid and gaseous phase, resulting in a lower particle emission, meaning that most of the contaminants remains in the ash at the bottom of the gasifier. Studies that used contaminated *Miscanthus*, *Sida hermaphrodita*, and *Spartina pectinata* demonstrated that the syngas obtained is comparable with the one obtained from biomass without contamination [48].

Another example is the gasification in entrained flow gasifier. Using higher temperatures, it had promoted the volatilization of almost all the Cd, Pb, and Zn, while less volatile metals were only partially affected. The volatilized metals later recondensed onto the fly ash or got captured by slag, reducing the risk of being released in the environment. Cui et al. studied the gasification of *Sedum alfredii* for energy production and heavy metal stabilization, having focused the behaviour of Cd, Pb, and Zn during the process and the environmental properties of the resulting biochar. The biomass was tested under a lab scale fixed bed gasifier, using $N_2$ and $CO_2$, in different temperatures ranging from 600°C to 900°C, aiming to truly isolate the effect of the temperature and the oxidizing agent on the metal's volatilization and biochar formation. It was observed that temperature highly affected the metals volatilization while the oxidizing agent affects the metals destination, since when $CO_2$ was used, biochar with a more developed structure was formed, capturing a significant part of the volatilized metals [49].

Apart from the solid and gaseous fraction, gasification can also produce liquid effluents during the gas cooling. Studies performed in downdraft gasifiers identified contamination in this fraction of the gasifier that if not treated can represent risks to the environment. The level of contamination also affects the environmental risk of the biomass valorisation. Most plants accumulate the contaminants in the root system. Plants with high translocation factor may need the implementation of cleaning strategies and residues managements to be able to be gasified without environmental risks [50].

Thermodynamically modelling studies shows that in biomass gasification, pressure influence in metal volatilization as much as temperature, as observed by Jiang et al. which showed that increasing gasification pressure shift the solid to phase transitions to higher temperatures, effectively reducing the release of toxic metals into the gas phase. It was also indicated in the study that highly volatile elements such as As, Cd, Zn, and Pb tends to transform in gaseous phase at relatively low temperatures (lower than 1000°C), whereas metals such as Ni, Cu, Mn, and Co volatilize in the range from 1000°C to 1200°C. In contrast less volatile elements such as Cr, Al, Fe, and Mg, remain in the solid phase even at high temperatures [51].

An emerging technique is the utilization of subcritical water gasification (SCWG), a promising alternative particularly when high moisture content makes traditional gasification routes unsuitable. In SCWG systems, biomass is converted using water above its critical point (374°C and 22.1 MPa), where supercritical water acts as a nonpolar reaction medium that efficiently dissolves organic compounds while strongly limits the solubility of inorganic salts and metal species. The results observed by Dutzi et al. indicates the possibility of obtained high concentration hydrogen gas while also achieving high metals immobilization (Zn, Pb, Cd, Mn, and Cr) being a suitable candidate for the combined phytoremediation + energy production [52]. This technique also showed to be efficient when studied by Su et al. that used contaminated *Sedum plumblizincicola* and observed that 99% of the contaminants were retained inn the biochar fraction through adsorption and mineral complexation under supercritical conditions [53].

Hydrothermal carbonization is also a promising thermochemical pathway for the valorization of contaminated biomass. Recent studies showed that more than 98% of metals such as Cu, Zn, Cd, and Pb can be retained in the solid fraction of the resulted char fraction, with metals undergoing to more stable chemical forms, including elementals, oxidized, carbonate, or organometallic species. The stabilization efficiency strongly affected by temperature and residence time, highlighting the dual-purpose of the technology combining biomass conversion and metals immobilization [54].

Apart from the process efficiency and contaminant control, the environmental performance of thermochemical valorization routes for contaminated biomass has been increasingly studied using life cycle analysis (LCAs) approach. LCAs studies indicated that the overall sustainability of the contaminated biomass valorization strongly depends on the biomass choice and the conversion path. Vocciante et al. demonstrated that when compared to conventionl remediation techniques such as soil excavation and landfilling, phytoremediation coupled with biomass valorization results in substantially lower global warming potential, even when considered auxiliary inputs, such as fertilizers [55].

Alshehri and collaborators analysed the conversion of contaminated biomass in terms of LCA and cost-benefit analysis (CBA), observing that nowadays, most studies focused in remediation efficiency and greenhouse gas impacts while post-remediation aspects – such as ecosystem services and the end of the managements of contaminants are frequently neglected [56].

Within the scope of the project Phy2Climate it was identified that the conversion of contaminated biomass into biofuels and biochar can deliver net environmental benefits when compared with conventional remediation techniques. In the study, Silma et al. highlight that thermochemical conversion paths are suitable both for energy recovery and contaminant concentration, which should make the correct treatment easier. Regarding the environmental viability, some factors are usually not well taken into consideration, such as biomass logistics, energy integration, and the fate of the co-products across the entire life cycle [57].

Todde et al. demonstrated that industrial hemp cultivated on heavy metal contaminated soils and valorized through thermochemical paths can present primary energy savings in the order of 25.3 to 36.8 $GJ.ha^{-1}$ and avoided emissions of 579 and 641 $CO_2e.ha^{-1}$, depending on the scenario's specification. These results truly demonstrated that coupling phytoremediation with bioenergy significantly improves the sustainability of the remediation [58]. The energetic valorization of the biomass provides the largest benefits to mitigate energy related global warming emissions and ozone layer depletion, when comparing this technology with composting or mulching. However the mobilization of some contaminants into fly ash can increase the toxicity risks for the surrounding ecosystem and humans [59].

### 12.3.2 Environmental Sustainability of the Biochemical Processes Valorization of the Biomass Harvested from Contaminated Soils

The study of Black et al. [60] applied LCA methodology to assess the sustainability of the lignocellulosic bioethanol chain from the cultivation of willows used in the phytoremediation of heavy metal contaminated soils contaminated in the context of the United Kingdom. The study shows that willow grown in heavy metal contaminated soils is a high quality raw material to produce second-generation bioethanol, combining phytoremediation capacity with reduced greenhouse gas emissions compared to fossil fuels. The results seem also to indicate that the environmental performance of the system is highly sensitive to agricultural management—with GHG emissions ranging from 8.14 to 16.68 kg $CO_2$ eq./ODT depending on fertilizer use – and that technical optimization of the process allows for a 50% increase in the energy efficiency of pretreatment, resulting in an increase in the total yield of biomass conversion to ethanol of around 5%. Nevertheless, the study also highlighted some critical limitations of using willow in contaminated soils: 1) the accumulation of heavy metals in the ashes from the post-processing phase; 2) the high transpiration rates of willow, which can deplete local water resources; and 3) the biochemical complexity of the process, which, despite being efficient, produces inhibitors during the thermal or acid pretreatment phases.

The study conducted by Witters et al. [61] examined the energy production and $CO_2$ reduction potential of three energy crops—willow (*Salix* spp.), silage corn (*Zea mays* L.), and rapeseed (*Brassica napus* L.)—grown on soil contaminated with Cd, Zn, and Pb in the Campine region (Belgium and The Netherlands).The work studied the conversion of corn for silage into biogas through anaerobic digestion. The authors used small scale batch tests (14 days) in which they compared the biogas production potential of corn grown in contaminated soils versus uncontaminated soils. The study concluded that there were no significant differences in biogas production between contaminated corn (215 ± 23 $Nm^3$/Mg) and the reference (194 ± 4 $Nm^3$/Mg), which points to the viability of the biochemical process even in the presence of metals. According to the authors, it appears that metals are concentrated entirely in the digestate and not in the biogas. Among the studied crops, corn showed the best overall performance, with a yield of 20 Mg dm $ha^{-1}$ $y^{-1}$ combined with high biogas production, achieving the highest net $CO_2$ reduction (14,214 kg $ha^{-1}$ $y^{-1}$) and net energy production (148,289 MJ $ha^{-1}$ $y^{-1}$). The thermochemical conversion of willow biomass (co-combustion) showed a calorific value of 19.92 GJ/Mg dm (not different from willow cropped in non-contaminated soils) and biomass yields ranging from 6 to 15.6 Mg dm $ha^{-1}$ $y^{-1}$. The process resulted in a reduction of up to 9,134 kg $CO_2$ $ha^{-1}$ $y^{-1}$. Rapeseed was mechanically processed to obtain the oil for biodiesel production (after transesterification). Despite returning a net energy value of up to 35,640 MJ $ha^{-1}$ $y^{-1}$, rapeseed has the lowest environmental mitigation potential among the crops studied, with a maximum reduction of 3,165 kg $CO_2$ $ha^{-1}$ $y^{-1}$. This study shows that phytoremediation technology avoids the energy and $CO_2$ costs of conventional methods, allowing the identification of the crop that maximizes energy production and reduces emissions per hectare.

The study conducted by Andersson-Sköld et al. [62] focused on the cultivation of corn, *Salix* spp. sunflower, wheat, miscanthus, alfalfa, and potatoes grown on contaminated land (brownfields) or in areas unsuitable for food production. The authors specifically mention heavy metal–contaminated soils sites, such as Micasasa (Copsa Mica) in Romania, where the soil is severely contaminated by heavy metal deposition: corresponding to an area of 900,000 ha, or about 0.4% of the country's total area. Seen from another perspective, the energy potential of that area could correspond to about 1% of the approximately 400 TWh of Romania's total annual energy consumption (in 2015). The article discusses the pros and cons of bioenergy production with biomass harvested from contaminated soils, noting that the total environmental impact depends on the entire cultivation chain, including the waste management phase, a key component in environmental and risk assessments. Anaerobic digestion and alcoholic fermentation processes are referred to as generators of a wet, carbon-rich solid residue (sludge or digestate) that, when properly managed, can be used to increase carbon levels in the soil (soil carbonization) and quickly provide nutrients. Although first-generation biofuels (such as wheat ethanol: 25 $gCO_2eq$/MJ) exhibit a considerably lower global warming potential (GWP) than gasoline (80 $gCO_2eq$/MJ), a major concern lies in the concentration of pollutants in the soil (through the application of sludge/digestate). The case study from Romania provides an illustration of this challenge and related risk potential. The uptake of contaminants in heavily contaminated areas with heavy metals by crops such as sunflower, wheat, and others (except corn) was above the reference values for grazing animals' purposes. The most critical outcome, though, is that the current use of these plants in the area, especially for feeding domestic livestock, poses a greater risk than using them for bioenergy. The development of standardized commercial methods for enriching and removing metals and nutrients from organic waste (before being placed in the soil or discarded) could increase the benefits in terms of sustainability. Moreover, using biodegradable organic waste (like digestate) could help capture carbon in the soil, with sequestration estimates ranging from 2 to 20 million tons of $CO_2eq$ per year.

The CERESiS project [63], focused on the remediation of heavy metal-contaminated soils through the application of energy crops, using plant based remediation to produce liquid biofuels. Regarding the biochemical process, the authors note that the conversion of contaminated material into ethanol may require additional chemical pre-processing (such as diluted acid or an alkaline catalyst) to decontaminate the feedstock: treatment with diluted acid can extract 90% of Mn and Zn. But, the pretreatment phase generates an effluent rich in heavy metals (hazardous). In addition, the impact of heavy metals on biochemical conversion can be both positive and negative: (i) the presence of Zn and Cd in biomass is detrimental to enzymatic hydrolysis; (ii) Cr in biomass is beneficial to enzymatic hydrolysis. However, Cd and Cr demonstrated a positive impact on simultaneous saccharification and fermentation (SSF), with ethanol yield from Cd samples increasing by 8–12%.

Abreu et al. [18] examined the cultivation of *Cynara cardunculus*, *A. donax*, *Miscanthus × giganteus*, sorghum, and *Salix viminalis* on marginal, degraded, and contaminated soils. This is a comprehensive review article that evaluates the production of biodiesel from cardoon and also emphasizes biochemical processes, including: (i) anaerobic digestion for biogas/biomethane production (e.g., cardoon, sorghum); (ii) alcoholic fermentation for bioethanol (e.g., cardoon, giant reed, paulownia, willow, poplar); (iii) production of other biofuels such as biobuthanol and biohydrogen. Regarding the bioethanol production, the review points to paulownia cultivation as offering high production potential, between 10,779 and 13,300 L bioethanol $ha^{-1}year^{-1}$. Willow is also mentioned by the authors as having the potential to achieve a conversion yield of 310 L ethanol.$t^{-1}$ (dry biomass). Alkaline pretreatment of residual hemp hurds biomass increased the biochemical methane potential by up to 28.5%. Furthermore, high volumes of biomethane can be obtained from cultivated cardoon compared to the wild varieties. Fermentation is considered a viable option for contaminated bioenergy feedstocks, even though the presence of metals can inhibit the process. Some metals, such as Zn, Ni, and As, have no effect on enzymic saccharification. On the other hand, metals like Pb, act as strong inhibitors, whereas Cd and Cu are moderate inhibitors. This study highlights the potential for bioenergy crops to reduce greenhouse gas emissions and improve soil quality. Another crucial point observed in the study is that biodiesel production from cardoon can have negative GHG emissions (–1.5 g $CO_2$eq $MJ^{-1}$) thanks to the positive credits provided by by-products (such as bagasse used for animal feed or biofertilizer).

Eranki and Dale [64] evaluated the cultivation of cellulosic raw materials from low-maintenance perennial grasses, including elephant grass and miscanthus, on marginal soils, such as land not used for conventional crops, abandoned and degraded agricultural land. The approach involved conducting a comparative life cycle assessment (LCA) study between distributed and centralized biomass processing systems, and the functional unit was described as a 5,000 tons per day biorefinery, focused on the production of lignocellulosic bioethanol. Ammonia fiber expansion process (AFEX) is the main biochemical pretreatment used, and relevant co-products include electricity (generated by burning lignin-rich waste) as well as animal feed. Distributed processing systems appear to generate 3.7% fewer GHG emissions than centralized systems on average (with a densified biomass), and perennials such as switchgrass exhibit the greatest potential for soil carbon sequestration. The distributed system consistently has a higher net energy yield (NEY) than the centralized system if combined with landscapes dominated by high-yield perennial grasses. The yield of perennial grasses is a key parameter that greatly affects the results of NEY (changes from 15 to 50%) and net carbon emission reduction (changes from 20 to 65%).

Vigil et al. [65] examined the sustainability of phytoremediation of Pb contaminated soils in sites where the metal concentration exceeds the legal reference value (Asturias, Northern Spain). They used the forage plant *Melilotus alba*, which is a hyperaccumulator suitable for phytoextraction. A comparative cradle-to-grave LCA was conducted to evaluate the environmental performance of the complete remediation system, whose functional unit was adopted as one hectare of contaminated soil, and whose analysis focused on the complete estimated remediation time, starting from the initial state of the soil until reaching the target concentration of 70 mg Pb.$kg^{-1}$ DM (after 32 years). The Baseline scenario used biomass as raw material for anaerobic co-digestion to produce biogas (60% methane) and digestate. Biogas was then cryogenically upgraded to produce synthetic natural gas (SNG) to suit the natural gas distribution network through a process that removes $CO_2$. Removed $CO_2$ is then captured and sent to a soft drink producer, thus avoiding its emissions into the atmosphere. Digested material is partially used on site to provide the nitrogen needed for the next growing season, with the remainder sold as a substitute for urea (fertilizer). Baseline scenario (phytoremediation with anaerobic co-digestion and SGN production) was the only one to show a net environmental benefit. The positive impact of replacing natural gas is enough to offset the large impacts caused by anaerobic co-digestion and cryogenic upgrading, with the latter mainly due to the electricity and gas burned to heat the process. Co-production of digestate is also a positive factor, avoiding the impacts associated with mineral fertilizer production. Concerning the sustainability of phytoremediation, they note that the distance from the site to the treatment center (up to 300 Km) is not a critical factor to the sustainability of the treatment scheme. However, due to the transport of soil, distance is a key factor when comparing phytoremediation with conventional off-site techniques (excavation).

The study by Lynch and Satrio [66] assessed the use of switchgrass both as a high yield feedstock for bioenergy and as a phytoextraction agent in soils contaminated with Pb of areas contaminated by industrial or mining activities. LCA analyzed the environmental impacts of producing bioethanol, bioenergy, and biochemicals from switchgrass grown on contaminated soil, and compared the system to a fossil fuel based refinery combined with *ex-situ* lead extraction. Complete remediation system is designed to remediate the soil in 3.5 years. Biorefinery system uses a combination of biochemical and thermochemical processes to make sugars and bio-oil. Biomass is then hydrolyzed (acid and enzymatic) into a sugar solution, followed by fermentation to make bioethanol. Switchgrass biomass is first dried and ground, then subjected to acid hydrolysis with 4% phosphoric acid and enzymatic hydrolysis. Through these processes, 65% of the cellulose and 35% of the hemicellulose are then converted to C6 and C5 sugars, respectively, transferring 97% of the Pb to the liquid solution. Then, the Pb is separated by precipitation using calcium carbonate ($CaCO_3$). The sugar solution, now free of lead, is sent to a fermenter where saccharification and simultaneous co-fermentation occur. The sugar monomers (C6 and C5) are converted into bioethanol with yields of 92.5% and 85%, respectively, followed by distillation for purification. Lynch and Satrio presented a combined system (phytoremediation + biorefinery) that is more environmentally viable than the conventional system, with total GHG emissions being three times lower. Bioethanol has the highest yield in the system, yielding 123.9 GJ/ha over the total remediation period (3.5 years). Energy costs for Pb extraction (including calcium carbonate and separation) resulted in a lower energy return on investment (EROI) (1.53) compared to clean biomass (2.54), however the system remains environmentally viable (EROI > 1).

Todde et al. [58] evaluated the cultivation of industrial hemp (*C. sativa* L.) for phytoremediation of heavy metal-contaminated soils. Research was conducted at two sites in Sardinia, Italy: S1 (low contamination by heavy metals, mainly Pb and Zn) and S2 (high contamination by Pb, Zn, Cd, and Co). Given that contaminated digestate cannot be used for agricultural purposes in Italy, it was assumed that it would be dehydrated (using heat from cogeneration) and then incinerated to generate electricity, this being the most sustainable option. In a biochemical focus scenario, raw hemp biomass was harvested and silage at the anaerobic digestion plant (ADP). Both the residual straw and silage underwent mechanical pretreatment followed by anaerobic digestion under mesophilic conditions, converting organic matter into biogas (mainly methane). Biogas is then burned in a cogeneration (CHP) unit to generate electricity (sent to the national grid) and heat. Surplus heat from the CHP is used to dehydrate contaminated digestate. The scenario fresh biomass for anaerobic digestion followed by digestate incineration, proved to be efficient in terms of energy and the environment, especially due to the high biomass production. However, the lower the yields the higher the environmental impacts per kg of dry matter. Regarding the results per hectare, primary energy savings were estimated at 36.8 GJ $ha^{-1}$, avoiding the emission of 641 kg $CO_2$ eq per hectare remediated. Electricity production results from the anaerobic digestion subsystem (using fresh biomass) were 0.73 kWh $kg^{-1}$ (dry matter). Due to the recovery of contaminated biomass in anaerobic digestion subsystem, the energy saved (compared to electricity from the national grid) was

6.25 MJ $kg^{-1}$ (dry matter) of primary energy, and the emission of 191 g $CO_2$ eq $kg^{-1}$ (dry matter) was avoided. Considering the overall results of the study, on average, the supply chain enabled energy savings of approximately 7.82 MJ $kg^{-1}$ (dry matter) and avoided the emission of 137 g $CO_2$ eq $kg^{-1}$ (dry matter).

Mata et al. [67] performed a gate-to-gate LCA to assess the environmental impacts of bioethanol production from corn stover grown on heavy metal contaminated soils for phytoremediation purposes in Estarreja, Portugal. The study was focused on the optimization of the biochemical conversion process, comparing specifically the use of four different acids ($H_2SO_4$, $HNO_3$, HCl, and $CH_3COOH$) for the essential pre-treatment step. Due to their optimized processes, sulfuric and hydrochloric acids showed better environmental performance, while nitric acid had the greatest impact. Authors recommend that renewable energy should be used in the process to reduce fossil fuel consumption.

## 12.4 CONCLUSIONS

Phytoremediation of heavy metal-contaminated soils using industrial crops emerges as a promising and multifunctional approach to address soil degradation while contributing to renewable energy production and broader sustainability goals. As discussed throughout this chapter, industrial crops combine tolerance to heavy metal stress with high biomass productivity, enabling the stabilization or gradual removal of contaminants while restoring key ecosystem functions such as soil structure, carbon sequestration, and erosion control. When cultivated on contaminated lands, these systems avoid competition with food production and offer a viable alternative to conventional, often costly, remediation techniques.

Although industrial crops typically accumulate lower concentrations of metals compared to hyperaccumulator species, their substantial biomass production enables significant contaminant removal per unit area, especially when applied in phytoextraction or phytostabilization approaches. In addition to site remediation, converting contaminated biomass through thermochemical and biochemical processes improves the overall sustainability of these systems. Technologies such as combustion, pyrolysis, gasification, anaerobic digestion, and fermentation can produce bioenergy while simultaneously concentrating, stabilizing, or facilitating the recovery of heavy metals in controlled settings, thereby supporting circular economy objectives.

Life cycle assessment findings consistently show that integrating phytoremediation with bioenergy generation can surpass traditional remediation methods regarding greenhouse gas reductions, net energy performance, and resource use efficiency. However, the environmental outcomes of such combined systems are strongly dependent on site-specific and contextual conditions, including the degree of contamination, crop type, agricultural practices, conversion technologies employed, and the management of residues and emissions. Potential challenges—such as metal volatilization during high-temperature treatments, depletion of water and soil nutrients, and the safe handling of contaminated by-products—require careful mitigation through suitable technological solutions and robust regulatory oversight.

Overall, phytoremediation with industrial crops represents a sustainable phytomanagement strategy capable of transforming contaminated lands from environmental liabilities into productive assets. Future research should focus on improving plant–soil–microbe interactions, optimizing biomass conversion technologies, and developing integrated assessment frameworks that fully capture long-term environmental, economic, and social benefits. When properly designed and managed, these systems can play a meaningful role in land restoration, climate change mitigation, and the transition toward a more resilient and circular bioeconomy.

## ACKNOWLEDGMENTS

This work was supported by Fundação para a Ciência e Tecnologia (FCT)/MECI through the MEtRICs research unit (UID/04077/2025, https://doi.org/10.54499/UID/04077/2025, Mechanical Engineering and Resource Sustainability Center) and by national funds from FCT, I.P., under the scope of the projects UID/50006/2025, UID/PRR/50006/2025 and LA/P/0008/2020 of the Associated Laboratory for Green Chemistry—LAQV REQUIMTE (https://doi.org/10.54499/UID/50006/2025, https://doi.org/10.54499/UID/PRR/50006/2025 and https://doi.org/10.54499/LA/P/0008/2020).

## REFERENCES

1. Kuang, Y. Y., Chen, X., Zhu, C. 2024. Characteristics of soil heavy metal pollution and health risks in Chenzhou City. *Processes*, 12 (3): 623. https://doi.org/10.3390/pr12030623.
2. Testa, G., Corinzia, S. A., Cosentino, S. L., Ciaramella, B. R. 2023. Phytoremediation of cadmium-, lead-, and nickel-polluted soils by industrial hemp. *Agronomy*, 13 (4): 995. https://doi.org/10.3390/agronomy13040995.
3. Barbosa, B., Costa, J., Boléo, S., Duarte, M. P., Fernando, A. L. 2016. Phytoremediation of inorganic compounds. In: Ribeiro AB, Mateus EP, Couto N (Eds) *Electrokinetics Across Disciplines and Continents: New Strategies for Sustainable Development*, Springer International Publishing, Switzerland, 469 p., pp. 373–400. https://doi.org/10.1007/978-3-319-20179-5_19.
4. Agnello, A. C., Bagard, M., van Hullebusch, E. D., Esposito, G., Huguenot, D. 2016. Comparative bioremediation of heavy metals and petroleum hydrocarbons co-contaminated soil by natural attenuation, phytoremediation, bioaugmentation and bioaugmentation-assisted phytoremediation. *Science of the Total Environment*, 563–564: 693–703. https://doi.org/10.1016/j.scitotenv.2015.10.061.
5. Do Nascimento, C. W. A., Xing, B. 2006. Phytoextraction: A review on enhanced metal availability and plant accumulation. *Scientia Agricola*, 63 (3): 299–311. https://doi.org/10.1590/S0103-90162006000300014.
6. Papazoglou, E. G., Fernando, A. L. 2017. Preliminary studies on the growth, tolerance and phytoremediation ability of sugarbeet (*Beta vulgaris* L.) Grown on heavy metal contaminated soil. *Industrial Crops and Products*, 107: 463–471. https://doi.org/10.1016/j.indcrop.2017.06.051
7. Limmer, M., Burken, J. 2016. Phytovolatilization of organic contaminants. *Environmental Science and Technology*, 50: 6632–6643. https://doi.org/10.1021/acs.est.5b04113.

8. Ali, A., Khan, E., Sajad, M. A. 2013. Phytoremediation of heavy metals: Concepts and applications. *Chemosphere*, 91 (7): 869–81. https://doi.org/10.1016/j.chemosphere.2013.01.075.
9. Gomes, L., Costa, J., Moreira, J., Cumbane, B., Abias, M., Santos, F., Zanetti, F., Monti, A., Fernando, A. L. 2022. Switchgrass and giant reed energy potential when cultivated in heavy metals contaminated soils. *Energies*, 15: 5538. https://doi.org/10.3390/en15155538.
10. Testa, G., Ciaramella, B. R., Fernando, A. L., Kotoula, D., Scordia, D., Gomes, L. A., Cosentino, S. L., Alexopoulou, E., Papazoglou, E. G. 2024. Harnessing lignocellulosic crops for phytomanagement of contaminated soils: A multi-country study. *Plants*, 13: 2671. https://doi.org/10.3390/plants13192671
11. Fernando, A. L., Rettenmaier, N., Soldatos, P., Panoutsou, C., 2018. Sustainability of perennial crops production for bioenergy and bioproducts. In: Alexopoulou, E. (Ed.), Perennial Grasses for Bioenergy and Bioproducts. Elsevier, London, pp. 245–283. https://doi.org/10.1016/B978-0-12-812900-5.00008-4.
12. Pires, J. R. A., Souza, V. G. L., Gomes, L. A., Coelhoso, I. M., Godinho, M. H., Fernando, A. L. 2022. Micro and nanocellulose extracted from energy crops as reinforcement agents in chitosan films. *Industrial Crops and Products*, 186: 115247. https://doi.org/10.1016/j.indcrop.2022.115247.
13. Dias, C., Mersmann, L., Pires, J. R. A., Fernando, A. L. 2025. From lignocellulosic biomass to nanolignin: Sustainable solutions for a greener future. *Industrial Crops and Products*, 238: 122286. https://doi.org/10.1016/j.indcrop.2025.122286.
14. Scordia, D., Papazoglou, E. G., Kotoula, D., Sanz, M., Ciria, C. S., Pérez, J., Co-sentino, S. L. 2022. Towards identifying industrial crop types and associated agronomies to improve biomass production from marginal lands in Europe. *GCB Bioenergy*, 14: 710–734. https://doi.org/10.1111/gcbb.12935.
15. Barbosa, B., Fernando, A. L. 2021. Ecological and ecosystem engineering for economic environmental revitalization. In: Prasad, M.N.V. (Ed) *Handbook of Ecological and Ecosystem Engineering*. John Wiley & Sons Ltd, UK, pp. 25–46. https://doi.org/10.1002/9781119678595.ch2
16. Blanco, H. 2016. Growing dedicated energy crops on marginal lands and ecosystem services. *Soil Science Society of America Journal*, 80 (4): 845–858. https://doi.org/10.2136/sssaj2016.03.0080
17. Cao, X., Dong, Q., Mao, L., Yang, X., Wang, X., Zou, Q. 2025. Enhanced phytoextraction technologies for the sustainable remediation of cadmium-contaminated soil based on hyperaccumulators: A review. *Plants*, 14 (1): 115. https://doi.org/10.3390/plants14010115.
18. Abreu, M., Silva, L., Ribeiro, B., Ferreira, A., Alves, L., Paixão, S. M., Gouveia, L., Moura, P., Carvalheiro, F., Duarte, L. C., Fernando, A. L., Reis, A., Gírio, F. 2022. Low indirect land use change (ILUC) energy crops to bioenergy and biofuels: A review. *Energies*, 15 (12): 4348. https://doi.org/10.3390/en15124348.
19. Barbosa, B., Costa, J., Boléo, S., Duarte, M. P., Fernando, A. L. 2015. Phytoremediation of heavy metal-contaminated soils using the perennial energy crops *Miscanthus* spp. and *Arundo donax* L. *BioEnergy Research*, 8: 1500–11. https://doi.org/10.1007/s12155-015-9688-9.
20. Ali, A., Anwar, S., Abbas, M., Batool, S. W., Zainab, S., Bajwa, S. 2025. Biomass crops for sustainable energy in Europe: Contributions to bioenergy, environment, and energy security. *Journal of Plant Interactions*, 20 (1): 2564839. https://doi.org/10.1080/17429145.2025.2564839.
21. Kabata-Pendias, A. 2011. *Trace Elements in Soils and Plants*. 4th ed. Boca Raton: CRC.

22. Guo, Z. H., Miao, X. F. 2010. Growth changes and tissues anatomical characteristics of giant reed (*Arundo donax* L.) in soil contaminated with arsenic, cadmium and lead. *Journal of Central South University of Technology*, 17: 770–777. https://doi.org/10.1007/s11771-010-0555-8
23. Yang, X., Feng, Y., He, Z., Stoffella, P. 2005. Molecular mechanisms of heavy metal hyperacumulation and phytoremediation. *J. Trace Elem Med Biol*, 18: 339–353. https://doi.org/10.1016/j.jtemb.2005.02.007.
24. United Nations Framework Convention on Climate Change (UNFCCC). 2015. The Paris Agreement. Accessed: Nov. 2025. https://unfccc.int/process-and-meetings/the-paris-agreement
25. UNFCCC. 2023. Outcome of the first global stocktake, Draft decision -/CMA.5 (FCCC/PA/CMA/2023/L.17), Dec. 2023. Accessed: Nov. 2025. https://unfccc.int/sites/default/files/resource/cma2023_L17_adv.pdf
26. European Union. 2022. Regulation (EU) 2021/1119 (European Climate Law) and REPowerEU Plan, COM (2022) 230 final. Accessed: Nov. 2025. https://eur-lex.europa.eu/legal-content/EN/TXT/?uri=CELEX:32021R1119
27. Baidoo, E. B., Tulashie, S. K., Alale, E. M., Munumkum, C. A., Acquah, D., Agudah, P. Q., Asante, P. A. 2025. Revolutionizing bioenergy production: A review on sustainable biomass feedstock. *Biomass and Bioenergy*, 201: 108113. https://doi.org/10.1016/j.biombioe.2025.108113.
28. Mignogna, D., Szabó, M., Ceci, P., Avino, P. 2024. Biomass energy and biofuels: Perspective, potentials, and challenges in the energy transition. *Sustainability*, 16: 7036. https://doi.org/10.3390/su16167036.
29. European Commission. 2012. Indirect Land Use Change (ILUC), European Commission – Press Corner, MEMO/12/787, Brussels, Belgium, Oct. 17, 2012. Accessed: Nov. 2025. https://ec.europa.eu/commission/presscorner/detail/en/memo_12_787.
30. Mukherjee, S., Leri, A. C., Bandaranayaka, C., Vázquez-Núñez, E., Barros, R., Khan, A. H. A., Zhou, P., Zhang, T., Bernal, M. P., Clemente, R., Bolan, N. 2025. Sustainable management of post-phytoremediation biomass. *Energy, Ecology and Environment*, 10: 675–709. https://doi.org/10.1007/s40974-025-00364-w.
31. Bao, C., Cao, Y., Zhao, L., Li, X. 2025. Biofuel production from phytoremediated biomass via various conversion routes: A review. *Energies*, 18 (4): 822. https://doi.org/10.3390/en18040822.
32. Espada, J. J., Rodríguez, R., Delgado, A., Vicente, G., Bautista, L. F. 2024. Assessing environmental sustainability of phytoremediation to remove copper from contaminated soils. *Sustainability*, 16: 2441. https://doi.org/10.3390/su16062441.
33. Xue, Z. C., Liu, J., Zhang, C. M., Lv, P., Jiao, Z. Y., Li, S. Z. 2025. Unlocking bioenergy potential in heavy metal polluted soils: Synergistic advanced solid-state fermentation and combustion for valorization of sweet sorghum. *BMC Plant Biology*, 25: 1784. https://doi.org/10.1186/s12870-025-08005-4.
34. Sui, H., Jiang, H., Li, G., Chen, J., Cheng, W., Chen, H. 2025. Combustion of phytoremediation biomass for recycling utilization: Thermal behavior, kinetic analysis, and release of heavy metal. *Industrial Crops & Products*, 230: 121047. https://doi.org/10.1016/j.indcrop.2025.121047.
35. Zha, J., Huang, Y., Zhu, Z., Yu, M., Clough, P. T., Yan, Y., Dong, L., Cheng, H. 2021. Dynamic transformations of metals in the burning solid matter during combustion of heavy metal-contaminated biomass. *ACS Sustainable Chemistry & Engineering*, 9 (20): 7063–7073. https://doi.org/10.1021/acssuschemeng.1c01048.
36. Espada, J. J., Rodríguez, R., Gari, V., Salcedo-Abraira, P., Bautista, L. F. 2022. Coupling phytoremediation of Pb-contaminated soil and biomass energy production: A comparative life cycle assessment. *Science of the Total Environment*, 840: 156675. https://doi.org/10.1016/j.scitotenv.2022.156675.

37. Dinh, T., Kovács, H., Dobó, Z. 2024. The fate of noble metals and rare earth elements during pelletized biomass combustion. *Heliyon*, 10: e23546. https://doi.org/10.1016/j.heliyon.2023.e23546.
38. Ma, R., Wang, H., Sun, Z., Gu, X. 2025. Release and ecological risks of heavy metals during coal combustion in coal-fired power plants. *Applied Sciences*, 15 (21): 11360. https://doi.org/10.3390/app152111360.
39. Zeng, K., Li, R., Minh, D. P., Weiss-Hortala, E., Nzihou, A., He, X., Flamant, G. 2019. Solar pyrolysis of heavy metal contaminated biomass for gas fuel production. *Solar Energy*, 193: 71–81. https://doi.org/10.1016/j.solener.2019.09.017.
40. Park, K.-B., Chae, D.-Y., Fini, E. H., Kim, J.-S. 2024. Pyrolysis of biomass harvested from heavy-metal contaminated area: Characteristics of bio-oils and biochars from batch-wise one-stage and continuous two-stage pyrolysis. *Chemosphere*, 355: 141715. https://doi.org/10.1016/j.chemosphere.2024.141715.
41. Amato, D., Giudicianni, P., Grottola, C. M., Stanzione, F., Ragucci, R. 2025. Effect of bonding and speciation of pb on contaminated biomass pyrolysis. *Energy & Fuels*, 39 (25): 15310–15319. https://doi.org/10.1021/acs.energyfuels.5c02798.
42. Özkan, A., Günkaya, Z., Banar, M., Kulaç, A., Yalçın, G., Taşpınar, K., Altay, A. 2015. Pyrolysis of zinc contaminated biomass from phytoremediation. *Anadolu University Journal of Science and Technology A – Applied Sciences and Engineering*, 16 (3): 385–393. https://doi.org/10.18038/btda.70683.
43. Han, Z., Guo, Z., Zhang, Y., Xiao, X., Peng, C. 2018. Potential of pyrolysis for the recovery of heavy metals and bioenergy from contaminated *Broussonetia papyrifera* biomass. *BioResources*, 13 (2): 2932–2944. https://doi.org/10.15376/biores.13.2.2932-2944.
44. Debela, F., Thring, R. W., Arocena, J. M. 2012. Immobilization of heavy metals by co-pyrolysis of contaminated soil with woody biomass. *Water, Air, & Soil Pollution*, 223 (3): 1161–1170. https://doi.org/10.1007/s11270-011-0934-2.
45. Voglar, J., Prašnikar, A., Moser, K., Carlon, E., Schwabl, M., Likozar, B. 2024. Pyrolysis of industrial hemp biomass from contaminated soil phytoremediation: Kinetics, modelling transport phenomena and biochar-based metal reduction. *Thermochimica Acta*, 742: 179899. https://doi.org/10.1016/j.tca.2024.179899.
46. He, J., Strezov, V., Zhou, X., Kumar, R., Kan, T. 2020. Pyrolysis of heavy metal contaminated biomass pre-treated with ferric salts: Product characterisation and heavy metal deportment. *Bioresource Technology*, 313: 123641. https://doi.org/10.1016/j.biortech.2020.123641.
47. Ritz, M., Dossow, M., Mörtenkötter, H., Spliethoff, H., Fendt, S. 2025. Experimental investigation of heavy metal release in entrained-flow biomass gasification. *Fuel*, 387: 134379. https://doi.org/10.1016/j.fuel.2025.134379.
48. Werle, S., Ziółkowski, Ł, Bisorca, D., Burnete, D., Pogrzeba, M., Krzyżak, J., Ratman-Kłosińska, I. 2017. Heavy metal contaminated energy crops as a local energy carrier: Fixed bed gasification process of the pretreated feedstock. *Chemical Engineering Transactions*, 61: 1393–1398. https://doi.org/10.3303/CET1761230.
49. Cui, X., Shen, Y., Yang, Q., Kawi, S., He, Z., Yang, X., Wang, C.-H. 2018. Simultaneous syngas and biochar production during heavy metal separation from Cd/Zn hyperaccumulator (*Sedum alfredii*) by gasification. *Chemical Engineering Journal*, 347: 543–551. https://doi.org/10.1016/j.cej.2018.04.150.
50. Ancona, V., Barra Caracciolo, A., Campanale, C., De Caprariis, B., Grenni, P., Uricchio, F., Borello, D. 2019. Gasification treatment of poplar biomass produced in a contaminated area restored using plant assisted bioremediation. *Journal of Environmental Management*, 239: 137–141. https://doi.org/10.1016/j.jenvman.2019.03.019.
51. Jiang, Y., Ameh, A., Lei, M., Duan, L., Longhurst, P. 2016. Solid–gaseous phase transformation of elemental contaminants during the gasification of biomass. *Science of the Total Environment*, 563–564: 724–730. https://doi.org/10.1016/j.scitotenv.2015.11.017.

52. Dutzi, J., Boukis, N., Sauer, J. 2024. Supercritical water gasification of heavy metal contaminated plants with focus on separation of heavy metal contaminants. *Biomass and Bioenergy*, 182: 107059. https://doi.org/10.1016/j.biombioe.2024.107059.
53. Su, W., Liu, P., Cai, C., Ma, H., Jiang, B., Xing, Y., Liang, Y., Cai, L., Xia, C., Le, Q. V., Sonne, C., Lam, S. S. 2021. Hydrogen production and heavy metal immobilization using hyperaccumulators in supercritical water gasification. *Journal of Hazardous Materials*, 402: 123541. https://doi.org/10.1016/j.jhazmat.2020.123541.
54. Wang, J., Zhang, S., Wei, C., Hou, H., Song, G., Cao, L., Zhang, J. 2025. Hydrothermal carbonization of heavy metal-contaminated biomass: Migration, transformation, and ecological stability changes of metals. *International Journal of Molecular Sciences*, 26 (6): 2551. https://doi.org/10.3390/ijms26062551.
55. Vocciante, M., Caretta, A., Bua, L., Bagatin, R., Franchi, E., Petruzzelli, G., Ferro, S. 2019. Enhancements in phytoremediation technology: Environmental assessment including different options of biomass disposal and comparison with a consolidated approach. *Journal of Environmental Management*, 237: 560–568. https://doi.org/10.1016/j.jenvman.2019.02.104.
56. Alshehri, K., Gao, Z., Harbottle, M., Sapsford, D., Cleall, P. 2023. Life cycle assessment and cost-benefit analysis of nature-based solutions for contaminated land remediation: A mini review. *Heliyon*, 9: e20632. https://doi.org/10.1016/j.heliyon.2023.e20632.
57. Simla, T., Korus, A., Petela, K., Stanek, W., Ortner, M., Szlęk, A. 2025. Phy2Climate: Life cycle assessment of phytoremediation combined with biofuel production. *Renewable Energy*, 247: 122979. https://doi.org/10.1016/j.renene.2025.122979.
58. Todde, G., Carboni, G., Marras, S., Caria, M., Sirca, C. 2022. Industrial hemp (*Cannabis sativa* L.) for phytoremediation: Energy and environmental life cycle assessment of using contaminated biomass as an energy resource. *Sustainable Energy Technologies and Assessments*, 52: 102081. https://doi.org/10.1016/j.seta.2022.102081.
59. Brás, I., Silva, E., Raimondo, R., Saetta, R., Mignano, V., Fabbricino, M., Ferreira, J. 2024. Valorisation of forest and agriculture residual biomass—The application of life cycle assessment to analyse composting, mulching, and energetic valorisation strategies. *Sustainability*, 16 (2): 630. https://doi.org/10.3390/su16020630.
60. Black, M. J., Whittaker, C., Hosseini, S. A., Diaz-Chavez, R., Woods, J., Murphy, R. J. 2011. Life cycle assessment and sustainability methodologies for assessing industrial crops, processes and end products. *Industrial Crops and Products*, 34: 1332–1339. https://doi.org/10.1016/j.indcrop.2010.12.002.
61. Witters, N., Mendelsohn, R. O., Van Slycken, S., Weyens, N., Schreurs, E., Meers, E., Tack, F., Carleer, R., Vangronsveld, J. 2011. Phytoremediation, a sustainable remediation technology? Conclusions from a case study. I: Energy production and carbon dioxide abatement. *Biomass and Bioenergy*, 35 (11): 4545–4554. https://doi.org/10.1016/j.biombioe.2011.08.016.
62. Andersson-Sköld, Y., Hagelqvist, A., Crutu, G., Blom, S. 2014. Bioenergy production on contaminated land—Opportunities and risks. *Journal of Environmental Management*, 145: 230–239. https://doi.org/10.1080/17597269.2014.996728.
63. Giannopoulos, D., Katsifis, I. 2022. D1.7: Integrated solution pathways (CERESiS: Contaminated land remediation through energy crops for soil improvement to liquid biofuel strategies; Grant Agreement No. 101006717). National Technical University of Athens.
64. Eranki, P. L., Dale, B. E. 2011. Comparative life cycle assessment of centralized and distributed biomass processing systems combined with mixed feedstock landscapes. *GCB Bioenergy*, 3: 427–438. https://doi.org/10.1111/j.1757-1707.2011.01096.x.
65. Vigil, M., Marey-Pérez, M. F., Martinez Huerta, G., Álvarez Cabal, V. 2015. Is phytoremediation without biomass valorization sustainable? Comparative LCA of landfilling vs. anaerobic co-digestion. *Science of the Total Environment*, 505: 844–850. https://doi.org/10.1016/j.scitotenv.2014.10.047.

66. Lynch, M. N., Satrio, J. A. 2022. Utilization of grassy biomass grown in heavy-metal contaminated soil as feedstock for bioenergy production: An LCA study. *IOP Conference Series: Earth and Environmental Science*, 1034 (1): 012022. https://doi.org/10.1088/1755-1315/1034/1/012022.
67. Mata, T. M., Rodrigues, S., Caetano, N. S., Martins, A. A. 2022. Life cycle assessment of bioethanol from corn stover from soil phytoremediation. *Energy Reports*, 8: 468–474. https://doi.org/10.1016/j.egyr.2022.01.059.

# 13 Recent Advances in Dansyl-Based Fluorescent Probes for Heavy Metal Sensing and Quantification in Chemical and Biological Environments

*Frederico Duarte, Inês Pereira-Gomes, Hugo Miguel Santos, José Luis Capelo-Martínez, Elisabete Oliveira, and Carlos Lodeiro*

A fluorescent molecular sensor is a (supra)molecular moiety that can undergo an interaction with an analyte, which causes a change in its own photophysical properties, effectively transforming that interaction into a measurable signal. Among various applications, fluorescent molecular sensors can be used for the detection of metallic cations, anions, neutral molecules, and gases.[1,2] These systems allow non-destructive and real-time visual detection of the analyte with high sensitivity, low cost, and experimental simplicity.[3] Fluorophores can be distinguished into three classes: the first and second classes differ depending on whether the interaction occurs through collision or binding with the analyte, while the third class requires special attention to its design, since the receptor may or may not be linked to the fluorophore by a spacer.[1]

Several recognition mechanisms arise depending on the interaction between the fluorophore and the target analyte and can be divided into five types, namely, photo-induced electron transfer (PET), intramolecular charge transfer (ICT), fluorescence resonance energy transfer (FRET), excited state intramolecular proton transfer (ESIPT), and aggregation-induced emission (AIE).[4,5]

Fluorescent sensors that operate through PET often rely on three integrated components: a fluorophore, which can function as either an electron donor or acceptor, a recognition unit, and a spacer that maintains these two elements at an appropriate electronic distance. In the absence of analyte, excitation of the fluorophore promotes electronic communication with the receptor, enabling PET.[4–8] This electron transfer

 DOI: 10.1201/9781003632092-14

process causes the suppression of emission, giving the probe a weak or completely "off" fluorescence background. Once the analyte binds to the receptor, PET becomes energetically or kinetically unfavorable, allowing the fluorophore to regain its emissive pathway. The resulting fluorescence enhancement provides a clear signal for analyte detection.

PET processes can occur in two directions, giving rise to a-PET and d-PET modes.[4–8] In a-PET systems, the receptor possesses a higher-energy highest occupied molecular orbital (HOMO) than the fluorophore. Upon excitation of the fluorophore, its HOMO becomes singly occupied, providing an opportunity for an electron from the receptor's HOMO to transfer into this position. This transfer blocks radiative relaxation and quenches fluorescence. In d-PET systems, the receptor's lowest unoccupied molecular orbital (LUMO) lies between the HOMO and LUMO of the fluorophore. Upon excitation, the electron promoted to the fluorophore's LUMO is diverted into the lower-lying LUMO of the receptor, again disabling emission. Binding of the target analyte alters the electronic structure of the receptor in a way that prevents PET, restoring the fluorophore's radiative decay pathway. This produces a strong turn-on response which forms the basis for analyte recognition.

Considering a donor–acceptor molecular system, where there is integration of an electron-donating group which is conjugated to an electron-withdrawing group, there is a driving force to display ICT behavior.[4,5] In these systems, the donor contributes most of the HOMO electron density, while the acceptor dominates the LUMO. When the molecule absorbs light, electrons are promoted in a way that induces charge toward the acceptor, leading to a noticeable ICT excited state. Given that this process is highly influenced by changes in electronical distribution, an analyte that interacts with the donor or the acceptor site can alter the HOMO-LUMO energies. This results in a shift of the ICT strength alongside its photophysical response which typically appears as changes in the emission wavelength (red or blue shifts) and fluorescence quantum yield. These optical modulations form the basis for detecting analytes with D-A type sensors.[4,5] If an analyte coordinates to the acceptor site of an ICT probe, it lowers the energy of the acceptor-centered LUMO, causing the emission band to shift to longer wavelengths. In contrast, if an analyte interacts with the donor portion, it stabilizes the HOMO, usually leading to a blue shift in the fluorescence. When these spectral changes are sufficiently large, the intensities at the emission maxima of the free and bound probe can be quantitatively compared, enabling precise determination of analyte concentration.

In systems whose design involves the presence of two separate chromophores within the same molecule, these units can exchange energy through a nonradiative pathway. In such cases, the excited-state energy of one unit (the donor) is transferred to another unit (the acceptor) through long-range dipole–dipole interactions. This mechanism is known as Förster resonance energy transfer (FRET).[4,5,9–11] The effectiveness of FRET depends essentially on three factors: the distance between the donor and acceptor, the overlap between the donor's emission spectrum and the acceptor's absorption band, and the relative orientation of their transition dipoles. FRET-based sensors typically employ fluorescent donor and acceptor units so that both the emission from the donor and the emission resulting from the sensitization of the acceptor can be monitored. These dual signals allow ratiometric detection, improving accuracy in analyte measurement. In such sensors, the analyte may bind to either the donor or the acceptor site of the probe. This binding event alters the probe's structure, which in turn changes either the spacing between the two chromophores or the way their dipoles are oriented. Because these structural changes directly influence the efficiency of FRET, the resulting variation in the optical output provides a reliable method for identifying and quantifying the analyte.[4,5,9–11]

Several fluorescent probes contain structural features that allow a proton to transfer internally from one part of the molecule to another, similar to what occurs in keto-enol tautomerism. These systems can undergo excited-state intramolecular proton transfer (ESIPT), a process that relies on a proton moving from a donor site to an acceptor site only after the molecule has been promoted to its excited state.[4,5,12] In most ESIPT-active molecules, the enol form dominates in the ground state, usually because it is stabilized by intramolecular hydrogen bonding. Upon excitation, the acidity and basicity of the donor and acceptor groups change significantly, allowing rapid conversion of the enol form to the keto form via ESIPT. The resulting keto tautomer emits at a significantly longer wavelength than the enol form, producing a characteristic dual-band emission profile. When an analyte binds to an ESIPT-capable chromophore, it can disrupt or weaken the proton-transfer pathway in the excited state. This suppression of ESIPT leads to enhanced emission from the enol form and a corresponding decrease in the emission from the keto form. Because these two emission bands respond in opposite directions, ESIPT-based probes are well suited for ratiometric sensing and are particularly powerful for detecting small analytes by comparing the relative intensities of the enol and keto emissions.[4,5,12]

Since Tang and colleagues first reported AIE in 2001,[13] this photophysical phenomenon has become highly prominent in fluorescence-based analyte sensing. AIE molecules are typically faint or non-emissive as isolated molecules because their flexible structures or freely rotating, propeller-like substituents enable efficient nonradiative relaxation of their excited states.[4,5] When these molecules begin to group, whether due to increased concentration or solvent-driven factors, such as polarity, viscosity, or electrostatic attraction to oppositely charged hosts, they switch to strong fluorescence. Aggregation restricts internal motions and limits conformational freedom, reducing nonradiative decay and directing excited-state energy into radiative emission, which substantially enhances fluorescence. This behavior is the opposite of what is observed for traditional fluorophores, which are bright in dilute solutions but lose emission efficiency upon aggregation, a process known as aggregation-caused quenching (ACQ). ACQ often hinders their utility in sensing, where relatively high dye concentrations are commonly needed. AIE-active probes avoid this drawback entirely since instead of being quenched, they become strongly emissive upon aggregation. For this reason, AIE molecules have become powerful tools in sensing, as the AIE effect naturally counters the limitations imposed by ACQ.[4,5]

## 13.1 FLUOROPHORES: THE DANSYL UNIT

Fluorophores can generally be divided into three categories: organic fluorophores, nanoparticle-based fluorophores, and biological fluorophores. Nanoparticle fluorophores, such as quantum dots, are recognized for their brightness and photostability, although some concerns about biocompatibility and long-term toxicity can be raised.[14] Biological fluorophores, including fluorescent proteins, are useful for tracking within living systems. At the same time, they often offer limited spectral range and responsiveness.[15] Organic fluorophores, by contrast, are very versatile, since their chemical structures can be designed to cover a wide spectral range, achieve high quantum yields, and respond sensitively to environmental changes.[16]

Within this category lies a notable diversity of molecular families: fluorescein, rhodamine, coumarin, cyanine, BODIPY, and dansyl (Figure 13.1), among others, each providing distinct photophysical behaviors that can specifically be tuned for an application.[17–19] For example, fluorescein derivatives are known for their brightness and reversible pH sensitivity, rhodamines for their stability and sharp signal changes through spirolactam ring-opening, coumarins for their strong donor properties in energy transfer systems, and cyanines for extending detection into the near-infrared region.[17] Among these examples, dansyl chloride (DNS-Cl) stands out as a widely used and consistent fluorophore due to its accessible conjugation, predictable photophysical properties, environmental sensitivity, and compatibility with a broad range of biochemical applications.[3,20] Due to these reasons, this family of compounds was chosen as central piece in this work.

DNS-Cl is one of the earliest and most influential fluorophores introduced into biochemical research, recognized for its sensitivity to microenvironmental changes and its strong reactivity toward soft metal centers.[3,20] First described by Weber in 1952,[21] DNS-Cl was initially employed to generate fluorescent conjugates of albumin, and this work established its broad applicability in protein chemistry. Since

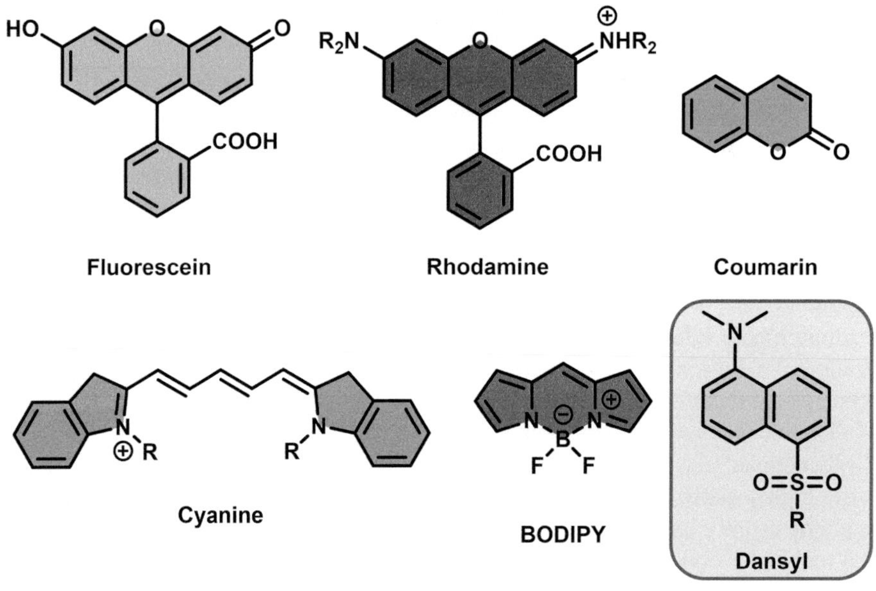


**FIGURE 13.1** Molecular structures of some common fluorophores used as fluorescent molecular sensors.

then, DNS-Cl has been widely used as a versatile reagent for labeling terminal amino groups, particularly in studies where fluorescence polarization measurements are required. Its widespread adoption was favored by its photophysical characteristics, including a fluorescence lifetime of approximately 10 ns, which is well suited for time-dependent fluorescence techniques.[3,20]

DNS-derived fluorophores exhibit distinct absorption behavior in the 275–360 nm region, corresponding to n–p* and p–p* transitions of the DNS group. They can be efficiently excited around 350 nm, a wavelength range where proteins typically do not absorb, allowing selective monitoring of the dansyl label without interference from intrinsic protein fluorescence.[3,20] The emission of the dansyl moiety is strongly influenced by solvent polarity, yielding significant Stokes shifts and notable changes in quantum yield as microenvironmental conditions vary. These properties make DNS-based fluorophores exceptionally sensitive reporters of local environmental changes.[3,20]

The historical development of DNS-Cl represents a major conceptual shift in fluorescence spectroscopy. Weber's work emphasized the importance of the molecular probe in determining the quality and interpretability of fluorescence data.[3] His contributions laid the foundation for modern fluorescence methods, demonstrating that fluorophores such as DNS-Cl could be used to extract important information about the behavior of proteins and their interactions. His insistence that effective fluorescence spectroscopy depends foremost on the choice and design of suitable probes continues to guide contemporary research.

## 13.2 DANSYL-BASED FLUORESCENT PROBES FOR DETECTION OF METAL IONS

Heavy metal ions are a significant environmental and public health concern due to their toxicity, persistence, and bioaccumulative nature. Safe and accessible water is essential for human health; however, contamination with metal ions remains prevalent, occurring from both natural processes and anthropogenic activities, such as industrial discharge, mining, and inappropriate wastewater treatment.[20,22,23] While certain metals, including iron, zinc, copper, and cobalt, are essential trace elements for biological processes, their excess or deficiency can lead to severe physiological disorders.[20,24,25] In contrast, nonessential heavy metals such as mercury, lead, and cadmium pose serious threats to ecosystems and human health, even at low concentrations.[20,22,25] Monitoring and detecting metal ions in environmental and biological systems are therefore of utmost importance.

Traditional analytical techniques, such as atomic absorption spectroscopy, inductively coupled plasma mass spectrometry (ICP-MS), and electrochemical methods, offer high sensitivity but are often costly, time-consuming, and require specialized equipment and expertise.[26–28] In contrast, fluorescent probes have surfaced as powerful tools for metal ion detection, offering advantages such as high sensitivity, rapid response, simplicity, and the potential for in situ and real-time monitoring.[25,29,30] Fluorescent sensors use the mechanisms discussed previously to generate measurable optical signals upon binding to metal ions.[25,30] Their application is particularly important for mercury and copper ions due to the severe environmental and health consequences associated with these metals.

Copper is an essential trace element, playing critical roles in enzymatic catalysis, mitochondrial respiration, melanin synthesis, neurotransmitter production, and antioxidant defense.[24,27] However, excessive $Cu^{2+}$ exposure can occur due to contaminated water, industrial waste, or soil leaching, leading to oxidative stress, gastrointestinal disturbances, liver and kidney toxicity, and neurodegenerative diseases, such as Wilson's and Alzheimer's disease.[24,27] This dual nature, essential at low concentrations but toxic at higher levels, requires careful monitoring of copper in environmental and biological systems. The choice of ligands or functional groups is important to ensure strong and specific coordination with $Cu^{2+}$ ions, which induces measurable changes in fluorescence or absorption.[24,27] Furthermore, fluorescent probes enable real-time monitoring of $Cu^{2+}$ in complex matrices, such as drinking water, biological fluids, and environmental samples, making them highly valuable for environmental protection and biomedical applications.[24,27]

Mercury is a highly toxic, nonessential metal that presents severe risks to both human health and ecosystems. It exists in various chemical forms, including elemental mercury, inorganic mercury salts, and organic mercury compounds, such as methylmercury, which is particularly toxic due to its lipophilicity and ability to cross biological barriers, such as the blood-brain barrier.[20,23,28,29,31] Mercury is released into the environment primarily through coal combustion, battery production, electronic waste, and other industrial activities, ultimately accumulating in water systems and the food chain.[20,23,31] Prolonged exposure to $Hg^{2+}$ can lead to neurological disorders, kidney damage, endocrine disruption, reproductive toxicity, and developmental defects, as exemplified by historical incidents such as Minamata disease in Japan.[20,23,28,29,31]

Fluorescent probes have become an essential tool for detecting $Hg^{2+}$ due to their high sensitivity, rapid response, and suitability for *in situ* monitoring. These probes rely on selective interactions with $Hg^{2+}$, often exploiting the strong affinity of $Hg^{2+}$ for sulfur- or nitrogen-containing functional groups, such as thiols, amines, or sulfonyl groups.[20,29] Strategies include "turn-on" sensors that become fluorescent upon $Hg^{2+}$ binding, quenching-based systems, and multifunctional materials that allow simultaneous detection and removal of $Hg^{2+}$ from aqueous media.[20,28,29] The integration of fluorescent detection with adsorption offers the dual advantage of monitoring and mitigating mercury pollution, a critical requirement given the global scale of contamination.[23,31]

Overall, fluorescent probes offer a versatile and effective approach for the detection of heavy metal ions, addressing the limitations of traditional analytical methods. The development of probes that combine high sensitivity, selectivity, and rapid response remains a critical area of research. This literature review focuses on strategies, mechanisms and applications of dansyl-based probes from 2016 onward for the detection of $Cu^{2+}$, $Hg^{2+}$ and $Cd^{2+}$. Although, an acknowledgment should be given to key pre-2016 studies for the detection of metal ions,[32–46] anions,[47–49] small molecules,[50,51] amino acids[52] and the use of dyad approaches[53–55] for the detection of the mentioned analytes.

### 13.2.1 Fluorescent Probes for $Cu^{2+}$ Detection

Copper is prevalent both in plants and animals occupying the third position as the most plentiful element in the human body usually in the form of divalent copper ($Cu^{2+}$).[56] Its presence is crucial for the normal functioning of the central nervous system; it is also tied as catalyst cofactor for electron transport being also able to act as a catalyst for protein production and the production of reactive oxygen species (ROS) essential for programmed cell death.[57] Excess concentrations of this metal ion leads to an overflow production of these reactive species that degrade biomacromolecules and is considered the basis process in which grave diseases emerge, such as neurodegenerative syndromes.[58,59] The World Health Organization (WHO) and the U.S. Environmental Protection Agency (US EPA) bodies have stipulated 31.5 mM an acceptable limit of $Cu^{2+}$ ion in drinking water and its presence in human blood should range between 15.7 and 23.6 mM.[59]

As a contribution towards the detection of $Cu^{2+}$, Liu *et al* designed a new dansyl-based compound **1** (Figure 13.2) composed of a dipeptide, i.e. glycyl-L-tyrosine with the intention of render water solubility and the presence of multiple N and O atoms enabling coordination with metal ions.[60] The assessment of **1** sensitivity towards metal ions was implemented in HEPES buffer solution at pH 7.4 against a variety of metal ions. The strong emission observed at 558 nm is only influenced significantly when $Cu^{2+}$ is present, promoting a 90% fluorescence quenching with an associated decrease of the fluorescence quantum yield from 4.70% to 0.08% after the addition of 5.0 equiv. $Cu^{2+}$. Moreover, the assigned absorption band with a peak at 337 nm gradually blue-shifts and broadens after the addition of $Cu^{2+}$, indicative of complexation and selective responsiveness given that no obvious shifts in absorbance were found for the other metal ions. The fluorescence titration allowed the determination of the detection limit (0.69 mM) which enables its potential applicability in real samples since it is far below the value suggested by the US EPA. Job's plot experiment and electrospray ionization mass spectrometry (ESI-MS) analysis concluded a 1:1 biding ratio of 1 and $Cu^{2+}$ with an association constant value of $6.47 \times 10^4$ $M^{-1}$ in the aqueous solution. Anti-interference studies showed no significant influence of other metals on compound **1** ability to selectively sense $Cu^{2+}$, although, when tested in the presence of several amino acids $_{L}$-His and $_{L}$-Glu showed competitive coordination. The influence of the pH demonstrated that this system can be applied in a pH range of 6 to 12. Finally, compound **1** has been used successfully to detect $Cu^{2+}$ in HepG2 cells with low toxicity in the range of 5 mM to 100 mM and the use of $Na_2EDTA$ demonstrated recovery of fluorescence (Figure 13.2a).

On top of the findings previously discussed and the lack of theoretical calculations, Li *et al* decided to perform time-dependent density functional theory (TDDFT) to provide a better understanding of the fundamental principles and mechanisms that rule in ion recognition.[56] Comparison between compound **1** and its copper(II) complex showed that the oscillator strength of the maximum emission peak of the fluorescence mechanism is large for **1** while a contrast is observed for the complex, reaching nearly zero. Thus, the predicted radiative rate $k_r$ of **1** was several orders of magnitude larger, yet the predicted internal conversion rate $k_{ic}$ was similar for both. Furthermore, due to the heavy atom effect, the simulated intersystem crossing

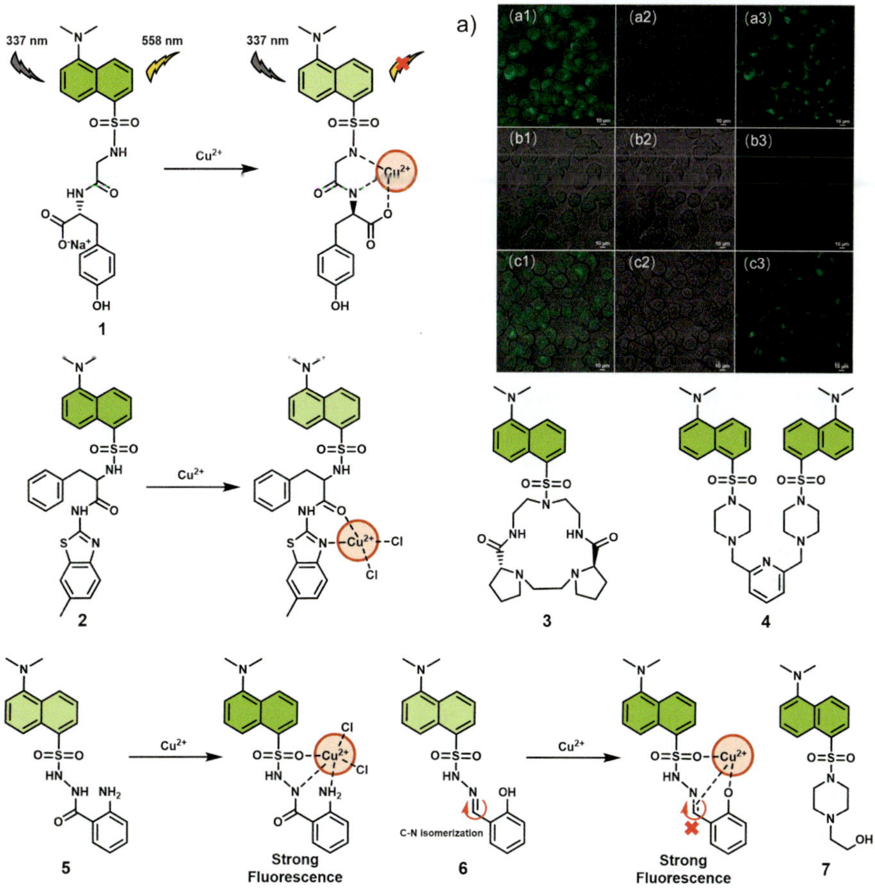


**FIGURE 13.2** Structures of compounds 1–7 and proposed detection mechanisms for $Cu^{2+}$. (a) Confocal laser scanning microscopy (CLSM) images of HepG2 cells: (A) Cells incubated with 1; (b) Cells incubated with 1 and then with $Cu^{2+}$; (c) cells incubated with 1 and then with $Cu^{2+}$ and subsequently treated with $Na_2EDTA$. (a1, b1, c1): Merged images; (a2, b2, c2): Bright-field images; (a3, b3, c3): Fluorescent images. (Adapted from ref.[60] copyright 2021, Elsevier.)

rate $k_{isc}$ of the complex was considerably higher than that of **1**. This information allowed them to obtain a calculated fluorescence quantum yield of 0.16% for **1** and $10^{-6}$% for the complex. The hole-electron pair analysis of the complex showed that the $S_0 \rightarrow S_1$ is attributed to metal-to-ligand charge transfer. The density-of-state diagram shows that the HOMO includes contributions from the copper atom's s orbital. This s orbital mixing lowers the energy of the frontier orbital and reduces the degree of overlap between the HOMO and the LUMO.

Lee's group have fabricated an unnatural benzothiazole conjugated phenylalanine bearing dansyl moiety **2** (Figure 13.2).[57] The sensing capabilities of compound **2** were tested against alkali, alkali-earth, and transition metal ions in DMSO:water

(1:1, v/v, pH 7.4). The selective $Cu^{2+}$ detection was marked by a change from light green to light brown color at the naked eye due to the coordination between the ligand and the metal ion. The absorption profile when $Cu^{2+}$ is added is characterized by a small bathochromic shift and increased intensity of the 307 nm maximum absorbance value attributed to the n-p* transition, while the fluorescence response is associated with a 95% quenching that suffers a hypsochromic shift from 539 to 534 nm when titrated. The interference studies revealed that the fluorescence remained strong in the presence of various metal ions but was completely quenched upon the addition of $Cu^{2+}$, indicating a strong affinity for this ion and suggesting its selective discrimination even in the presence of competing metals. The fluorescence lifetime measurements showed that **2** is subject to dynamic quenching since the increasing concentration of $Cu^{2+}$ was associated with a consistent decrease in its value. The $^{1}H$ NMR titration revealed that the NH protons were not involved in the binding of $Cu^{2+}$, while the IR spectrum analysis demonstrated the presence of metal-oxygen and metal-nitrogen bonds. Finally, given that the limit of detection (LOD) was found to be 3.0 mM, compound **2** was applied for the determination of $Cu^{2+}$ in ordinary water samples being able to recover up to 99.92% of the spiked $Cu^{2+}$ ion.

Wang and co-workers developed compound **3** (Figure 13.2), a novel L-proline-based heterocyclic ligand functionalized with a dansyl fluorophore for copper detection.[61] In an aqueous solution (DMSO:$H_2O$ = 9:1, HEPES buffer, pH 7.4), the free ligand exhibits strong fluorescence, but the addition of one equivalent of $Cu^{2+}$ quenches the emission to just 11.4% of its original intensity. Other biologically relevant metal ions, including $Zn^{2+}$, $Ni^{2+}$, and $Pb^{2+}$, cause negligible quenching, with only $Fe^{3+}$ producing a minor decrease. High-resolution ESI mass spectrometry confirmed the formation of a 1:1 ligand-$Cu^{2+}$ complex, and the binding constant was determined to be $1.32 \times 10^4$ $M^{-1}$. The detection limit of 1.39 μM is well below the WHO guideline. Importantly, the sensor was successfully applied in living HeLa cells, where intracellular fluorescence was efficiently quenched upon $Cu^{2+}$ exposure, demonstrating its promise for bioimaging applications.

Further examples of the design of dansyl-based compounds incorporating peptides moieties and associated detection parameters for $Cu^{2+}$ are summarized in Table 13.1. Alongside simpler examples such as DGly and GG-Dns,[62,63] Wang's group developed compounds DNH to DP5 (Table 13.1) through solid-phase peptide synthesis (SPPS) technology.[64–70] All compounds presented solubility in water and were found to exhibit a "turn off" of the fluorescence in the presence of $Cu^{2+}$.

Compounds DNH, DAH, DAHK and DP5 (Table 13.1) also showed some level of quenching towards $Hg^{2+}$ but not to the full extent observed for $Cu^{2+}$. Aside from compound L (Table 13.1), the remaining fluorophores were successfully used for bioimaging applications including different cell lines and zebrafish larvae demonstrating negligible biological toxicity. Furthermore, as a way of improving user-friendliness and portability, compounds DNH, DAH, DAHK and L were embedded in test trips showing the absence of emission only when submerged in $Cu^{2+}$ solution while DNH and DAH were further explored for the semi-quantitative detection of $Cu^{2+}$ in solution using a color recognition app on the smartphone to obtain Red, Green, Blue (RGB/values. The linear equations were fitted as R/B = – 0.1358X + 0.9539

**TABLE 13.1**
**Comparison of Dansyl-Based Compounds Composed of Peptide Moieties for the Detection of $Cu^{2+}$**

| Compound | Peptide Composition | Recognition Unit | Limit of Detection (nM) | pH Range | Ref |
|---|---|---|---|---|---|
| DGly | (Gly) | –NH and carbonyl groups | $1.32 \times 10^3$ | 6–8 | 62 |
| GG-Dns | (Gly–Gly) | –NH and –OH groups | 290 | 6–13 | 63 |
| DNH | (Asn–His–$NH_2$) | –NH group | 19.6 | 7–12 | 64 |
| DAH | (Ala–His–$NH_2$) | –NH group | 18.0 | 6–11 | 65 |
| DSD | (Ser–Asp–$NH_2$) | –NH and carbonyl groups | 2.4 | 7–8 | 66 |
| DAHK | (Ala–His–Lys–$NH_2$) | –NH and carbonyl groups | 38.9 | 4–12 | 67 |
| DGGH | (Gly–Gly–His–$NH_2$) | –NH group | 37.9 | 6–12 | 68 |
| L | (Ala–Ser–Arg–His–$NH_2$) | –NH and –OH groups | 12.04 | 6–12 | 69 |
| DP5 | (Ser–Pro–Gly–His–Gly–$NH_2$) | –NH and –OH groups | 23.5 | 6–12 | 70 |
| P | (Asp–Asp–Gly–Gly–Glu–Glu) | –OH groups | 47 | 7–12 | 71 |

($R^2 = 0.9749$) and G/B = − 0.0603X + 1.1361 ($R^2 = 0.9888$) with associated limits of detection of 0.49 mM and 0.25 mM for DNH and DAH, respectively. Finally compound P (Table 13.1) incorporating six amino acids into the dansyl unit was reported by Hu *et al.* that recognizes $Cu^{2+}$ without interference of $Hg^{2+}$ and other metal ions which showed good biocompatibility and low biotoxicity for the imaging of HeLa cells and zebrafish.

Beyond peptide-appended dansyl probes, Zhou *et al.* designed compound **4** (Figure 13.2) through the nucleophilic substitution of 2,6-bis(bromomethyl)pyridine and N-dimethyl-5-(piperazin-1-ylsulfonyl)naphthalen-1-amine, a five-step synthetic route resulting in a 49% overall yield.[59] HEPPS buffer system was used to evaluate the selectivity of sensor **4** in physiological pH window. Although some slight changes in fluorescence were observed for most of the metal ions tested, only $Cu^{2+}$ was able to considerably attenuate the emission at the peak of 540 nm leading to a 40-fold decrease in fluorescence being hardly affected when performed in competition experiments. This system was found to be functional between pH 5-14 with a detection limit of 0.25 mM and formed a 1:1 complex formation with $Cu^{2+}$. Furthermore, the successful bioimaging application, supported by living-cell experiments, demonstrated that **4** possesses high cell-membrane permeability and strong capability for $Cu^{2+}$ detection.

Sahoo's group reported compound **5** (Figure 13.2) through a simple one pot reaction between dansyl chloride and 2-aminobenzohydrazide in the presence of a base.[72] Compound **5** has a fluorescence maximum wavelength at 438 nm when excited at 335 nm in DMSO:$H_2O$ (1:1, v/v). Upon it was tested with several metal ions, only $Cu^{2+}$ was able to enhance, and blue shift its original emission to 474 nm with also the appearance of a new shoulder at 411 nm. The 1:1 biding stoichiometry was supported by HRMS data, and the LOD was estimated to be 0.98 μM. Probe **5** was used for the determination of $Cu^{2+}$ concentration in water samples from tap and river being able to measure the concentration of spiked $Cu^{2+}$ between the range of 92-112%. A similar compound (**6**, Figure 13.2) with a quantum yield of 0.51 was obtained by Nasomphan *et al.* that exhibited also a "turn-on" fluorescence response towards $Cu^{2+}$ with an associated blue shift from 547 to 479 nm.[73] In this case, $Hg^{2+}$ and $Fe^{2+}$ would cause significant degree of fluorescence quenching. A LOD of 0.5 μM was achieved at the physiological pH 7.2 while exciting at 366 nm. The proposed mechanism suggests that the rigid chelation inhibits the non-radiative pathway for a C=N isomerization. Given the absence of cytotoxicity **6** was applied for the detection $Cu^{2+}$ in Vero cells.

A simple probe **7** (Figure 13.2) was constructed by Jiang *et al.* using 2-(piperazin-1yl)ethanol moiety as receptor unit and its photophysical properties were evaluated in $CH_3CN$-$H_2O$ (95:5, v/v).[74] The emission would suffer a red-shift from 540 to 557 nm only in the presence of $Cu^{2+}$. The Job's plot confirmed the formation of 1:1 stoichiometry and further supported by DFT calculations which suggest that $Cu^{2+}$ coordinates with the O atom of the sulfonyl group and the N atom of the sulfur. The detection limit falls below the value that is present in the US EPA guidelines being 0.11 μM. This compound was applied to two-photon confocal fluorescence imaging for A549 cells, a technique that uses two photons of lower energy that provide advantages regarding lower autofluorescence levels, low damage induced by light and higher tissue penetration depth.

Considering more sophisticated systems, Zhao *et al.* prepared a new biosensor by covalently linking DNS-Cl with amine groups of the Copper resistance operon coded protein C (CopC) and FRET behavior was implemented by exciting the DNS-Cl through the emission of the tryptophan residue (Trp83) on CopC.[75] The selectivity was performed in 30 mM $Na_2CO_3$/$NaHCO_3$ buffer solution (pH 9.3) and no metal ion studied aside from $Cu^{2+}$ was able to quench the emission with no obvious interference when in the presence of other metal ions and amino acids. The mechanism involves the excitation at 280 nm of the Trp83 of CopC that in turn fluoresces at 320 nm. This emission overlaps with the absorption band of DNS-Cl, enabling efficient FRET and producing a strong secondary emission at 505 nm. The detection mechanism arises when copper ions coordinate with specific residues of CopC (His1, His91, Glu27, and Asp89). These processes quench the tryptophan fluorescence at 320 nm, thereby reducing the excitation of DNS-Cl and leading to a marked decrease in the 505 nm emission. Sensitivity tests showed a linear response over the range of 0.04 to 11 μM, with a detection limit of 6.81 ± 0.07 nM. Finally, the biosensor was successfully applied to real-world samples, including tap water and industrial wastewater, with results in agreement with ICP-MS measurements.

Biswas' group reported the synthesis of a dansyl anchored hafnium-based fluorescent metal-organic framework (MOF) that demonstrated high stability in the pH range of 4 to 10.[76] The sensing of $Cu^{2+}$ was achieved in 10 Mm HEPES buffer at pH 7.4 reveling strong emission at 428 nm with gradual decrease in emission. This system maintained a quenching efficiency of $Cu^{2+}$ between 78 to 84% given the absence of interference from other metals when they are present with a LOD of 229 nM. The MOF was applied for the detection of $Cu^{2+}$ in different water specimens (tap water, river water, distilled water and lake water) with its recovery percentage found to fall between 96.3 and 103.9%. Time-resolved photoluminescent measurements confirmed that the fluorescence lifetime of the probe remained unchanged in the presence of $Cu^{2+}$, indicating static quenching while powder X-ray diffraction patterns showed no structural changes after $Cu^{2+}$ treatment, ruling out trans-metalation with the Hf(IV) ions of the framework. Instead, $Cu^{2+}$ engages in weak complexometric interactions with oxygen and nitrogen atoms of the linker, as evidenced by X-ray photoelectron spectroscopy (XPS), which revealed shifts in O 1s and N 1s binding energies while signals for Hf, C, and S remained unchanged. Selectivity arises because $Cu^{2+}$, with its high charge density and smaller ionic radius, exhibits stronger complexation than other competing ions, while only minimal interference was observed from paramagnetic species like $Fe^{2+}$ and $Cr^{3+}$.

### 13.2.2 Fluorescent Probes for $Hg^{2+}$ Detection

Mercury (Hg) represents one of the most hazardous heavy metals and an impactful global pollutant due to its extreme toxicity, persistence, and ability to bioaccumulate in living organisms.[20,77,78] Mercury released into aquatic systems can be transformed by microorganisms into highly toxic organic forms, which bioaccumulate in fish and other organisms and pose serious risks to humans through dietary exposure. Owing to the extreme toxicity of $Hg^{2+}$, the US EPA has defined a maximum contaminant level of 10.0 nM (2.0 ppb) in drinking water.[20,79]

Wang's group reported probe **8** (Figure 13.3) based on dansyl and tyrosine using SPPS.[80] The metal ion selectivity experiment was carried out in HEPES buffer solution at pH 7.4 displaying a "turn-on" response to $Hg^{2+}$ with no relevant effect upon exposure to the remaining metal ions. The presence of 5 equiv. of coexisting metal ions did not display the inability for $Hg^{2+}$ to maintain the increase of the fluorescence with an associated blue shift from 545 to 515 nm. Job's plot suggests a 2:1 (L:M) stoichiometric complexation which promotes the presence of monomer-excimer mechanism causing the 30 nm shift in emission. The NMR experiments revealed the strong contribution of the sulfonamide group and the phenol moiety on the binding with $Hg^{2+}$. This system showed a detection limit of 22.65 nM and its reversibility was demonstrated with EDTA. pH was found to limit the detection of $Hg^{2+}$ between 7 to 10 for maximum enhancement of emission. Finally, cell viability was maintained at 90% when using **8** until 100 mM and was able to serve as a fluorescent probe for $Hg^{2+}$ in HeLa cells and HK2.

Contribution of Zhang *et al.* involved the synthesis of a dansylated tryptophan **9** (Figure 13.3) through SPPS.[77] Among different metal ions tested in HEPES buffer (pH 7.4), **9** only $Hg^{2+}$ was able to cause any meaningful change in the emission causing an enhancement of fluorescence accompanied by a 54 nm blue shift from 550 to 496 nm and no interference was detected by other metals (Figure 13.3a). The CHEF effect was elucidated through fluorescence lifetime since an increase from 3.17 to 5.91 ns was observed with a related increase of the quantum yield from 6.31 to 10.41%. The detection limit was found to be as low as 6.53 nM and a was functional between pH 7 to 12. A 2:1 complexation with $Hg^{2+}$ was proposed and evidence suggested the involvement of the sulfonamide and the amide groups in the interaction with $Hg^{2+}$. The portability aspect of detecting $Hg^{2+}$ was demonstrated through a smartphone color recognition APP showing good linearity in the range of 10 to

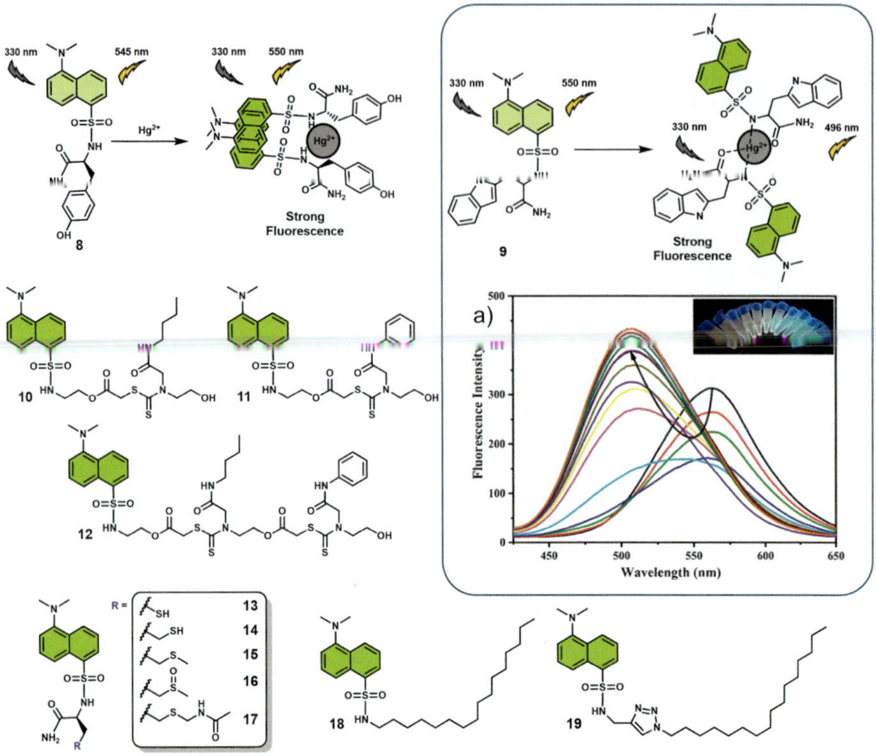


**FIGURE 13.3** Structures of compounds 8–19 and proposed detection mechanisms for $Hg^{2+}$. (a) Fluorescence spectra of 9 with excitation wavelength of 330 nm upon addition of increasing concentrations of $Hg^{2+}$ in HEPES buffer at pH 7.4. Inset picture is the variation of color excited at 365 nm. (Adapted from ref.[77] copyright 2024, Elsevier.)

70 mM with a corresponding linear regression equation of $R/B = -0.0171X + 1.3832$ ($R^2 = 0.9969$). Finally, the potential monitoring of $Hg^{2+}$ was tested in real water samples and presented good recoveries in the range of 6 to 10 mM when spiked.

Implementation of three or more amino acids as the backbone of the design of dansyl based compounds are gathered in Table 13.2.

Xue *et. al* reported compound L (Table 13.2) that integrated two dansyl units into a lysine backbone using a serine aminoacid.[81] The chemosensor would adopt a 2:1 stoichiometry with $Hg^{2+}$ causing a quenching of the emission in aqueous solution. The MTT assay showed low cytotoxicity in HK2 normal cells when tested until 80 mM, which promoted monitoring of $Hg^{2+}$ ions in living LNCap cells.

The dansyl-peptide probe DGRK (Table 13.2) was synthesized via SPPS for selective $Hg^{2+}$ detection in water and displayed strong fluorescence at 560 nm.[82] Upon $Hg^{2+}$ binding, DGRK formed a 1:1 non-fluorescent complex. Fluorescence titration revealed a linear response up to 1 μM $Hg^{2+}$ with a detection limit of 33.6 nM. In live-cell assays (RKO and GES-1), DGRK showed low cytotoxicity (>90% viability at 80 μM) and efficient cell permeability, enabling imaging of intracellular $Hg^{2+}$.

**TABLE 13.2**
**Comparison of Dansyl-Based Compounds Composed of Peptide Moieties for the Detection of $Hg^{2+}$.**

| Compound | Peptide Composition | Recognition Unit | Limit of Detection (nM) | pH Range | Ref |
|---|---|---|---|---|---|
| L | (DNS-Ser-Lys-Ser-DNS) | –OH groups | 7.59 | 6–12 | 81 |
| DGRK | (DNS-Gly-Arg-Lys-$NH_2$) | –NH groups | 33.6 | 6–12 | 82 |
| DKA | (DNS-Ala-Lys-Ala-DNS) | Carbonyl groups | 13.4 | 5–10 | 78 |
| DWPH | (DNS-Trp-Pro-His-$NH_2$) | –NH and carbonyl groups | 24.6 | 7–11 | 79 |
| D-P4 | (DNS-Glu-Cys-Glu-Trp-$NH_2$) | –NH and -SH groups | 23.0 | 8–11 | 83 |
| D-P4′ | (DNS-His-Pro-Gly-Asp-$NH_2$) | –Sulfonamide and imidazole | 61.0 | 7–12 | 84 |
| KEEP | ((Dansyl-Glu-Glu)$_2$-Lys-$NH_2$) | –NH and carbonyl groups | 40.0 | 7–11 | 85 |

For portable detection, paper-based test strips gave clear fluorescence quenching with $Hg^{2+}$ under 365 nm UV light, and smartphone-assisted RGB analysis allowed semi-quantitative detection ($R/B = -0.1609\,X + 1.98$, $R^2 = 0.9773$) with an LOD of 0.23 µM.

Closely related to the previous work, Wang's group also designed compounds DKA and DWPH (Table 13.2) that this time a "turn-on" with an associated blue shift of the emission was noticed.[78,79] For both compounds the initial emission observed at 560 nm would weaken while at 510 nm an increase was observed only in the presence of $Hg^{2+}$. DKA showed very low cytotoxicity to Hacat cells up to 80 mM and its ability to measure $Hg^{2+}$ levels in live cells was demonstrated. The use of a smartphone APP that reads RGB values was used for the semi-quantitative analysis of $Hg^{2+}$ providing good linearity ($R/G = -0.0439\ X + 0.7249$, $R^2 = 0.9799$) with a detection limit of 2.5 mM. In turn, DWPH was found to be under a PET phenomenon from the lone pair electrons of the imidazole that would quench the emission and could only be restored through a CHEF mechanism initiated by $Hg^{2+}$. HGC-27 cells were used to assess the cytotoxicity which had a survival rate above 90% when incubated up to 80 mM of DWPH. The monitoring of $Hg^{2+}$ in zebrafish larvae was validated. The use of the smartphone APP revealed good linearity ($R/G = 0.6412X + 1.1330$ ($R^2 = 0.9855$)) with a detection limit of 0.45 mM.

Pang *et. al* developed D-P4 (Table 13.2) that manifested a turn on response toward $Hg^{2+}$ in HEPES solution (pH 7.1).[83] While excited at 330 nm, the band located at 550 nm would blue shift 45 nm to 505 nm and an increase of fluorescence intensity was observed, clear evidence of CHEF effect. The presence of Trp residue allowed the monitoring of the emission of D-P4 using the 295 nm wavelength which evidenced a fluorescence emission peak at 337 nm and 550 nm. With the addition of $Hg^{2+}$ the band at 337 nm decreased due to the folding of D-P4 that would shorten the distance between the Trp residue (donor) and the dansyl unit (acceptor). This translated in an increase of the fluorescence intensity at 505 nm suggesting a FRET effect and another pathway for the sensing of $Hg^{2+}$. Fluorescence lifetime decay profiles suggest a decrease from 2.80 to 2.77 ms at 337 nm emission when exciting at 295 nm while the fluorescence lifetime at 505 nm emission peak increased from 7.43 to 11.18 ms.

Furthermore, D-P4 was found to recover $Hg^{2+}$ in the range of 0.1 to 1.5 mM from river water while the visual detection was assessed in agarose gel flakes and test paper.

Compound KEEP (Table 13.2) was designed by Li *et. al* as a fluorescent probe capable of producing distinct fluorescence responses depending on the mercury form.[85] When binding $Hg^{2+}$, the probe undergoes fluorescence quenching, giving a clear "turn-off" signal. Conversely, interaction with organic mercury compounds ($CH_2Hg^{2+}$ and $PhHg^+$) enhances fluorescence emission, generating a "turn-on" response. Quantitative studies revealed high sensitivity, with a detection limit as low as 40 nM for $Hg^{2+}$, and micromolar detection limits for $CH_3Hg^+$ (5.32 μM) and $PhHg^+$ (0.79 μM). This enhancement is attributed to steric hindrance introduced by the organic groups, which restricts non-radiative relaxation pathways and thus strengthens fluorescence. Among them, $PhHg^+$ produces a stronger enhancement than $CH_3Hg^+$ due to greater steric effects.

Anna Jose and co-workers have designed three water soluble sequence-defined oligo-dithiocarbamate dansyl derivatives **10** to **12** (Figure 13.3) for the detection of $Hg^{2+}$.[86] Compound **10** was used as model compound showing in aqueous medium an absorption and emission maxima of 330 and 560 nm with an associated quantum yield of 0.138. A blue shift of 60 nm was observed when $Hg^{2+}$ is present and no interference from other metal ions were detected. This sensor adopts a 1:1 binding stoichiometry and has a LOD of 9.3 nM. Comparable LOD value (8.1 nM) was demonstrated by substitution of the butyl moiety with the phenyl ring (**11**). The possibility of increasing the receptor units for $Hg^{2+}$ was accomplished by increasing the chain length. Sensor **12** that incorporates two dithiocarbamate units yielded an LOD of 60 pM. Visual detection of $Hg^{2+}$ was made possible through paper strips and fluorescence microscopy which was validated in MCF-7 and HeLa cancer cell lines.

Guided by the principles demonstrated in earlier studies on the dansyl-methionine chemosensor, Andrea Cusano's group developed new dansyl-amino acids incorporating methionine and cysteine (cys) derivatives for Hg²+ ion recognition in aqueous buffer (**13** to **17**, Figure 13.3).[87] Sensors containing only a thioether moiety, **15** and **17**, exhibited a turn-on fluorescence response to $Hg^{2+}$ in the 0–120 μM range. In contrast, chemosensors containing oxidized methionine or free cysteine/homocysteine showed a turn-off response under the same conditions. The most sensitive probe **15** had a LOD of 140 nM, which exceeds the EPA limits for drinking water. To enhance sensitivity, a portable optical fiber-based experimental setup was tested, improving the LOD by ~100-fold to approximately 5 nM, well below safe human exposure limits. However, the sensors are not suitable for detecting $Hg^{2+}$ in marine water due to interference from high chloride concentrations.

The self-assembly properties originated from the reaction between dansyl chloride and hexadecyl amine enabled Gadiyaram et al. to integrate the compound **18** (Figure 13.3) into sphere-like phospholipid nanovesicles with an average particle size of 102.9 nm at biological pH.[88] The positioning of the dansyl unit on the vesicle surface motivated studies on its interaction with metal ions. In the presence of $Hg^{2+}$, the emission at 525 nm was significantly quenched, corresponding to a LOD of 6.41 μM. Fluorescence lifetime measurements supported this observation, showing a decrease from the initial 13.24 ns, which was attributed to $Hg^{2+}$-induced aggregation of the vesicles, further confirmed by scanning electron microscopy (SEM) imaging.

A similar approach was taken by Kim's group who fused the dansyl unit with a cetyl chain bridged by a triazole (**19**, Figure 13.3).[89] Probe **19** was found to emit strongly in the green region with an emission maximum at 500 nm in aqueous solution containing 0.5% DMSO. Addition of $Hg^{2+}$ would lead to a shift of the emission to 476 nm at its peak with an associated quenching. A five-membered ring between the $Hg^{2+}$ center and the N donors of the sulfonamide and the triazole in a 1:1 stoichiometry was the proposed recognition site. The sensing of $Hg^{2+}$ was associated with the initial induced self-aggregation of the probe, since dynamic light scattering (DLS) data revealed the presence of nanoaggregates with 120 nm average diameter while transmission electron microscopy (TEM) showed diameters between 20–50 nm in the solid state. The critical aggregation concentration (CMC) was 7.86 μM and presented a sensing response between pH 4 to 11. MTT assay demonstrated 90% cell viability until 40 μM for 24h in HepG-2 liver cancer cells with further studies supported the fluorescence imaging of intracellular $Hg^{2+}$ using this probe.

Uttam and co-workers reported a lower rim dansyl-functionalized calix[4]arene (DanC4A) as a selective fluorescent probe for mercury ions ($Hg^{2+}$) in solution, solid state, and living cells.[90] DanC4A exhibited strong emission at 510 nm in THF, which was quenched by nearly 80% upon $Hg^{2+}$ addition, with a detection limit of 0.56 μM and a 1:1 binding stoichiometry confirmed by Job's plot. XPS studies revealed that both sulfur atoms from thiophene and nitrogen atoms from tertiary amines participated in $Hg^{2+}$ coordination. DanC4A was biocompatible in HeLa cells (>70% viability up to 100 μg/mL) and intracellular fluorescence imaging confirmed quenching upon $Hg^{2+}$ uptake. To address solubility and reusability issues, DanC4A was encapsulated into ZIF-8, forming DanC4A@ZIF-8 hybrid MOFs, maintaining its fluorescence. These hybrid materials not only enabled confocal $Hg^{2+}$ sensing but also achieved up to 95% removal of $Hg^{2+}$ from aqueous solutions. In contrast, Oguz *et al.*[91] designed water-soluble sulfonated dansyl-calixarenes for live-cell $Hg^{2+}$ imaging in SW-620 cells, achieving non-toxicity up to 200 μM and PET-based fluorescence quenching with detection limits in the $10^{-5}$ M range, thereby emphasizing biological compatibility rather than environmental remediation.

### 13.2.3 Fluorescent Probes for $Cd^{2+}$ Detection

Cadmium (Cd) is a naturally occurring metallic element mainly found as cadmium sulfide and cadmium carbonate. It is widely applied in metallurgy, electroplating, pigments, phosphate fertilizers, nickel–cadmium batteries, semiconductors, and other industrial fields. However, $Cd^{2+}$ is a highly toxic and bioaccumulative heavy metal that poses serious threats to the environment and human health. Anthropogenic activities such as metal smelting, sewage discharge, fertilizer use, and industrial waste disposal have led to the long-term accumulation of cadmium in soil and water, from where it enters the food chain and accumulates in organisms.[92–95]

In the human body, $Cd^{2+}$ binds to proteins containing sulfhydryl and hydroxyl groups, forming cadmium-thionein complexes that accumulate mainly in the liver and kidneys. It disrupts cellular processes by damaging membranes, inhibiting enzymes, altering DNA, and generating reactive oxygen species (ROS), leading to oxidative stress. Acute exposure can cause gastrointestinal symptoms, while chronic exposure is associated with kidney, liver, and bone damage, as well as cancers such as lung, prostate, and bladder cancer.[96–98] Recognized as one of the most dangerous global pollutants, cadmium's maximum contaminant levels in drinking water are limited to 5.0 ppb (EPA) and 3.0 ppb (WHO). Owing to its chemical similarity to $Zn^{2+}$, achieving selective and sensitive detection of $Cd^{2+}$ in environmental and biological systems remains an important research priority.[92–95]

Notably, Peng Wang's group have developed methods for cadmium detection using dansyl chromophores.[92–95] Figure 13.4 gathers probes **20** to **23** synthesized by

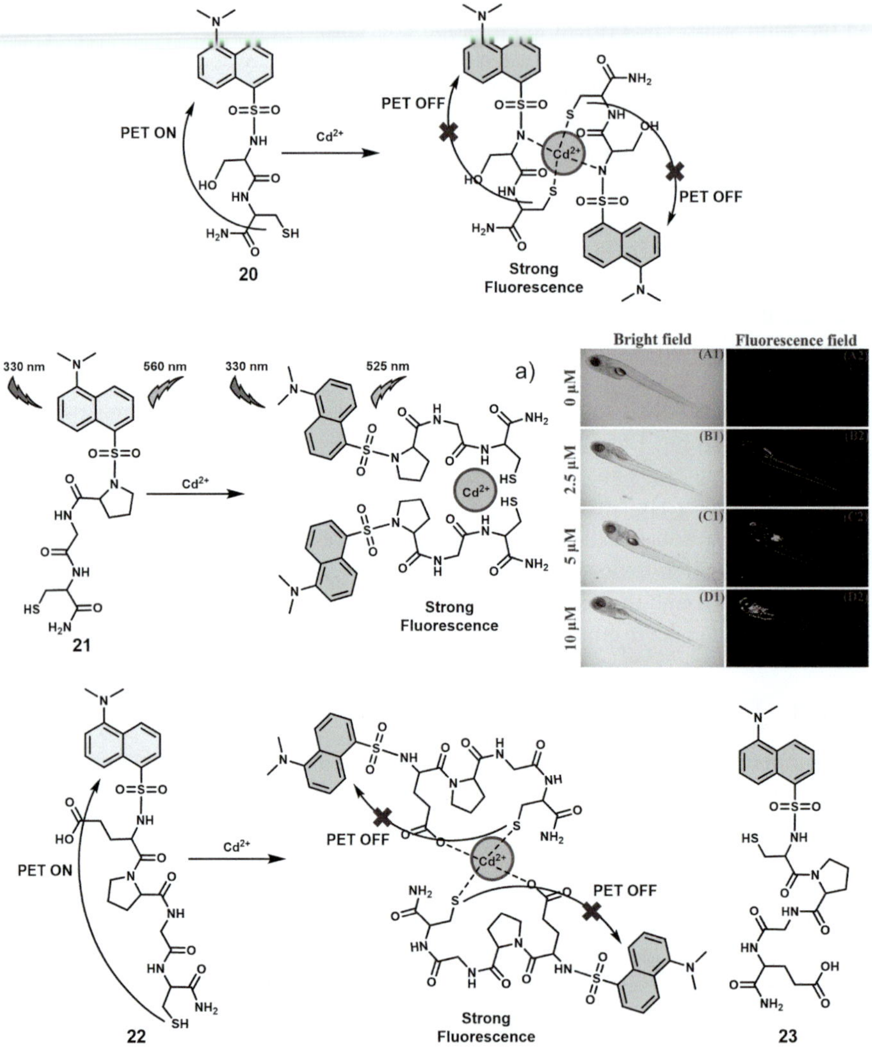


**FIGURE 13.4** Structures of compounds 20–23 and proposed detection mechanisms for $Cd^{2+}$. (a) Confocal images of zebrafish larvae treated with 21 for 1 h (A1-A2) and further treated with 2.5 µM (B1-B2), 5 µM (C1-C2) and 10 µM (D1-D2) of $Cd^{2+}$ for 1 h. (Adapted from ref. [93] copyright 2024, Elsevier.)

SPPS technology. Compound **20**, a dipeptide conjugated with a dansyl group (DNS-Ser-Cys-$NH_2$), exhibits a fluorescence 'turn-on' response at 545 nm upon addition of $Cd^{2+}$, with the quantum yield increasing from 0.068 to 0.154 in aqueous solution.[92] Binding studies demonstrate that **20** coordinates in 2:1 stoichiometry with $Cd^{2+}$ by the sulfhydryl group with an associated LOD of 13.8 nM. The sensing mechanism involves a PeT process from the sulfhydryl group of cysteine to dansyl unit quenching the initial emission that is restored upon addition of $Cd^{2+}$ exhibiting some interference only to $Cu^{2+}$. Given that the optimal response to $Cd^{2+}$ was observed between pH 7 to 10, cytotoxicity experiments were carried out using living H2K cells revealing no toxicity until 80 μM of **20**. The same cell line helps validate the usefulness of **20** on the detection of $Cd^{2+}$ in biological applications.

Probe **21** (Figure 13.4) is a dansyl labelled tripeptide (DNS-Pro-Gly-Cys-$NH_2$) that shows a fluorescence peak at 560 nm in aqueous solution that is blue shifted to 525 nm with a 22-fold increase upon exposure to $Cd^{2+}$.[93] Adopting a 2:1 (L:M) stoichiometry through coordination of the sulfur atom in the sulfhydryl group, the probe exhibited a calculated detection limit of 12.4 nM for $Cd^{2+}$. Fluorescence decay curves suggest an increase from 2.095 ns to 18.724 ns when $Cd^{2+}$ is present allied with an increase of the quantum yield from 6.14% to 51.35%. Moreover, the optimal pH for the detection of $Cd^{2+}$ was found to be between 7 and 11. Cadmium(II) detection using **21** (>95% cell viability at 80 μM, SW620 cells) in living cells was demonstrated using SW620 cells showing the increase of intracellular fluorescence which was also evident in zebrafish models (Figure 13.4a). The use of **21** for recovery of $Cd^{2+}$ in diverse water samples reach values from 95.3% to 108.0%. Finally, using a Colour Picket App it was possible to validate a smartphone-assisted sensing system showing excellent linear relationship between G and B value ($G/B = 0.2430X + 0.6452$, $R^2 = 0.9792$) achieving a detection limit of 0.29 μM of $Cd^{2+}$.

The dansyl-conjugated tetrapeptide **22** (DNS-Glu-Pro-Gly-Cys, Figure 13.4) exhibited a 5-fold enhancement in fluorescence intensity upon interaction with $Cd^{2+}$ ions in aqueous solution, accompanied by an increase in quantum yield from 0.0857 to 0.1769.[94] Notably, the probe demonstrated high selectivity toward $Cd^{2+}$, showing minimal interference from other metal ions. Experimental studies suggest that **22** adopt a 2:1 stoichiometry with $Cd^{2+}$ through the O atom of Glu and the S atom of Cys which prevents the initial PeT process from the sulfhydryl group to the dansyl unit. Fluorescence lifetime measurements revealed an increase from 9.32 ns to 17.36 ns in the presence of $Cd^{2+}$. Moreover, the calculated LOD was demonstrated to reach 45 nM in optimal pH values between 7 to 10. MTT assays revealed a survival rate above 90% to 200 μM of **22** in HeLa cells which were used to validate the imaging of endogenous $Cd^{2+}$ ions.

A similar response toward $Cd^{2+}$ was found for compound **23** (DNS-Cys-Pro-Gly-Glu-$NH_2$, Figure 13.4) which adopts a 2:1 (L:M) stoichiometry with the initial quenching caused by PeT process being reverted after complexation.[95] The LOD for this system was found to be 93 nM. Moreover, the coordination process occurred through the O atoms of the Glu and sulfhydryl of Cys which induced a change in the fluorescence lifetime from 9.86 ns to 18.78 after complexation. Cytotoxicity experiments were conducted using HeLa cells (>90% cell viability at 200 μM) and an additional cell line (H2K) was used to demonstrate the usefulness of **23** as a probe for detecting $Cd^{2+}$ in living cells.

### 13.2.4 Fluorescent Probes for Dual Metal Ion Detection

While most dansyl-based probes have been developed for the selective detection of individual metal ions, several systems have shown the ability to recognize two different metal ions simultaneously. These dual-responsive sensors often display distinct optical or fluorescence changes depending on the ion present, enabling differential or even sequential detection within the same molecular framework. This unique behavior typically comes from multiple coordination sites or flexible binding modes associated with the fluorophore.[99–101] The versatility of such systems not only broadens the sensing scope of dansyl-based probes but also provides valuable insights into their underlying recognition mechanisms. The following section summarizes representative examples of these dual-metal sensing systems and their characteristic fluorescence responses.

A notable contribution to this area was made by Wang's group, who developed the dual-responsive dansyl probes **24** to **28** for metal ion detection (Figure 13.5).[102–106] Compound **24** is a dansyl-appended dipeptide (DNS-Gly-Cys-$NH_2$) that in HEPES buffer solution displays a "turn on" response to $Cd^{2+}$ with a change in the emission peak from 560 to 515 nm while $Cu^{2+}$ detection is characterized by "turn off" effect.[102] For both metal ions, the binding modes would originate a 2:1 complex (L:M) and the calculated detection limits were 14.5 and 26.3 nM for $Cd^{2+}$ and $Cu^{2+}$, respectively. The proposed mechanism involves a monomer-excimer mechanism in aqueous solution upon complexation. Since the optimal sensing conditions were between pH 7 to 12, cell viability was performed in H2K cells (>90% cell viability at 80 μM) while fluorescence imaging was carried out in LNCaP cancer cells demonstrating the successful implementation of **24** for sensing $Cd^{2+}$ and $Cu^{2+}$ in living cells.

The dansyl unit modified tripeptide **25** (DNS-Gly-Cys-Ala-$NH_2$, Figure 13.5) acts as a dual sensor exhibiting a fluorescence enhancement by $Ag^+$ and fluorescence quenching by $Cu^{2+}$ in HEPES buffer solution (pH 7.4).[103] With an emission maximum at 560 nm, a 6.38-fold enhancement was observed upon addition of $Ag^+$ revealing also a blue shift to 520 nm. Compound **25** showed good anti-interference for both metals which assemble in a 1:1 stoichiometric ratio through one -NH of sulfonamide, the -SH group and two O atoms. The optimized detection of $Ag^+$ and $Cu^{2+}$ could be done in pH 5 to 11 and 7 to 12 with associated LODs of 7.4 nM and 5.9 nM, respectively. Lifetime measurements revealed an increase from 2.8 ns to 6.9 ns with complexation with $Ag^+$ and a static quenching was elucidated for $Cu^{2+}$ give the almost unchanged fluorescence lifetime upon its addition. Quantum yield experiments demonstrated an increase from 5.98% to 14.80% with the addition of $Ag^+$. Given these results a logic gate was built where the ON state was present only with $Ag^+$ input alone. Cytotoxicity studies were carried out with **25** through the MTT assay in SW620 cells (>90% cell viability at 80 μM) while confocal fluorescence images support the capacity of **25** to detect both metals in living cells and zebrafish larvae. Compound **25** was further used for recovery experiments in three real water samples (91.6% to 107.9%), three tea samples (94.5% to 109.5%) and real food samples (92.3% to 107.7%). Finally, semi-quantitative detection of each metal was carried out using a smartphone through color recognition software which

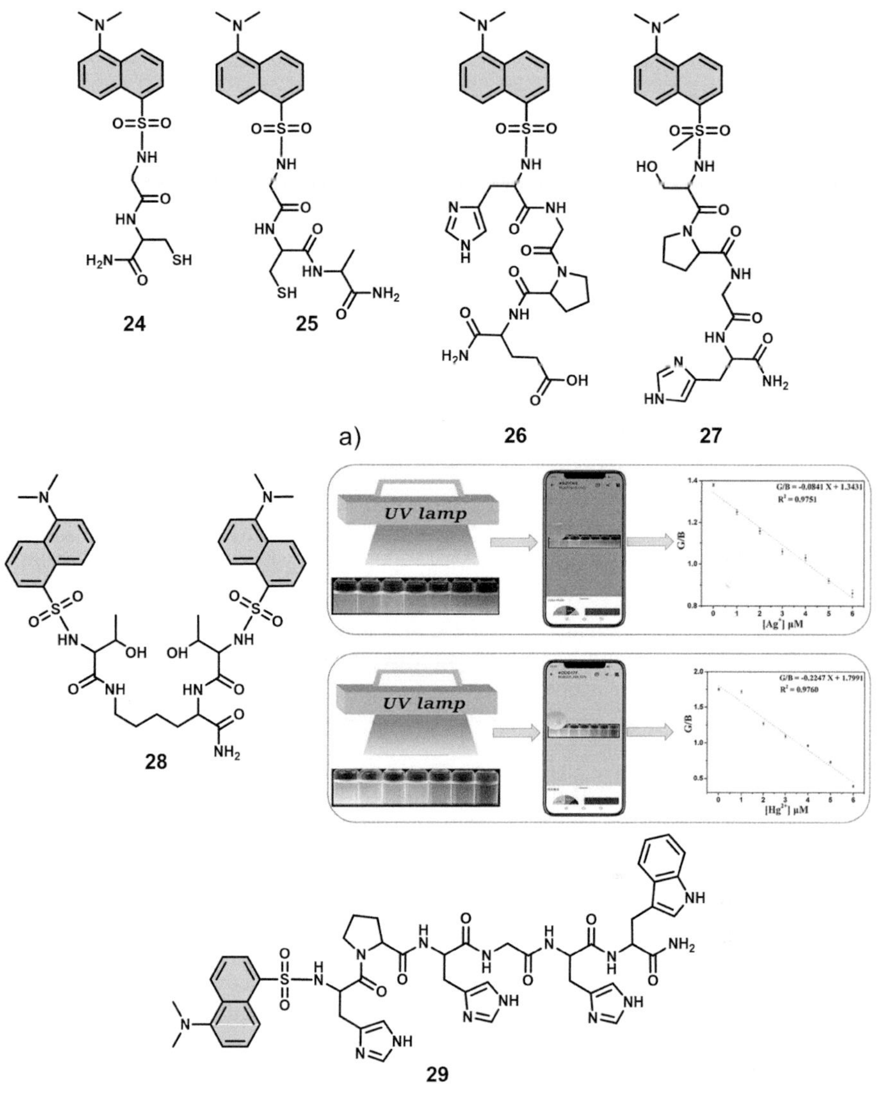


**FIGURE 13.5** Structures of compounds 24–29 for the dual detection of several metal ions (a) Smartphone-assisted RGB responses for the recognition of $Ag^+$ and $Hg^{2+}$. (Adapted from ref.[106] copyright 2025, Elsevier.)

provided good linear relationships for $Ag^+$ (G/B = 0.0357X + 0.9864, $R^2$ = 0.9944) and $Cu^{2+}$ (G/B = −0.0284 X + 0.9783, $R^2$ = 0.9954) with accompanying LODs of 0.92 and 1.37 μM.

Probe **26** is assembled by conjugating the dansyl unit with a tetrapeptide (DNS-His-Pro-Gly-Glu-$NH_2$, Figure 13.5) which manifest a fluorescent "turn on" response to $Zn^{2+}$ and a fluorescence quenching to $Cu^{2+}$ in aqueous solution.[104] In both cases, a blue shift was observed from 545 to 515 nm, with $Zn^{2+}$ inducing an increase of the quantum yield from 0.115 to 0.168, while other metal ions did not induce any meaningful interference. A 2:1 (L:M) stoichiometric ratio was found for both metal ions coordinating by the O atom of Glu and the N atom of His while LOD calculations demonstrated values of 4.9 nM and 15 nM to $Zn^{2+}$ and $Cu^{2+}$, respectively. The proposed mechanism relates to the initial PeT process arising from the electron-donating ability of the imidazole groups to the dansyl unit, which is suppressed upon coordination with $Zn^{2+}$. Since the optimal pH values for the detection of both metals were between 7 to 12, intracellular fluorescence imaging began with cytotoxicity experiments in HeLa cells demonstrating a survival rate above 95% up to 160 μM of **26** with confocal microscopy supporting its use for the detection of $Zn^{2+}$ and $Cu^{2+}$ in living cells.

Similar results have been obtained for the dansyl-based tetrapeptide **27** (DNS-Ser-Pro-Gly-His-$NH_2$, Figure 13.5) which also revealed a "turn-on" behavior $Zn^{2+}$ and a fluorescence quenching for $Cu^{2+}$ in HEPES buffer solution.[105] The increase of fluorescence was associated with a blue shift from 560 to 530 nm due to the blocking of the PeT process initially imposed between the imidazole and the dansyl unit. The 2:1 (L:M) stoichiometry was proposed for both metal ions through the coordination of the hydroxyl group from Ser and imidazole group of His. Both metals could be detected in the pH range of 6 to 12 with the LOD values being 13.2 nM and 4.1 nM for $Zn^{2+}$ and $Cu^{2+}$, respectively. Cell viability was evaluated through the MTT assay in HK2 cells (88.145% cell viability at 80 μM), HeLa cells (>90% cell viability at 80 μM) and LNCaP cells (>90% cell viability at 80 μM) while the detection of $Zn^{2+}$ and $Cu^{2}$ by fluorescence imaging was carried out using LNCaP cells suggesting positive results.

Probe **28** integrates two dansyl units into a tripeptide backbone (DNS-Thr-Lys-Thr-DNS, Figure 13.5) which can detect $Ag^{+}$ in a "turn on" manner and $Hg^{2+}$ through fluorescence quenching in aqueous solution.[106] The detection of $Ag^{+}$ is associated with a blue shift in emission from 553 to 531 nm translating into an increase of the quantum yield from 12.25 to 19.94% while the fluorescence lifetime decay saw an increase from 3.67 to 9.12 ns. A static quenching was assigned to $Hg^{2+}$ given the small change in the lifetime of **28** after complexation tested as 3.26 ns. ESI-HRMS and Job's plot suggest a 2:1 (L:M) stoichiometry for both metals through four O atoms of Thr. The underlying LOD values were determined to be 37.1 nM and 26.7 nM for $Ag^{+}$ and $Hg^{2+}$, respectively. pH dependent studies revealed that **28** has an optimal detection of $Ag^{+}$ between 7 to 12 and 5 to 10 for $Hg^{2+}$. MTT assays showed biocompatibility of **28** to LLC cells (>80% cell viability at 80 μM) while fluorescence imaging in the same cell line and zebrafish larvae provided the results to demonstrate the detection of $Ag^{+}$ and $Hg^{2+}$ in biological systems. Probe **28** was further used for the recovery detection in water samples (lake, purified and tap water) ranging from 97.6% and 107.6% and four food samples (green, black and jasmine tea and watermelon juice) which saw recoveries from 90.6 to 106.8%. Lastly, with the aid of a Color Recognizer application good linear relationships were obtained (Figure 13.5a) regarding the concentration of $Ag^{+}$ ($G/B = 0.0841X + 1.3431$, $R^2 = 0.9751$) and $Hg^{2+}$ ($G/B = 0.2247X + 1.7991$, $R^2 = 0.9760$) achieving LOD values of 0.37 μM and 0.21 μM, respectively.

Notomista's group synthesized the peptide DNS-His-Pro-His-Gly-His-Trp-$NH_2$ (**29**, Figure 13.5) that exhibited a "turn-on" effect to $Zn^{2+}$ and a strong quenching for $Cu^{2+}$.[107] Excitation at 295 nm revealed two peaks at 370 nm and 527 nm attributed to the indole ring and the dansyl unit. Upon the addition of $Zn^{2+}$ the peak at 370 nm decreased while the peak at 527 nm increased and shifted to 515 nm, suggesting a FRET effect. It was found that the peptide binds to $Zn^{2+}$ in a 2:1 stoichiometry promoting an increase in the quantum yield from 0.04983 to 0.1933. Interference studies revealed that $Ni^{2+}$ and $Co^{2+}$ were able to displace $Zn^{2+}$ to some degree with $Cu^{2+}$ totally quenching the emission indicating a higher affinity for dH3w. In this case $Cu^{2+}$ promotes a 1:1 stoichiometric ratio. The determined LOD values were 5.2 nM and 5.5 nM for $Zn^{2+}$ and $Cu^{2+}$, respectively, within the optimal pH of 7. Finally, peptide **29** was used for the staining of HeLa cells which did not show negative effects in cell viability up to 50 μM.

Our contribution to the detection metal ions and the design of new applications using dansyl compounds was accomplished through compounds **30** to **41** (Figure 13.6).[108–112] Compounds **30** to **36** were designed to compare the influence of a disulfide group versus a single sulfur bridge on metal ion detection, while also examining the effect of different spacers or direct linkage, as seen in compounds **30** and **31**, as well as the presence of a more complex system, such as the thiophene-based compound **36**. Compounds **37** and **38** are examples of tetradansylated compounds bearing cyclen and cyclam cores which provide higher molar absorptivity and signal (Figure 13.6a) while the implementation of different alkyl side chains may modulate physical properties seen in compounds **39** to **41**.

The photophysical properties of each compound were studied in a diverse set of solvents to elucidate the solvatochromic behavior using the Kamlet-Taft equation for quantitatively characterize the solute-solvent interactions. Given the low solubility in water, aggregation studies were performed using THF or acetonitrile with different water fractions providing insights into the modulation of the emission culminating in the presence of AIE in high water content. The presence of aggregates was supported by DLS measurements and the enhanced emission properties were subject to quantum yield and fluorescence lifetime experiments. Except for compound **30**, which was non-emissive due to a disulfide bridge directly attached to the dansyl unit, and compound **36**, whose high structural rigidity precluded its use as a sensor, all derivatives reliably detected $Cu^{2+}$ and $Hg^{2+}$ through a "turn off" response in acetonitrile. Table 13.3 gathers the association constants considering the 1:1 stoichiometry alongside the LOD for comparison of compounds **31** to **39**.

Furthermore, integration of these compounds in polymeric thin films revealed the possibility to construct materials with tunable emission since depending on the polymer used the resulting thin film would acquire a different emission peak. The obtained materials were then applied for the development of temperature smart materials exhibiting good linearity in a wide range of temperatures. Finally, compound **34** was used for the synthesis of polymeric microparticles using styrene-butadiene-styrene polymer through an oil-in-water approach as method to avoid solubility restrictions while **41** was found useful to stabilize platinum microparticles employing ethanol as the reducing agent.

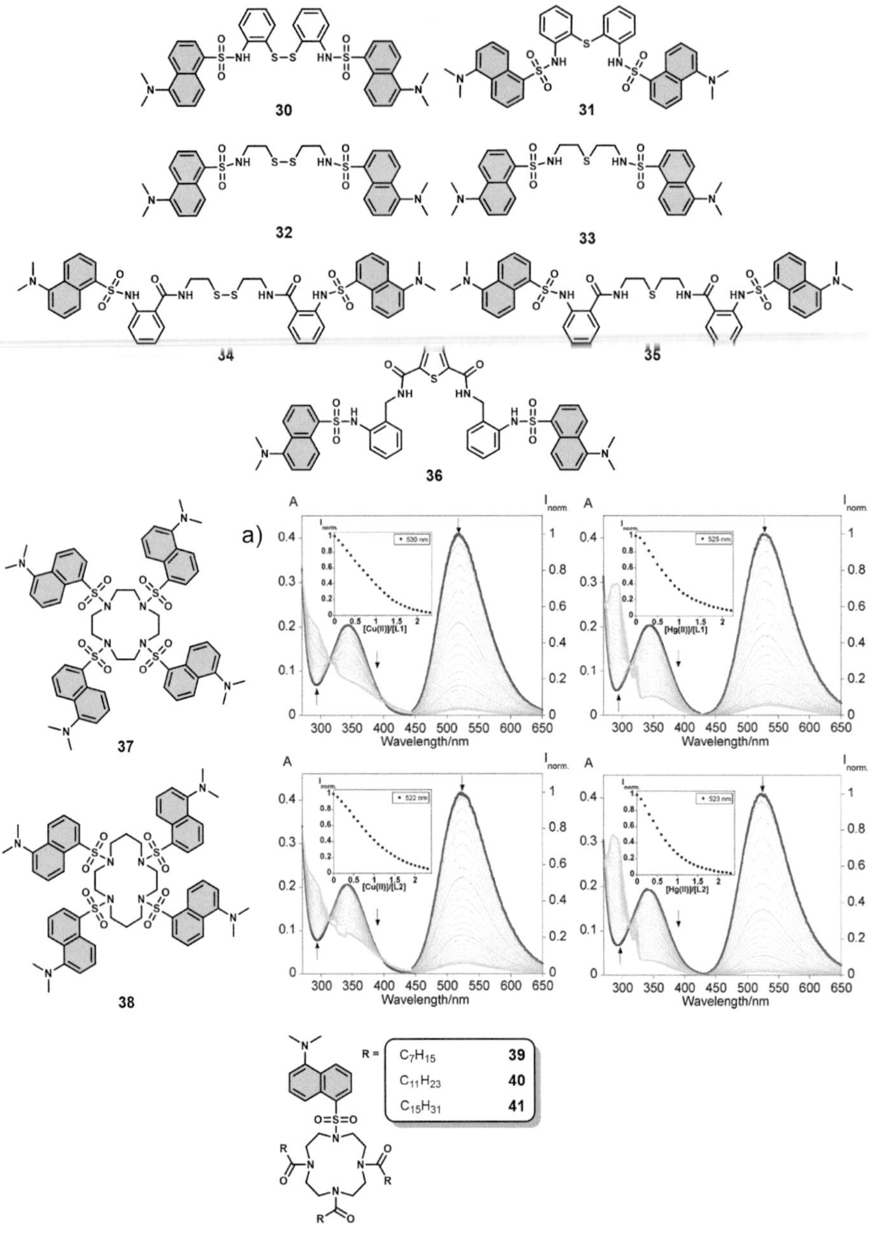


**FIGURE 13.6** Structures of compounds 30–41 for the dual detection of $Cu^{2+}$ and $Hg^{2+}$. (a) Spectrophotometric and spectrofluorimetric titrations of compounds 37 and 38 with increased additions of $Cu^{2+}$ and $Hg^{2+}$ in acetonitrile. The inset represents the emission as a function of the number of equivalents of metal. (Adapted from ref.[111] copyright 2024, Elsevier.)

**TABLE 13.3**
**Comparison of Dansyl-Based Compounds Reported by Us for the Detection of $Cu^{2+}$ and $Hg^{2+}$ in Acetonitrile**

| Compound | Metal Ion | $LogK_{ass}$ | Limit of Detection (μM) | Ref |
|---|---|---|---|---|
| **31** | $Cu^{2+}$ | 5.73 ± 0.01 | 4.0 | 108 |
| | $Hg^{2+}$ | 5.80 ± 0.02 | 2.0 | |
| **32** | $Cu^{2+}$ | 4.59 ± 0.01 | 6.0 | 109 |
| | $Hg^{2+}$ | 6.38 ± 0.06 | 2.5 | |
| **33** | $Cu^{2+}$ | 5.00 ± 0.04 | 6.0 | 110 |
| | $Hg^{2+}$ | 6.38 ± 0.02 | 4.0 | |
| **34** | $Cu^{2+}$ | 5.11 ± 0.06 | 7.5 | 109 |
| | $Hg^{2+}$ | 5.00 ± 0.01 | 8.0 | |
| **35** | $Cu^{2+}$ | 4.82 ± 0.03 | 6.0 | 110 |
| | $Hg^{2+}$ | 5.29 ± 0.06 | 7.0 | |
| **37** | $Cu^{2+}$ | 5.613 ± 0.002 | 2.9 | 111 |
| | $Hg^{2+}$ | 5.104 ± 0.002 | 3.5 | |
| **38** | $Cu^{2+}$ | 5.870 ± 0.003 | 2.8 | 111 |
| | $Hg^{2+}$ | 5.047 ± 0.002 | 2.0 | |
| **39** | $Cu^{2+}$ | 4.73 ± 0.01 | 2.0 | 112 |
| | $Hg^{2+}$ | 3.10 ± 0.01 | 40 | |

## 13.3 DANSYL-BASED DYAD SYSTEMS

Small organic dyes are widely used in bioimaging, biosensing, medical diagnostics, and environmental monitoring due to their bright fluorescence and versatility. However, many suffer from narrow Stokes shifts, leading to self-quenching and interference from excitation light, and intensity-based measurements can be affected by environmental factors, probe localization, and optical path variations.[113] Ratiometric fluorescence, which measures emission at two wavelengths and calculates the intensity ratio, provides a more reliable approach. While single-fluorophore systems can achieve ratiometric responses via ICT, multi-fluorophore constructs transfer energy from a donor to an acceptor, generating larger pseudo-Stokes shifts and dual-emission signals with built-in correction for environmental effects.[113] Dansyl derivatives are particularly suited for these architectures, as their high quantum yields, environmental sensitivity, and easy modification allow them to act as donors. Conjugating dansyl with other chromophores can tune absorption and emission, enhance photostability, and improve energy transfer efficiency. Exploring these donor-acceptor strategies with dansyl derivatives thus holds significant promise for the search of robust and tunable fluorescent probes. This section examines the principles, strategies, and recent developments in conjugating dansyl with other chromophores, emphasizing how such designs expand the functionality and applicability of fluorescent probes in sensing, imaging, and molecular diagnostics.

Our contribution to this field was introduced by the design of four naphthalimide-derived ligands, two of which are engineered as dansyl-naphthalimide dyads (**40** and **41**, Figure 13.7).[114] These dyads were designed to exploit intramolecular electron transfer processes, enabling dual responsiveness toward Hg(II) and Cu(II) ions. Photophysical characterization revealed that the dyads (**40** and **41**) display distinct dual-emission behavior compared to single naphthalimide analogues. In acetonitrile at 298 K, they absorb around 425 nm and emit at 515 nm, with quantum yields ranging from 7.5% to 33% depending on solvent. Solvatochromism was evident, with emission shifting from green to yellow as solvent polarity increased. Metal

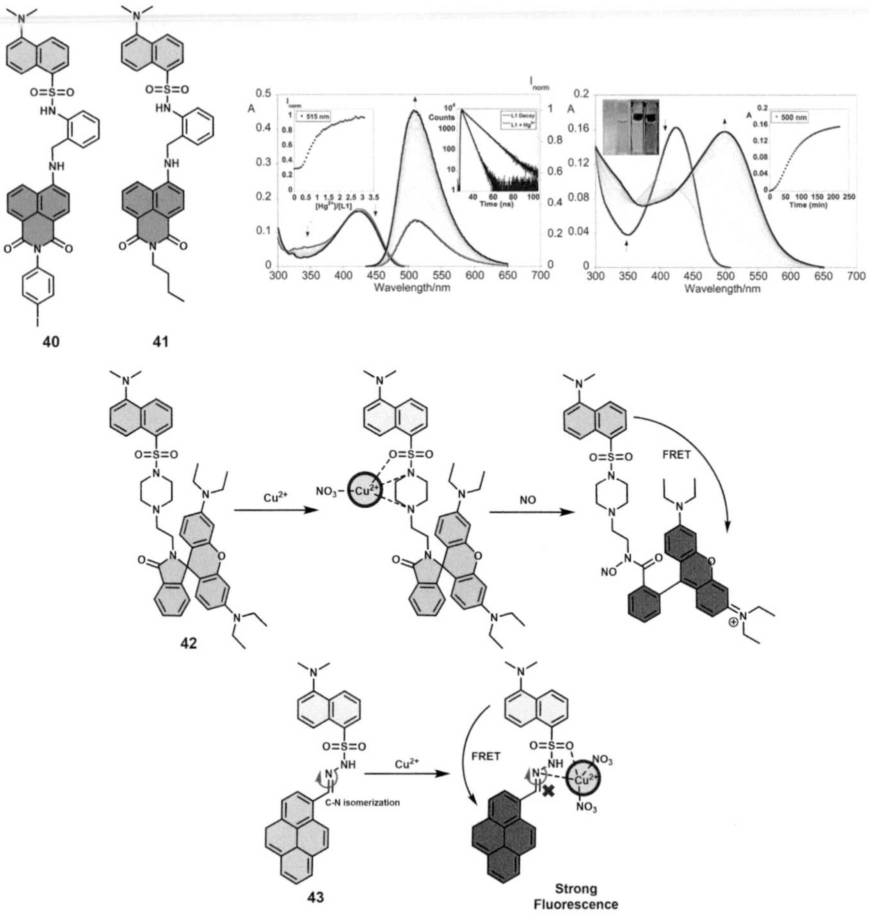


**FIGURE 13.7** Structures of dansyl dyads integrating naphthalimide (40 and 41), rhodamine (42), and pyrene (43). UV–vis and luminescent titration of compound 40 (left) upon continuous additions of $Hg^{2+}$. The inset represents the emission of [Hg2+]/[L1] at 515 nm (left) and the time-resolved fluorescence decay profile of 40 before and after the titration with $Hg^{2+}$ions (right). UV–vis time continuous spectra of L1 (left) after the addition of 7 eq. of $Cu^{2+}$. The inset represents The images under natural light and UV light (left) and the time-dependent response of the absorbance at 500 nm (right). (Adapted from ref.[114] copyright 2025, Elsevier.)

ion sensing highlighted the unique role of the dyads. Upon addition of $Hg^{2+}$, both **40** and **41** demonstrated CHEF, a rare "turn-on" response given mercury's heavy atom effect (Figure 13.7a). Quantum yields increased two- to threefold, and fluorescence lifetimes extended to ~9 ns. In contrast, $Cu^{2+}$ induced quenching and a bathochromic shift of absorption bands, accompanied by visible color changes from yellow to pink (Figure 13.7a). Time-dependent studies revealed ligand-mediated reduction of $Cu^{2+}$ to $Cu^{+}$, supported by EPR and XPS data, establishing the dyads as non-redox-innocent systems capable of electron transfer. Mechanistically, the dansyl unit in the dyads initiates PeT, which is suppressed upon $Hg^{2+}$ coordination, restoring fluorescence. For $Cu^{2+}$, the dyads promote ligand-to-metal electron transfer, reducing $Cu^{2+}$ to $Cu^{+}$ and altering absorption profiles. This dual functionality, fluorescence "turn-on" for $Hg^{2+}$ and colorimetric response for $Cu^{2+}$, distinguishes the dyads from single naphthalimide analogues, which generally show quenching behavior. Catalytic relevance was demonstrated by incorporating an iodophenyl moiety as a substrate probe. In **40**, $Cu^{2+}$ reduction facilitated Ullmann-type amidation, highlighting the potential of these dyads as ancillary ligands in copper-catalyzed reactions. The dual responsiveness and catalytic activity underscore their versatility as both sensors and functional ligands.

Srivastava and co-workers reported the development of a rhodamine-dansyl conjugate (**42**, Figure 13.7) designed as a FRET-based ratiometric probe for nitric oxide detection mediated by $Cu^{2+}$ ions.[115] The optical behavior of the probe was examined in aqueous phosphate buffer (10 mM, pH 7.4) with acetonitrile as co-solvent. The system showed strong dependence on sequential chemical inputs since $Cu^{2+}$ binding first quenched the dansyl emission at 481 nm, and subsequent addition of NO triggered ring opening of the rhodamine spirolactam, producing a new emission band at 580 nm. This sequence-dependent response was accompanied by visible color changes, from colorless to pink under ambient light and from pale green to orange under UV illumination, allowing naked-eye detection. Spectral studies confirmed that the probe-$Cu^{2+}$ complex exhibited high selectivity for NO over other reactive oxygen and nitrogen species. The detection limit for NO was determined to be approximately 1 nM, with a tenfold increase in quantum yield upon NO binding. FRET efficiency was calculated at 88%. The probe maintained stable performance across a biologically relevant pH range (6.5–9.0), confirming suitability for cellular applications. Metal ion interference experiments showed that **42** was selective for $Cu^{2+}$ among a panel of cations. Binding stoichiometry was established as 1:1 by Job's plot analysis, with a binding constant of $4.36 \times 10^4\ M^{-1}$. Cyclic voltammetry confirmed the easily achieved reduction of $Cu^{2+}$ to $Cu^{+}$ at 0.201 V vs Ag/AgCl, consistent with the proposed mechanism in which NO coordinates to $Cu^{2+}$ before inducing rhodamine ring opening. The probe was also evaluated as a molecular keypad lock device, where the correct input sequence, $Cu^{2+}$ followed by NO, produced the "unlock" state with strong emission at 580 nm. Alternative sequences, such as NO before $Cu^{2+}$ or $Cu^{2+}$ followed by EDTA, resulted in no signal, demonstrating sequence-dependent logic gate behavior. Biological experiments confirmed low cytotoxicity in HeLa and MCF-7 cells, with viability maintained after 24 h exposure to probe concentrations up to 40 μM. Confocal microscopy showed efficient cellular uptake and dual-channel emission. In HeLa cells, the probe–$Cu^{2+}$ complex alone produced weak green emission, while subsequent NO treatment generated strong signals in both green and red channels, consistent with the dual fluorophore activation mechanism.

Paul *et al.* designed a pyrene-dansyl dyad (**43**, Figure 13.7) as a multi-channel probe for $Cu^{2+}$ and $CN^-$ ions, with additional responsiveness to aqueous environments.[116] The dyad links a dansyl donor to a pyrene acceptor through an imine bridge. Compound **43** shows weak emission when excited at 290 nm, displaying only faint blue emission around 450 nm. This low intensity arises from PET between the hydrazone nitrogen and the pyrene ring, as well as cis-trans isomerization of the imine bond, both of which act as non-radiative decay pathways. Upon addition of $Cu^{2+}$ in aqueous acetonitrile, the emission is enhanced and shifted to green-cyan at around 490 nm. The mechanism involves chelation of $Cu^{2+}$ by the azomethine nitrogen and sulfonyl oxygen, which blocks PET and stabilizes the excited state. At the same time, the dansyl emission overlaps with pyrene absorption, enabling FRET. The combined CHEF-FRET-isomerization suppression pathway produces a "turn-on" fluorescence response with a detection limit of 37.75 nM. Increasing addition of $CN^-$ to **43** in acetonitrile exhibits a yellow color solution and cyan emission. The acidic -NH proton of the hydrazone establish hydrogen bonding with $CN^-$, which induces ICT and shifts the absorption band bathochromically to ~395 nm. The emission is initially quenched by PET which is then suppressed resulting in an increase of the intensity as the hydrogen bond is formed. The detection limit is 0.62 mM, with no interference from other common anions. As the proportion of water in acetonitrile increases up to 90%, **43** undergoes AIEE. Pyrene excimer formation and J-type aggregation produce a broad blue emission band, shifted relative to the monomer. Solid-state fluorescence is further red-shifted at around 515 nm compared to solution, consistent with intermolecular $\pi$–$\pi$ stacking. Plant cells of Tecoma stans and Peperomia pellucida incubated with $Cu^{2+}$ followed by the addition of **43** displayed bright green-cyan luminescence under UV excitation, confirming cell permeability and selective $Cu^{2+}$ tracking. In fetal bovine serum, **43** detected copper at 0.28 ppm, a concentration relevant to the diagnostic window for Menke's disease, showing potential as a biomarker. DNA binding studies with calf thymus DNA revealed hypochromism in absorption spectra, consistent with intercalation. Competitive binding assays showed that **43**-$Cu^{2+}$ could displace ethidium bromide, suggesting possible cytotoxic activity.

## 13.4 CONCLUSIONS AND PERSPECTIVES

The studies discussed in this chapter highlights the broad potential of dansyl-based and related fluorophores as adaptable platforms for sensing and functional material development. Future advances in this field will benefit from a deeper understanding of structure–property relationships, supported by systematic photophysical studies and complementary computational approaches. Such efforts are expected to enable more predictable control over solvatochromism, AIE, and ICT processes.

One of the most promising directions lies in advanced chemical and biological sensing. Dansyl fluorophores exhibit pronounced sensitivity to polarity, viscosity, pH, and microenvironmental changes, making them well suited for next-generation chemosensors and biosensors. Their integration into modular receptor architectures, including macrocycles, hydrogen-bonding motifs, and metal-binding units, may enable selective detection of ions, small molecules, and biologically relevant targets. In this context, ratiometric and lifetime-based sensing strategies are expected to gain increasing relevance, as they overcome the limitations of intensity-based measurements and allow more reliable analysis in complex matrices.

In parallel, dansyl derivatives are likely to continue evolving within biomedical imaging and diagnostic applications. Through molecular engineering, energy-transfer strategies, or incorporation into AIE systems, their optical properties may be further optimized for multichannel or functional imaging. Probes responsive to enzymatic activity, redox status, or local microviscosity may offer new opportunities for monitoring dynamic biological processes, particularly when combined with time-resolved fluorescence techniques.

The role of dansyl units in materials science is also expected to expand. Their incorporation into polymers, hydrogels, liquid-crystalline systems, and hybrid nanomaterials enables fluorescence-based reporting on molecular organization and environmental changes. Such systems hold promises for responsive materials, optical sensing platforms, and supramolecular assemblies, where dansyl fluorophores can serve as versatile reporters within self-organized architectures.

In environmental and food-related analysis, future developments may focus on portable and low-cost sensing platforms. The integration of dansyl-based probes into paper-based devices, microfluidic systems, or wearable formats could facilitate real-time monitoring of contaminants and pollutants, benefiting from the inherent sensitivity of the dansyl chromophore to subtle microenvironmental variations.

Finally, emerging opportunities lie at the intersection of translational and data-driven chemistry. The combination of dansyl fluorophores with machine learning, high-throughput screening, and advanced spectroscopic analysis may accelerate the discovery of new structure–property relationships and sensing concepts. Coupled with sustainable synthetic approaches, these efforts position dansyl-based compounds to evolve from classical fluorescent labels into multifunctional components across analytical, biomedical, and materials-oriented applications.

## REFERENCES

1. B. Valeur, Molecular Fluorescence: Principles and Applications, Wiley-VCH, Weinheim, 2002.
2. D. Wu, A. C. Sedgwick, T. Gunnlaugsson, E. U. Akkaya, J. Yoon and T. D. James, Fluorescent chemosensors: The past, present and future, *Chem. Soc. Rev*, **46** (2017) 7105–7123. DOI: 10.1039/c7cs00240h.
3. J. R. Lakowicz, Principles of Fluorescence Spectroscopy, 3rd ed., Springer, New York, 2010.
4. Y. Yang, F. Gao, Y. Wang, H. Li, J. Zhang, Z. Sun and Y. Jiang, Fluorescent organic small molecule probes for bioimaging and detection applications, *Molecules*, **27** (2022) 8421. DOI: 10.3390/molecules27238421.
5. D. K. Mal, H. Pal and G. Chakraborty, A comprehensive review on recent advances in fluorescence-based bio-analytes sensing, *TrAC Trends in Analytical Chemistry*, **171** (2024) 117493. DOI: 10.1016/j.trac.2023.117493.

6. H. Niu, J. Liu, H. M. O'Connor, T. Gunnlaugsson, T. D. James and H. Zhang, Photoinduced electron transfer (PeT) based fluorescent probes for cellular imaging and disease therapy, *Chem. Soc. Rev*, **52** (2023) 2322–2357. DOI: 10.1039/d1cs01097b.
7. L. Yang, J. Y. Niu, R. Sun, Y. J. Xu and J. F. Ge, The pH-influenced PET processes between pyronine and different heterocycles, *Org. Biomol. Chem*, **15** (2017) 8402–8409. DOI: 10.1039/C7OB02247F.
8. D. Escudero, Revising intramolecular photoinduced electron transfer (PET) from first-principles, *Acc. Chem. Res*, **49** (2016) 1816–1824. DOI: 10.1021/acs.accounts.6b00299.
9. J. A. Broussard, B. Rappaz, D. J. Webb and C. M. Brown, Fluorescence resonance energy transfer microscopy as demonstrated by measuring the activation of the serine/threonine kinase akt, *Nat. Protoc*, **8** (2013) 265–281. DOI: 10.1038/nprot.2012.147.
10. B. Zhang, G. Lyu, E. A. Kelly and R. C. Evans, Förster resonance energy transfer in luminescent solar concentrators, *Adv. Sci*, **9** (2022) 2201160. DOI: 10.1002/advs.202201160.
11. L. Wu, C. Huang, B. P. Emery, A. C. Sedgwick, S. D. Bull, X. P. He, H. Tian, J. Yoon, J. L. Sessler and T. D. James, Förster resonance energy transfer (FRET)-based small-molecule sensors and imaging agents, *Chem. Soc. Rev*, **49** (2020) 5110–5139. DOI: 10.1039/c9cs00318e.
12. D. Udhayakumari, P. Jerome, N. Vijay and T. Hwan Oh, ESIPT: An approach and future perspective for the detection of biologically important analytes, *J. Lumin*, **267** (2024) 120350. DOI: 10.1016/j.jlumin.2023.120350.
13. J. Luo, Z. Xie, Z. Xie, J. W. Y. Lam, L. Cheng, H. Chen, C. Qiu, H. S. Kwok, X. Zhan, Y. Liu, D. Zhu and B. Z. Tang, Aggregation-induced emission of 1-methyl-1,2,3,4,5-pentaphenylsilole, *Chem. Commun*, **37** (2001) 1740–1741. DOI: 10.1039/B105159H.
14. J. Krämer, R. Kang, L. M. Grimm, L. De Cola, P. Picchetti and F. Biedermann, Molecular probes, chemosensors, and nanosensors for optical detection of biorelevant molecules and ions in aqueous media and biofluids, *Chem. Rev*, **122** (2022) 3459–3636. DOI: 10.1021/acs.chemrev.1c00746.
15. N. W. Nkune, K. Moloudi, B. P. George and H. Abrahamse, An update on recent advances in fluorescent materials for fluorescence molecular imaging: A review, *RSC Adv*, **15** (2025) 22267–22284. DOI: 10.1039/d5ra03102h.
16. T. C. Lovell, B. P. Branchaud and R. Jasti, An organic Chemist's guide to fluorophores – Understanding common and newer non-planar fluorescent molecules for biological applications, *Eur. J. Org. Chem*, **27** (2024) e202301196. DOI: 10.1002/ejoc.202301196.
17. R. Gui and H. Jin, Organic fluorophores-based molecular probes with dual-fluorescence ratiometric responses to in-vitro/in-vivo pH for biosensing, bioimaging and biotherapeutics applications, *Talanta*, **275** (2024) 126171. DOI: 10.1016/j.talanta.2024.126171.
18. D. Ceballos-Ávila, I. Vázquez-Sandoval, F. Ferrusca-Martínez and A. Jiménez-Sánchez, Conceptually innovative fluorophores for functional bioimaging, *Biosens. Bioelectron*, **264** (2024) 116638. DOI: 10.1016/j.bios.2024.116638.
19. D. Udhayakumari, S. Ramasundaram, P. Jerome and T. H. Oh, A review on small molecule based fluorescence chemosensors for bioimaging applications, *J. Fluoresc*, **35** (2025) 3763–3787. DOI: 10.1007/s10895-024-03826-2.
20. I. Muhammad, T. Z. Ren, R. Han, X. Zhang, K. Dawood and A. Shahzad, Design strategies and analytical performance of dansyl fluorophore-based fluorescent sensors for mercury ion ($Hg^{2+}$) detection in water, *Crit. Rev. Anal. Chem* (2025) 1–15. DOI: 10.1080/10408347.2024.2449532.
21. G. Weber, Polarization of the fluorescence of macromolecules. I. Theory and experimental method, *Biochem. J*, **51** (1952) 145–155. DOI: 10.1042/bj0510145.
22. U. N. Ritu and V. Kumar, Heavy metal detection with organic moiety-based sensors: Recent advances and future directions, *ChemBioChem*, **25** (2024) e202400191. DOI: 10.1002/cbic.202400191.

23. M. Li, Q. Shi, N. Song, Y. Xiao, L. Wang, Z. Chen and T. D. James, Current trends in the detection and removal of heavy metal ions using functional materials, *Chem. Soc. Rev*, **52** (2023) 5827–5860. DOI: 10.1039/d2cs00683a.
24. T. Chopra, S. Sasan, L. Devi, R. Parkesh and K. K. Kapoor, A comprehensive review on recent advances in copper sensors, *Coord. Chem. Rev*, **470** (2022) 214704. DOI: 10.1016/j.ccr.2022.214704.
25. D. T. Quang and J. S. Kim, Fluoro- and chromogenic chemodosimeters for heavy metal ion detection in solution and biospecimens, *Chem. Rev*, **110** (2010) 6280–6301. DOI: 10.1021/cr100154p.
26. S. Chowdhury, B. Rooj, A. Dutta and U. Mandal, Review on recent advances in metal ions sensing using different fluorescent probes, *J. Fluoresc*, **28** (2018) 999–1021. DOI: 10.1007/s10895-018-2263-y.
27. M. Saleem, M. Rafiq, M. Hanif, M. A. Shaheen and S. Y. Seo, A brief review on fluorescent copper sensor based on conjugated organic dyes, *J. Fluoresc*, **28** (2018) 97–165. DOI: 10.1007/s10895-017-2178-z.
28. Z. H. Yuan, Y. S. Yang, P. C. Lv and H. L. Zhu, Recent progress in small-molecule fluorescent probes for detecting mercury ions, *Crit. Rev. Anal. Chem*, **52** (2022) 250–274. DOI: 10.1080/10408347.2020.1797466.
29. Q. Li and Y. Zhou, Recent advances in fluorescent materials for mercury(II) ion detection, *RSC Adv*, **13** (2023) 19429–19446. DOI: 10.1039/d3ra02410e.
30. L. Li, J. Wang, S. Xu, C. Li and B. Dong, Recent progress in fluorescent probes for metal ion detection, *Front. Chem*, **10** (2022) 875241. DOI: 10.3389/fchem.2022.875241.
31. X. Cheng, T. Luo, F. Chu, B. Feng, S. Zhong, F. Chen, J. Dong and W. Zeng, Simultaneous detection and removal of mercury (II) using multifunctional fluorescent materials, *Sci. Total Environ*, **905** (2023) 167070. DOI: 10.1016/j.scitotenv.2023.167070.
32. P. Wang, L. Liu, P. Zhou, W. Wu, J. Wu, W. Liu and Y. Tang, A peptide-based fluorescent chemosensor for multianalyte detection, *Biosens. Bioelectron*, **72** (2015) 80–86. DOI: 10.1016/j.bios.2015.04.094.
33. J. Huang, M. Tang, M. Liu, M. Zhou, Z. Liu, Y. Cao, M. Zhu, S. Liu and W. Zeng, Development of a fast responsive and highly sensitive fluorescent probe for $Cu^{2+}$ ion and imaging in living cells, *Dyes and Pigments*, **107** (2014) 1–8. DOI: 10.1016/j.dyepig.2014.02.022.
34. P. Wang, J. Wu, P. Zhou, W. Liu and Y. Tang, A novel peptide-based fluorescent chemosensor for measuring zinc ions using different excitation wavelengths and application in live cell imaging, *J. Mater. Chem. B*, **3** (2015) 3617–3624. DOI: 10.1039/C5TB00142K.
35. P. Wang, J. Wu, P. Su, C. Xu, Y. Ge, D. Liu, W. Liu and Y. Tang, Fluorescence "on-off-on" peptide-based chemosensor for the selective detection of $Cu^{2+}$ and $S^{2-}$ and its application in living cell bioimaging, *Dalton Trans*, **45** (2016) 16246–16254. DOI: 10.1039/C6DT03330J.
36. P. Wang, J. Wu, P. Su, C. Shan, P. Zhou, Y. Ge, D. Liu, W. Liu and Y. Tang, A novel peptide-based fluorescent probe for selective detection of $Fe^{3+}$ and its application in living cells, *J. Mater. Chem. B*, **4** (2016) 4526–4533.
37. Y. Bin Ruan, A. Depauw and I. Leray, Aggregation-induced emission enhancement upon $Al^{3+}$ complexation with a tetrasulfonated calix[4]bisazacrown fluorescent molecular sensor, *Org. Biomol. Chem*, **12** (2014) 4335–4341. DOI: 10.1039/C4OB00187G.
38. K. H. Jung, E. T. Oh, H. J. Park and K. H. Lee, Development of new peptide-based receptor of fluorescent probe with femtomolar affinity for $Cu^+$ and detection of $Cu^+$ in Golgi apparatus, *Biosens. Bioelectron*, **85** (2016) 437–444. DOI: 10.1016/j.bios.2016.04.101.
39. S. Zhou, Z. Q. Zhou, X. X. Zhao, Y. H. Xiao, G. Xi, J. T. Liu and B. X. Zhao, A dansyl-based fluorescence chemosensor for $Hg^{2+}$ and its application in complicated environment samples, *Spectrochim. Acta A Mol. Biomol. Spectrosc*, **148** (2015) 348–354. DOI: 10.1016/j.saa.2015.03.126.

40. M. G. Choi, J. Kim, J. M. Hong, I. J. Chang, S. Ahn and S. K. Chang, $Cu^{2+}$-selective ratiometric fluorescence signaling probe based on the hydrolysis of dansylhydrazine, *Tetrahedron Lett*, **57** (2016) 975–978. DOI: 10.1016/j.tetlet.2016.01.052.
41. L. N. Neupane and K. H. Lee, Selective and sensitive turn on detection of $Hg^{2+}$ in aqueous solution using a thioether-appended dipeptide, *Tetrahedron Lett*, **54** (2013) 5007–5010. DOI: 10.1016/j.tetlet.2013.07.003.
42. P. Srivastava, R. Ali, S. S. Razi, M. Shahid, S. Patnaik and A. Misra, A simple blue fluorescent probe to detect $Hg^{2+}$ in semiaqueous environment by intramolecular charge transfer mechanism, *Tetrahedron Lett*, **54** (2013) 3688–3693. DOI: 10.1016/j.tetlet.2013.05.014.
43. J. Isaad and A. El Achari, Azathia crown ether possessing a dansyl fluorophore moiety functionalized silica nanoparticles as hybrid material for mercury detection in aqueous medium, *Tetrahedron*, **69** (2013) 4866–4874. DOI: 10.1016/j.tet.2013.04.072.
44. L. J. Ma, J. Liu, L. Deng, M. Zhao, Z. Deng, X. Li, J. Tang and L. Yang, Selective and sensitive fluorescence-shift probes based on two dansyl groups for mercury(II) ion detection, *Photochem. Photobiol. Sci*, **13** (2014) 1521–1528. DOI: 10.1039/C4PP00094C.
45. S. Zhang, T. Yu, M. Sun, H. Yu, Z. Zhang, S. Wang and H. Jiang, Highly sensitive and selective fluorescence detection of copper (II) ion based on multi-ligand metal chelation, *Talanta*, **126** (2014) 185–190. DOI: 10.1016/j.talanta.2014.03.076.
46. L. J. Ma, W. Cao, J. Liu, M. Zhang and L. Yang, Chromium (III) ion-selective "turn-on" and ratiometric probe based on the asparagine bearing dansyl group in aqueous solutions, *Sens. Actuators B Chem*, **181** (2013) 782–786. DOI: 10.1016/j.snb.2013.01.095.
47. G. N. Wei, J. L. Zhang, C. Jia, W. Z. Fan and L. R. Lin, Selective recognition of sulfate anions in a 95% ethanol solvent with a simple neutral salicylaldehyde dansyl hydrazine Schiff base tuned by Brønsted–Lowry acid–base reaction, *Spectrochim. Acta A Mol. Biomol. Spectrosc*, **128** (2014) 168–175. DOI: 10.1016/j.saa.2014.02.088.
48. Y. A. Jeong, I. J. Chang and S. K. Chang, Discriminative sensing of sulfide and azide ions in solution by a nitrobenzoxadiazole-dansyl dyad, *Sens. Actuators B Chem*, **224** (2016) 73–80. DOI: 10.1016/j.snb.2015.10.013.
49. S. Areti, S. Bandaru, D. S. Yarramala and C. P. Rao, Optimizing the electron-withdrawing character on benzenesulfonyl moiety attached to a glyco-conjugate to impart sensitive and selective sensing of cyanide in HEPES buffer and on cellulose paper and silica gel strips, *Anal. Chem*, **87** (2015) 12396–12403. DOI: 10.1021/acs.analchem.5b04085.
50. R. Alam, T. Mistri, P. Mondal, D. Das, S. K. Mandal, A. R. Khuda-Bukhsh and M. Ali, A novel copper(II) complex as a nitric oxide turn-on fluorosensor: Intracellular applications and DFT calculation, *Dalton Trans*, **43** (2014) 2566–2576. DOI: 10.1039/C3DT52521J.
51. R. Qian, L. Ding, L. Yan and H. Ju, Fluorescence imaging for in situ detection of cell surface sialic acid by competitive binding of 3-(dansylamino)phenylboronic acid, *Anal. Chim. Acta*, **894** (2015) 189–198. DOI: 10.1016/j.aca.2015.08.054.
52. S. Areti, S. K. Verma, J. Bellare and C. P. Rao, Selenocysteine vs cysteine: Tuning the derivatization on benzenesulfonyl moiety of a triazole-linked dansyl connected glycoconjugate for selective recognition of selenocysteine and the applicability of the conjugate in buffer, in serum, on silica gel, and in HepG2 cells, *Anal. Chem*, **88** (2016) 7259–7267. DOI: 10.1021/acs.analchem.6b01518.
53. K. Kaur and S. Kumar, Dansyl-anthracene dyads for ratiometric fluorescence recognition of $Cu^{2+}$, *Dalton Trans*, **40** (2011) 2451–2458. DOI: 10.1039/C0DT01016B.
54. Z. Hu, J. Hu, Y. Cui, G. Wang, X. Zhang, K. Uvdal and H. W. Gao, A facile "click" reaction to fabricate a FRET-based ratiometric fluorescent $Cu^{2+}$ probe, *J. Mater. Chem. B*, **2** (2014) 4467–4472. DOI: 10.1039/C4TB00441H.

55. H. J. Lee, M. J. Cho and S. K. Chang, Ratiometric signaling of hypochlorite by the oxidative cleavage of a sulfonhydrazide-based rhodamine-dansyl dyad, *Inorg. Chem*, **54** (2015) 8644–8649. DOI: 10.1021/acs.inorgchem.5b01284.
56. H. Li, X. Wang, K. Yuan, L. Lv, K. Liu and Z. Li, Fluorescent mechanism of a highly selective probe for copper(II) detection: A theoretical study, *ACS Omega*, **8** (2023) 17171–17180. DOI: 10.1021/acsomega.3c01528.
57. R. Nagarajan, H. I. Ryoo, B. D. Vanjare, N. Gyu Choi and K. Hwan Lee, Novel phenylalanine derivative-based turn-off fluorescent chemosensor for selective $Cu^{2+}$ detection in physiological pH, *J. Photochem. Photobiol. A Chem*, **418** (2021) 113435. DOI: 10.1016/j.jphotochem.2021.113435.
58. H. Xu, Y. Xiao, Y. Liu and W. Sun, Research progress on naphthalimide fluorescent probes, *Adv. Sens. Res*, **3** (2023) 2300032. DOI: 10.1002/adsr.202300032.
59. M. Zhou, X. Wang, K. Huang, Y. Huang, S. Hu and W. Zeng, A fast, highly selective and sensitive dansyl-based fluorescent sensor for $Cu^{2+}$ ions and its imaging application in living cells, *Tetrahedron Lett*, **58** (2017) 991–994. DOI: 10.1016/j.tetlet.2017.01.090.
60. Y. Liu, B. Jiang, L. Zhao, L. Zhao, Q. Wang, C. Wang and B. Xu, A dansyl-based fluorescent probe for sensing $Cu^{2+}$ in aqueous solution, *Spectrochim. Acta A Mol. Biomol. Spectrosc*, **261** (2021) 120009. DOI: 10.1016/j.saa.2021.120009.
61. X. Wang, P. Xia and X. Huang, A dansyl-appended n-heterocycle for $Cu^{2+}$ and $S^{2-}$ recognition via a displacement mode, *Spectrochim. Acta A Mol. Biomol. Spectrosc*, **210** (2019) 98–104. DOI: 10.1016/j.saa.2018.11.017.
62. H. Cai, Y. Liang, L. Huang and J. Wang, Elay detection of $Cu^{2+}$ and bovine serum albumin by a dansyl derivative-based fluorescent probe, *Acta A Mol. Biomol. Spectrosc*, **277** (2022) 121281. DOI: 10.1016/j.saa.2022.121281.10.1016/j.saa.2022.121281.
63. Y. Wang, J. Zhou, L. Zhao and B. Xu, A dual-responsive and highly sensitive fluorescent probe based on a dansyl derivative for $Cu^{2+}$ and pH, *Dyes Pigments*, **180** (2020) 108513. DOI: 10.1016/j.dyepig.2020.108513.
64. P. Wang, X. Cao, S. Xue, Z. Wang, Y. Zhou and J. Wu, Peptide-based fluorescent probe with large Stokes shift for specific sequential detection of copper(II) ions and histidine, *Microchem. J*, **207** (2024) 111728. DOI: 10.1016/j.microc.2024.111728.
65. Y. Ren, M. Jia, S. Xiong, C. Pu, P. Wang and Y. An, Highly selective sequential detection of $Cu^{2+}$ and $S^{2-}$ using a large Stokes shift peptide-based fluorescent probe and its multi-application, *Microchem. J*, **216** (2025) 114732. DOI: 10.1016/j.microc.2025.114732.
66. Z. Guo, Q. Wang, D. Zhou, Y. An, P. Wang and F. Liao, A novel peptide-based fluorescent probe with a large Stokes shift for rapid and sequential detection of $Cu^{2+}$ and $CN^-$ in aqueous systems and live cells, *Spectrochim. Acta A Mol. Biomol. Spectrosc*, **264** (2022) 120257. DOI: 10.1016/j.saa.2021.120257.
67. P. Wei, L. Xiao, Y. Gou, F. He, D. Zhou, Y. Liu, B. Xu, P. Wang and Y. Zhou, Fluorescent "on–off–on" probe based on copper peptide backbone for specific detection of $Cu^{2+}$ and $H_2S$ in 100% aqueous medium and application in cell imaging, real water samples and test strips, *Microchem. J*, **182** (2022) 107848. DOI: 10.1016/j.microc.2022.107848.
68. Q. Wang, Z. Guo, D. Zhou, J. Wu, P. Wang, X. Yang and S. Wen, A novel fluorescent probe for highly selective and sensitive detection of sulfur ions in real samples and living cells based on the tripeptide-$Cu^{2+}$ ensemble system, *Microchem. J*, **169** (2021) 106612. DOI: 10.1016/j.microc.2021.106612.
69. P. Wei, L. Xiao, P. Hou, Q. Wang and P. Wang, A novel Cu(II)-assisted peptide fluorescent probe for highly sensitive detection of glyphosate in real samples: Real application in test strips and smartphone, *Anal. Bioanal. Chem*, **415** (2023) 5985–5996. DOI: 10.1007/s00216-023-04869-3.

70. Y. An, P. Wang and Z. Yue, A sequential and reversibility fluorescent pentapeptide probe for Cu(II) ions and hydrogen sulfide detections and its application in two different living cells imaging, *Spectrochim. Acta A Mol. Biomol. Spectrosc*, **216** (2019) 319–327. DOI: 10.1016/j.saa.2019.03.065.
71. Y. Hu, S. Quan, C. Zhao, J. Li, X. Sun and J. Xiao, An "on–off–on" fluorescent peptide probe for the specific detection of $Cu^{2+}$ and $S^{2-}$ in living cells and zebrafish, *New J. Chem*, **46** (2022) 7663–7668. DOI: 10.1039/D2NJ00408A.
72. A. Thangaraj, V. Bhardwaj and S. K. Sahoo, A multi-analyte selective dansyl derivative for the fluorescence detection of Cu(ii) and cysteine, *Photochem. Photobiol. Sci*, **18** (2019) 1533–1539. DOI: 10.1039/C9PP00080A.
73. W. Nasomphan, P. Tangboriboonrat and S. Smanmoo, Dansyl based "Turn-on" fluorescent sensor for $Cu^{2+}$ ion detection and the application to living cell imaging, *J. Fluoresc*, **27** (2017) 2201–2212. DOI: 10.1007/s10895-017-2161-8.
74. H. Jiang, Z. Li, Y. Kang, L. Ding, S. Qiao, S. Jia, W. Luo and W. Liu, A two-photon fluorescent probe for $Cu^{2+}$ based on dansyl moiety and its application in bioimaging, *Sens. Actuators B Chem*, **242** (2017) 112–117. DOI: 10.1016/j.snb.2016.11.033.
75. Y. Zhao, D. Chen, J. Yang and B. Yang, Visual and fast detection of trace copper ions using biosensor based on FRET, *Spectrochim. Acta A Mol. Biomol. Spectrosc*, **217** (2019) 101–106. DOI: 10.1016/j.saa.2019.03.082.
76. S. Mukherjee, K. Sarkar and S. Biswas, A fluorophore anchored MOF for fast and sensitive sensing of Cu(II) and 3-nitrotyrosine in a physiological medium, *Dalton Trans*, **52** (2023) 5597–5605. DOI: 10.1039/D3DT00660C.
77. L. Zhang, P. Zhao, S. Yu, J. Kong, L. Gao and L. Li, Dansyl-modified tryptophan fluorescence sensor for highly sensitive and selective detection of $Hg^{2+}$ in aqueous solutions, *Microchem. J*, **200** (2024) 110520. DOI: 10.1016/j.microc.2024.110520.
78. W. Deng, S. Li, M. Zhou, M. Zheng, P. Wang and Y. An, Ratiometric peptide-based fluorescent probe with large Stokes shift for detection of $Hg^{2+}$ and $S^{2-}$ and its applications in cells imaging and smartphone-assisted recognition, *Spectrochim. Acta A Mol. Biomol. Spectrosc*, **315** (2024) 124306. DOI: 10.1016/j.saa.2024.124306.
79. M. Zhou, M. Zheng, P. Wang and Y. An, A novel ratiometric peptide-based fluorescent probe for sequential detection of $Hg^{2+}$ and $S^{2-}$ ions and its application in living cells and zebrafish imaging, *Spectrochim. Acta A Mol. Biomol. Spectrosc*, **309** (2024) 123829. DOI: 10.1016/j.saa.2023.123829.
80. P. Wang, Y. An and J. Wu, Highly sensitive turn-on detection of mercury(II) in aqueous solutions and live cells with a chemosensor based on tyrosine, *Spectrochim. Acta A Mol. Biomol. Spectrosc*, **230** (2020) 118004. DOI: 10.1016/j.saa.2019.118004.
81. S. Xue, P. Wang and K. Chen, A novel fluorescent chemosensor for detection of mercury(II) ions based on dansyl-peptide and its application in real water samples and living LNCaP cells, *Spectrochim. Acta A Mol. Biomol. Spectrosc*, **226** (2020) 117616. DOI: 10.1016/j.saa.2019.117616.
82. P. Wei, L. Xiao, Y. Gou, F. He, P. Wang and X. Yang, A novel peptide-based relay fluorescent probe with a large stokes shift for detection of $Hg^{2+}$ and $S^{2-}$ in 100% aqueous medium and living cells: Visual detection via test strips and smartphone, *Spectrochimica Acta Part A: Molecular and Biomolecular Spectroscopy*, **285** (2023) 121836. DOI: 10.1016/j.saa.2022.121836.
83. X. Pang, J. Dong, L. Gao, L. Wang, S. Yu, J. Kong and L. Li, Dansyl-peptide dual-functional fluorescent chemosensor for $Hg^{2+}$ and biothiols, *Dyes and Pigments*, **173** (2020) 107888. DOI: 10.1016/j.dyepig.2019.107888.10.1016/j.dyepig.2019.107888.
84. S. Yu, L. Gao, R. Li, C. Fu, W. Meng, L. Wang and L. Li, Ultrasensitive mercury ion and biothiol detection based on dansyl-His-Pro-Gly-Asp-$NH_2$ fluorescent sensor, *Spectrochim. Acta A Mol. Biomol. Spectrosc*, **250** (2020) 119246. DOI: 10.1016/j.saa.2020.119246.

85. Y. Li, Z. Ren, Y. Ge, C. Di, J. Zhou, J. Wu and L. Jia, A novel peptide fluorescent probe based on different fluorescence responses for detection of mercury species and hydrogen sulfide, *Microchem. J*, **184** (2023) 108160. DOI: 10.1016/j.microc.2022.108160.
86. A. Jose, R. Sahadevan, M. Vijay, S. Sadhukhan and M. Porel, Dansyl-appended sequence-defined oligomers for selective ultrasensitive detection of $Hg^{2+}$ in water, paper strips, living cells and its efficient removal, *Sens. Actuators B Chem*, **380** (2023) 133335. DOI: 10.1016/j.snb.2023.133335.
87. A. Aliberti, P. Vaiano, A. Caporale, M. Consales, M. Ruvo and A. Cusano, Fluorescent chemosensors for $Hg^{2+}$ detection in aqueous environment, *Sens. Actuators B Chem*, **247** (2017) 727–735. DOI: 10.1016/j.snb.2017.03.026.
88. S. Gadiyaram, M. Aakshika Sree, N. Sharma and D. Amilan Jose, An amphiphilic dansyl based multianalyte sensor for the detection of $Hg^{2+}$, PPi, and TNP·, *Methods*, **223** (2024) 45–55. DOI: 10.1016/j.ymeth.2024.01.007.
89. L. He, H. Tao, S. Koo, G. Chen, A. Sharma, Y. Chen, I. T. Lim, Q. Y. Cao and J. S. Kim, Multifunctional fluorescent nanoprobe for sequential detections of $Hg^{2+}$ ions and biothiols in live cells, *ACS Appl. Bio Mater*, **1** (2018) 871–878. DOI: 10.1021/acsabm.8b00300.
90. B. Uttam, S. Polepalli, S. Sinha, A. Majumder and C. P. Rao, Selective sensing and removal of mercury ions by encapsulating a dansyl-appended calix[4]arene conjugate in a zeolitic imidazolate framework as an organic–inorganic hybrid nanomaterial, *ACS Appl. Nano Mater*, **5** (2022) 11371–11380. DOI: 10.1021/acsanm.2c02459.
91. M. Oguz, A. A. Bhatti, S. Karakurt, M. Aktas and M. Yilmaz, New water soluble $Hg^{2+}$ selective fluorescent calix[4]arenes: Synthesis and application in living cells imaging, *Spectrochim. Acta A Mol. Biomol. Spectrosc*, **171** (2017) 340–345. DOI: 10.1016/j.saa.2016.08.042.
92. P. Wang, Y. An and Y. Liao, A novel peptide-based fluorescent chemosensor for cd(II) ions and its applications in bioimaging, *Spectrochim. Acta A Mol. Biomol. Spectrosc*, **216** (2019) 61–68. DOI: 10.1016/j.saa.2019.03.014.
93. W. Deng, J. Liu, S. Zhao, Z. Liu, Y. Wang, P. Wang, Y. An and J. Wu, Highly selective peptide-based fluorescent turn-on probe with large Stokes shift for the detection of $Cd^{2+}$ and its applications in bio-imaging and smartphone-assisted recognition, *J. Mol. Struct*, **1321** (2024) 140146. DOI: 10.1016/j.molstruc.2024.140146.
94. P. Wang, D. Zhou and B. Chen, A fluorescent dansyl-based peptide probe for highly selective and sensitive detect $Cd^{2+}$ ions and its application in living cell imaging, *Spectrochim. Acta A Mol. Biomol. Spectrosc*, **207** (2019) 276–283. DOI: 10.1016/j.saa.2018.09.029.
95. P. Wang, L. Duan and Y. Liao, A retrievable and highly selective peptide-based fluorescent probe for detection of $Cd^{2+}$ and Cys in aqueous solutions and live cells, *Microchem. J*, **146** (2019) 818–827. DOI: 10.1016/j.microc.2019.02.004.
96. C. T. Shi, Z. Y. Huang, A. Bin Wu, Y. X. Hu, N. C. Wang, Y. Zhang, W. M. Shu and W. C. Yu, Recent progress in cadmium fluorescent and colorimetric probes, *RSC Adv*, **11** (2021) 29632–29660. DOI: 10.1039/d1ra05048f.
97. A. Akheel Ahmed, N. Mohan, U. Mithra, V. Vijayakumar and S. Sarveswari, Organic chemosensors: A comprehensive review of lead, cadmium, and arsenic ion detection strategies, *Results Chem*, **18** (2025) 102823. DOI: 10.1016/j.rechem.2025.102823.
98. W. Li, X. Zhang, X. Hu, Y. Shi, Z. Li, X. Huang, W. Zhang, D. Zhang, X. Zou and J. Shi, A smartphone-integrated ratiometric fluorescence sensor for visual detection of cadmium ions, *J. Hazard. Mater*, **408** (2021) 124872. DOI: 10.1016/j.jhazmat.2020.124872.
99. L. Yu, S. Wang, K. Huang, Z. Liu, F. Gao and W. Zeng, Fluorescent probes for dual and multi-analyte detection, *Tetrahedron*, **71** (2015) 4679–4706. DOI: 10.1016/j.tet.2015.04.115.

100. J. L. Kolanowski, F. Liu and E. J. New, Fluorescent probes for the simultaneous detection of multiple analytes in biology, *Chem. Soc. Rev*, **47** (2018) 195–208. DOI: 10.1039/C7CS00528H.
101. Y. Yue, F. Huo, F. Cheng, X. Zhu, T. Mafireyi, R. M. Strongin and C. Yin, Functional synthetic probes for selective targeting and multi-analyte detection and imaging, *Chem. Soc. Rev*, **48** (2019) 4155–4177. DOI: 10.1039/C8CS01006D.
102. P. Wang, J. Wu and C. Zhao, A water-soluble peptide fluorescent chemosensor for detection of $Cd^{2+}$ and $Cu^{2+}$ by two different response modes and its application in living LNcap cells, *Spectrochim. Acta A Mol. Biomol. Spectrosc*, **226** (2020) 117600. DOI: 10.1016/j.saa.2019.117600.
103. P. Wang, M. Jia and S. Xue, Rational design of a peptide-based fluorescent probe for selective recognition of $Ag^{+}$ and $Cu^{2+}$ ions and its multifunctional applications, *Luminescence*, **40** (2025) e70291. DOI: 10.1002/bio.70291.
104. P. Wang and J. Wu, Highly selective and sensitive detection of Zn(II) and Cu(II) ions using a novel peptide fluorescent probe by two different mechanisms and its application in live cell imaging, *Spectrochim. Acta A Mol. Biomol. Spectrosc*, **208** (2019) 140–149. DOI: 10.1016/j.saa.2018.09.054.
105. P. Wang, S. Xue and X. Yang, A novel peptide-based fluorescent chemosensor for detection of $Zn^{2+}$ and $Cu^{2+}$ through differential response and application in logic-gate and bioimaging, *Microchem. J*, **158** (2020) 105147. DOI: 10.1016/j.microc.2020.105147.
106. C. Pu, S. Li, X. Cao, Y. Ren, R. Bai, F. You and P. Wang, A novel dual-function peptide-based fluorescent probe with large Stokes shift for the simultaneous detection and quantification of Ag(I) and Hg(II) ions, *Spectrochim. Acta A Mol. Biomol. Spectrosc*, **336** (2025) 126032. DOI: 10.1016/j.saa.2025.126032.
107. G. Donadio, R. Di Martino, R. Oliva, L. Petraccone, P. Del Vecchio, B. Di Luccia, E. Ricca, R. Isticato, A. Di Donato and E. Notomista, A new peptide-based fluorescent probe selective for zinc(II) and copper(II), *J. Mater. Chem. B*, **4** (2016) 6979–6988. DOI: 10.1039/C6TB00671J.
108. I. Lourenço, F. Duarte, G. M. Dobrikov, A. Kurutos, I. Slavchev, J. L. Capelo-Martínez, H. M. Santos and C. Lodeiro, Design and evaluation of dansyl-derived chemosensors for disulfide-cleavage-triggered detection: Photophysical, metal sensing, and thermometric applications, *J. Mater. Chem. C*, **13** (2025) 6708–6719. DOI: 10.1039/D4TC04750H.
109. F. Duarte, G. Dobrikov, A. Kurutos, H. M. Santos, J. Fernández-Lodeiro, J. L. Capelo-Martinez, E. Oliveira and C. Lodeiro, Enhancing water sensing via aggregation-induced emission (AIE) and solvatofluorochromic studies using two new dansyl derivatives containing a disulfide bond: Pollutant metal ions detection and preparation of water-soluble fluorescent polymeric particles, *Dyes Pigments*, **218** (2023) 111428. DOI: 10.1016/j.dyepig.2023.111428.
110. F. Duarte, G. Dobrikov, A. Kurutos, J. L. Capelo-Martinez, H. M. Santos, E. Oliveira and C. Lodeiro, Development of fluorochromic polymer-doped materials as platforms for temperature sensing using three dansyl derivatives bearing a sulfur bridge, *J. Photochem. Photobiol. A: Chem*, **445** (2023) 115033. DOI: 10.1016/j.jphotochem.2023.115033.
111. I. Pereira-Gomes, F. Duarte, G. M. Dobrikov, I. Slavchev, A. Kurutos, J. L. Capelo-Martinez, H. M. Santos and C. Lodeiro, Tetra dansylamides substituted cyclen and cyclam macrocycles as fluorescent sensing probes for metal ions and temperature-responsive materials in doped polymers, *Dyes Pigments*, **232** (2025) 112461. DOI: 10.1016/j.dyepig.2024.112461.
112. F. Duarte, C. Cuerva, G. M. Dobrikov, A. Kurutos, N. G. Vassilev, J. Fernández-Lodeiro, J. L. Capelo-Martinez, H. M. Santos and C. Lodeiro, Engineering multifunctional cyclen fluorophores for metal ion sensing, self-assembly and nanocolloid applications, *Spectrochim. Acta A Mol. Biomol. Spectrosc*, **349** (2026) 127285. DOI: 10.1016/j.saa.2025.127285.

113. J. Fan, M. Hu, P. Zhan and X. Peng, Energy transfer cassettes based on organic fluorophores: Construction and applications in ratiometric sensing, *Chem. Soc. Rev*, **42** (2013) 29–43. DOI: 10.1039/C2CS35273G.

114. F. Duarte, D. Torres, G. M. Dobrikov, A. Kurutos, L. B. Maia, O. Bondarchuk, I. Slavchev, J. L. Capelo-Martínez, H. M. Santos and C. Lodeiro, A dual-responsive Dansyl–Naphthalimide system for Hg(II) CHEF detection and ligand-mediated Cu(II) reduction with catalytic relevance, *Dyes Pigments*, **243** (2025) 113088. DOI: 10.1016/j.dyepig.2025.113088.

115. P. Srivastava, M. Verma, S. Sivakumar and A. K. Patra, A smart FRET probe exhibiting a molecular keypad-lock device based on rapid detection of nitric oxide mediated by $Cu^{2+}$ ion, *Sens. Actuators B Chem*, **291** (2019) 478–484. DOI: 10.1016/j.snb.2019.04.093.

116. S. Paul, S. Mukherjee, D. Kundu, S. Nag, S. Bhuyan, N. Chandra Murmu and P. Banerjee, AIEE activated pyrene-dansyl coupled FRET probe for discriminating detection of lethal $Cu^{2+}$ and $CN^-$: Bio-imaging, DNA binding studies and prompt prognosis of *Menke's disease*, *Spectrochim. Acta A Mol. Biomol. Spectrosc*, **303** (2023) 123110. DOI: 10.1016/j.saa.2023.123110.

# 14 Metals in the Environment–Health Binomial

## *Tools for Understanding and Action*

*C. Morro-Blazquez, R. Callejón Fernández, C. Gomez-de las Heras, and J. L. Gomez-Ariza*

Metals are present in the environment, sometimes naturally, as a result of geological activity, leaching and fluvial transport, or atmospheric dynamics. However, alterations in natural metal levels are usually anthropogenic, resulting from human activities in industry, mining, and other productive activities. The transfer of metals from our environment to living beings (plants, animals, and humans) is extremely complex due to the involvement of numerous factors. Human interaction with metals occurs, usually, through what we eat and drink, our lifestyle, habitat, work, and other sources. Therefore, their presence is ubiquitous and their consequences significant. Above all, their impact on our health is crucial and occurs throughout our lives. Numerous findings demonstrate that serious and highly prevalent diseases, such as Alzheimer's disease (AD), lung cancer, and breast cancer, are associated with the influence of our environment and, therefore, with the presence of certain elements lifelong [1].

When we talk about environment, metals, and health, we are talking about a complex system in which metals play an important role, but integrated into complex equilibria, in which metals/elements adopt specific chemical forms that affect their function, their toxicity or essentiality, their ability to cross biological barriers, their mobility, and ultimately their beneficial or lethal function [2, 3]. In the biological context, metals are carefully checked by the organism machinery [4] being detected and transported to the timely biomolecules just in the demanded biological microenvironment [5].

Living organisms continuously extract essential elements by interacting with the environment, mainly from the gastrointestinal tract (linked to diet), through breathing, and through the lungs, although non-essential and toxic elements can also be involved in this interchange. Three main biochemical processes are involved in these elements equilibria: (1) the "pass" of necessary amounts of essential elements from the gastrointestinal tract into the bloodstream; (2) the translocation of necessary

 DOI: 10.1201/9781003632092-15

amounts of essential elements to the different organs by transport proteins; and (3) the import of elements into the organs through the biosynthesis of suitable metaloproteins in the interphase bloodstream-organ [3]. These endogenous metalloproteins reflect the dynamic interplay between the environment and living organisms [6] through the gastrointestinal tract, the cell membrane, and cell cytosol [7], which represent the "selectivity filters" that only certain element species can overcome before metalloprotein assemblies occurs [8].

Therefore, studying this intricate balance involving the metals/environment/health triad (MEH) also requires powerful tools that not only inform us about the presence of elements in the various matrices involved, but also about their chemical form (species), their binding to proteins (metalloproteomics), metabolites and lipids (metalometabolomics/metallolipidomics), as well as to other biomolecular families. The complexity of these systems also increases the complexity of the methodological arrangements used for their study, the omics: genomics, transcriptomics, epigenomics, proteomics, metabolomics, lipidomics, metallomics, and ionomics, among others. The individualized application of these omics represents the exciting present of these studies, but integrating data from different omics sources is challenging due to the heterogeneity and dimensionality of the data, and their integration: multi-omics [9] represents the new frontier in the study of metal–biomolecule interactions. Despite the advances, significant challenges remain, especially in the massive use of multi-omics data, particularly in their statistically rigorous, computationally efficient, and biologically meaningful integration [10]. The complexity of multi-omics includes data from human and animal models, at the molecular, cellular, tissue, organ and fluid level, as well as behavioral and even textual information [11], which requires the support of artificial intelligence [12].

Studying this MEH three-dimensional problem throughout life requires the application of multi-omics, to get information about the molecules and their interactions (LEVEL 1: *Molecular level*-ML), but also those tools and methodologies that allow us to "see the scenery" in which these processes occur, membranes, cell walls, fluids, organic conducts. Understanding the presence of biomolecules in living systems is essential (Level 1), but so is understanding the complexity of their natural environment and the impact this can have on their basic functions and properties. This presents an enormous challenge, like trying to reassemble a mirror shattered into a thousand fragments. However, we must strive to understand cellular and organic complexity as fully as possible. The intracellular environment is crowded and heterogeneous, with a high concentration of macromolecules, leaving little room for soluble molecules to move. The available space is irregular and heterogeneous, so a given molecule can only utilize a fraction of the free space, depending on its size, shape, and flexibility. Some macromolecules do not move, leading to "confinement": the restriction of solutes to smaller volumes due to the effective limits of diffusion. The combination of macromolecular crowding and confinement results in restricted translational diffusion both intracellularly and extracellularly [13–15] and therefore determine their behavior and interactions at the molecular/cellular/microscale level (LEVEL 2: *Microspatial level* [MSL]). These Level 2 alterations typically manifest over time in organs through tissue damage and/or the development of tumors (e.g., lung and breast cancer) or the formation of pathological molecular accumulations

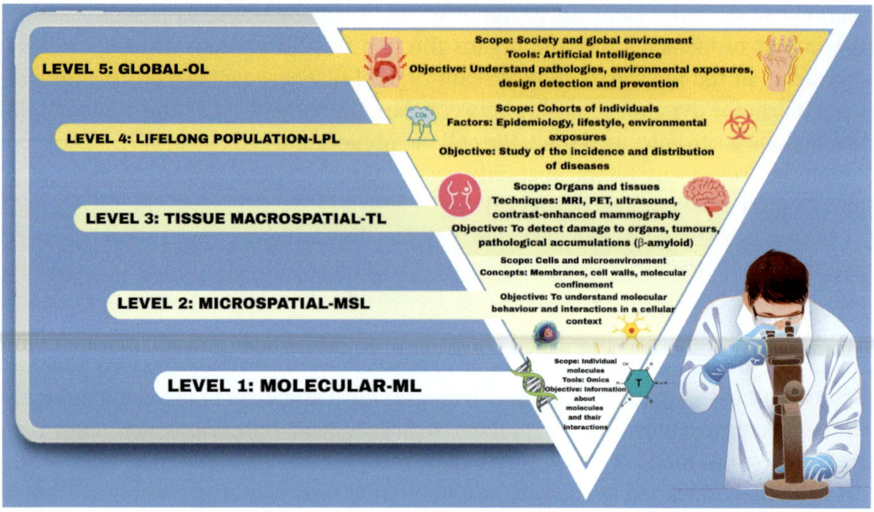


**FIGURE 14.1** Diagram of the methodological stages for the global study of the metals–environment–health triad (MEH).

(β-amyloid deposits in the case of AD). These alterations can be studied and diagnosed using imaging techniques such as magnetic resonance imaging (MRI), positron emission tomography (PET), ultrasound, and contrast-enhanced mammography (LEVEL 3: *Macrospatial tissue level* [TL]). However, since these are processes that occur over time, it is also necessary to study the epidemiological aspects of these pathologies by recruiting appropriate cohorts of individuals [16–18] associated with the environment and lifestyle (LEVEL 4: *Lifelong population level* [LPL]). The final understanding of the problem and its consequences requires a comprehensive approach supported by AI [19, 20] to achieve knowledge of these pathologies, their link to the environment and exposures, to establish procedures for early diagnosis, efficient treatments, and the design of disease detection campaigns (LEVEL 5: *Overall level* [OL]; Figure 14.1).

A thorough understanding of risk factors associated with MEH has important implications for public health, particularly in the prevention and early detection of highly prevalent diseases. Conclusive evidence of causal associations could inform long-term policy reforms in areas such as occupational health and safety, urban design, energy consumption and production, particulate emissions, and other human activities, with the aim of establishing appropriate control policies.

## 14.1 LEVEL 1: *MOLECULAR LEVEL*. TOOLS AND METHODOLOGICAL APPROACHES FOR THE STUDY OF BIOMOLECULES AND METAL–BIOMOLECULES INTERACTIONS

In the environment–living organisms interface we have to take into account the presence of elements such as As, Hg, Cd, and Pb, considered toxic or harmful, as well as their chemical species and the metallic biomolecules that arise from their interaction with living organisms. But also, essential or beneficial elements such as Cu, Zn, Mn, and Se, also in diverse chemical forms and molecular configurations. Measuring their presence, changes in chemical form, and molecular interactions in tissues, fluids, or living systems requires the design of appropriate analytical instrument configurations, methods, and sample treatments to provide the necessary information about their behavior and mechanisms of action (Figure 14.1).

In this regard, we can consider the studies and instrumental arrangements steps of progressive complexity that are discussed in subsequent subsections.

### 14.1.1 Step 1: Estimation of Elements' Presence

Analysis of essential (e.g., Zn, Cu, Mn, W, Mo, Se) or toxic (e.g., As, Cd, Hg) elements at trace or ultra-trace levels is mainly based on some atomic spectrometry techniques (e.g., AAS, AFS, ICP-OES, ICP-MS) after sample treatment to get metal extract [21]. This is the first step because, above all, we need measuring devices that can "see" and "count" the elements present in the system under study (e.g., tissue, fluid, extract), which bind to the molecules. These devices must be capable of measuring elements at concentrations up to ng/g and lower with high precision.

### 14.1.2 Step 2: Targeted Analysis of Element Chemical Species (Speciation)

As a second step, after the elements interact with biomolecules in the organism, it is necessary to identify and quantify these metallobiomolecules, considering that a single metal can give rise to different metallobiomolecules. Hence the need the coupling of chromatographic techniques with atomic detectors, where the metal acts as a "tag" for the atomic detector.

In this way have been proposed a first-generation of coupling liquid chromatography (anion or cation exchange chromatography) with ICP-OES, ICP-MS, AFS or other atomic detector, through a direct connection of the HPLC, in the case of ICP-OES and ICP-MS, or online pretreatment of chemical species eluted by the chromatograph for the online introduction into the AFS detector.

**Example Step 2-1:**

*Arsenic Speciation*

a. It is important to know the presence of chemical forms (species) of toxic elements as arsenic in living organisms used as food. This is the case of the anemone (*Anemonia sulcata*) after cooking. After extraction, the arsenic species were analyzed by anion exchange chromatography coupled to triple-quadrupole mass spectrometry (HPLC-ICP-QQQ-MS) [22]. The results showed the greater presence of methylated forms of As (monometylarsenic and dimethylarsenic) and arsenobetaine, which are less toxic, than more toxic inorganic forms of this element ($As^{III}$ and $As^{V}$). On the other hand, bioaccessibility of arsenic species in human after ingestion of cooked anemone was evaluated using the same instrumental approach in serum and urine samples collected along time after anemone ingestion [23]. The results showed that around 95% of total arsenic in cooked anemones is bioaccessible and 85% of the amount ingested was eliminated from the body by urine in 90 h.

b. Presence of arsenic species in the liver and kidney from the mice *Mus musculus* [24]. A similar approach was applied to mice exposed to As(III) (in the form of $As_2O_3$ in water) by oral administration. The following As species were quantified: DMA > MMA >> As(III). It was interesting the absence of arsenobetaine, final product of detoxification process of the very toxic inorganic arsenic, which was previously checked in marine organisms, such as the anemone. This is due to the inability of terrestrial organisms, particularly mammals, to metabolize arsenic to this completely innocuous form. Therefore, inorganic arsenic is biotransformed into methylated species (MMA and DMA) in the liver of mammals and finally excreted in the urine, as was previously reported by Suzuki et al. [25].

**Example Step 2-2:**

*Mercury Speciation*

Mercury speciation in aquatic organisms has an important environmental significance with potential repercussions in the food and health sectors. This is due to the toxicity and neurodegenerative action of the methylated form of this element. It is therefore important to know the presence of mercury ($Hg^{2+}$) and methylmercury ($MeHg^+$), especially in seafood samples, due to mercury methylation by water bacteria [26]. In this way, it has been proposed a method based on gas chromatography coupled to a home-modified AFS detector [27]. Mercury species were extracted in basic medium assisted by sonication and ethylated ($NaEt_4B$) for final extraction in hexane. The extract was analyzed with an instrumental arrangement including species separation by gas chromatography, followed by mercury species pyrolization and dragging of elemental mercury to the AFS detector. The procedure was applied to samples of bivalves, striped venus clam (*Chamelea gallina),* truncate donax (*Donax trunculus,* DT), and peppery furrow shell (*Scrobicularia plana*), from various areas of the Andalusian coast (southern Spain), with different types of productive activities: fishing, mining, and industrial. The results showed a higher concentration of methylmercury in the areas affected by mining activities.

### 14.1.3 Step 3: Targeted and Untargeted Analysis of Elements Interactions with Biomolecules (Metallomics)

Metal cofactors contributes to proteins up- and down-regulation expression in cellular compartments, cells or organisms, which are characterized non only by its genome, proteome or metabolome, but also by the metallome, their inorganic complement. The study of metallomes and their correlation with genomes, proteomes, and metabolomes are referred to as metallomics [5]. This omics considers metals, metalloids, or their chemical forms or bound to biomolecules, within a biological context, relating their presence to the genome, proteome, and metabolome, always from a systemic and global perspective.

### Example Step 3-1:

#### *Metallomic Study of Metals–Mice Interactions in Relation to Environmental Exposure*

The action of contamination, mainly metallic contamination, in natural areas has been addressed with a metallomic approach based on the use of size-exclusion chromatography coupled to ICP-MS, as an initial classification system of metallo-biomolecules by size, for subsequent analysis of the different molecular fractions classified by size ranges, using cationic or anionic chromatography with ICP-MS detection. This approach was applied to the free-living mouse (*Mus spretus*) used as bioindicator from Doñana National Park (southwest Spain) and the surroundings, affected by differential contamination from mining, industrial, and agricultural activities. The results showed the high presence of Mn, Cu, and Zn in liver bound to the antioxidant protein superoxide dismutase, and the toxic element Cd in kidney bound to the detoxifying protein metallothionein I. All these facts reflect the activation of detoxification mechanisms in exposed mice, and the potential of this methodology to study environmental problems [21]

### Example Step 3-2:

#### *Metallomic Study of Selenium in Human Serum, Plasma and Breast Milk*

To study the presence of selenoamino acids and selenoproteins in serum and plasma, due to the antioxidant and protective properties of this element, a metallomic methodology has been proposed based on the arrangement of several chromatographic columns interconnected by a column-switching valve: two size-exclusion columns stacked in series, followed by a heparin-sepharose affinity column for selenoprotein P (HEP-HP) and a blue-sepharose affinity column (BLU-HP) for selenoalbumin. All columns were configured using a six-port column-switching system. Final detection of selenium species was performed by ICP-MS. The operational flow in analytical pipeline is as follows: (1) Column-switching valve in series connect all the columns and mobile phase A (0.05 M ammonium acetate pH 7.4) elutes glutathione peroxidase (GPx), inorganic selenium species, and selenometabolites that are separated with suitable resolution due to the two stacked SEC columns, being detected by ICP-MS; (2) The valve is switched to position B, which deactivates the affinity column containing the SeAlb, passing mobile phase B (1.5 M ammonium acetate, pH 7.4) through the system for the elution and detection of SelP; (3) Valve change again to position A and mobile phase A elutes SeAlb ([28]; Figure 14.2).

This approach allows the quantification of glutathione peroxidase, seleno protein P and selenoalbumin, but inorganic selenium and selenoamino acids are quantified together. Therefore, when speciation of inorganic selenium is required, the setup described in Figure 14.2 has to be modified by adding a second 8-port valve and an anion-exchange chromatographic column, as well as a second auxiliary pump for the separation of inorganic selenium species, preceded by a 3-ml loop [28].

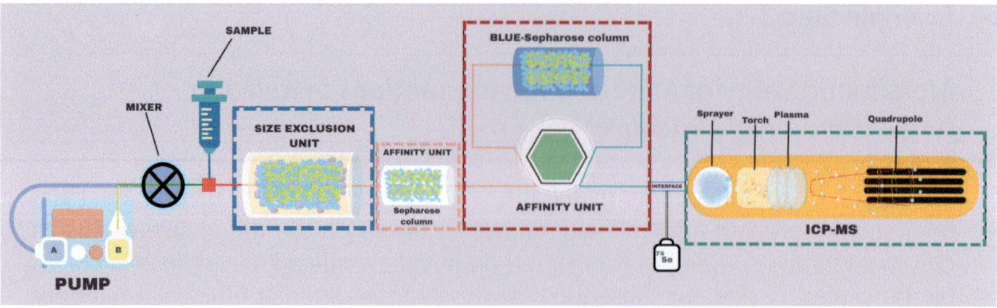


**FIGURE 14.2** Schematic diagram of multichromatography-column switching-ICP-MS arrangement for selenium containing proteins speciation in plasma and serum.

This seleno-metalomic instrumental approach has also been applied to serum samples from lung cancer (LC) and chronic obstructive pulmonary disease patients (COPD), to study the potential of extracellular glutathione peroxidase (eGPx), selenoprotein P (SELENOP), and selenoalbumin (SeAlb) as biomarkers of these diseases [16].

Finally, a similar instrumental arrangement has been used for the metallomics study of selenoprotein P in human breast milk [29].

### 14.1.4 Step 4: Untargeted Analysis of Metabolites Expression Under Metal Contamination and Pathological Health Processes (Metabolomics)

As we have previously seen, the action of toxic elements from the environment on living organisms is highly complex and involves numerous interactions with organs, tissues, fluids, and other components of their systemic equilibrium. These interactions can be studied by considering direct element–biomolecule equilibria (speciation, metallomics), and also indirectly, through the consequences that arise at the systemic level by the presence of metals. This results in changes in the expression of metabolites, proteins, enzymes, antibodies, and others. The changes that occur at the metabolic level, due to overexpression or inhibition of the metabolites that maintain vital equilibrium, are the focus of this approach to the problem: metabolomics. This also require a change in the instrumental tools focused in the study of biomolecules without destroying their identity, as is done in speciation and metallomics based on atomic detectors with the "*filter*" of chromatography. In metabolomics, the analytical techniques primarily used for biomolecules identification is mass spectrometry (MS).

Several illustrative examples can be cited about the application of metabolomics to environmental and health issues. These are discussed in subsequent text.

#### Example Step 4-1:

#### *Metabolomic Study of Mice Under Arsenic Exposure*

Given the environmental and foodborne nature of arsenic (As), it is necessary to delve deeper into the molecular mechanisms associated with its toxicity.

Metabolomics is a valuable tool in this regard and has been applied in laboratory exposure experiments of mice (*Mus musculus*) to inorganic arsenic. Direct infusion MS has been applied to different organs and fluids (e.g., liver, kidney, plasma). The metabolomic study revealed alterations in various metabolic pathways related to energy (e.g., glycolysis, Krebs cycle), amino acids, choline, methionine cycle, and degradation of membrane phospholipids (cell apoptosis), analyzing the extracts obtained from different organs with (1:1, *v/v*) methanol:water mixture (polar extract) and the resulting pellet with (2:1, *v/v*) chloroform:methanol mixture (non-polar extract) for direct infusion in high resolution mass spectrometer (DI-ESI-QTOF-MS) [30].

### Example Step 4-2:

#### *Metabolite Profiling to Identify Altered Metabolic Pathways in AD*

Given recent findings linking AD to a systemic pathology, not exclusively tied to the central nervous system (CNS), the study of metabolic alterations related to the onset of the disease has gained interest. Gas chromatography-MSGC-MS was applied to serum samples from AD patients and healthy controls after proteins precipitation and extraction with 1:1 (*v*/v) methanol/ethanol mixture. The extracts were injected after derivatization in the GC-MS system. The results revealed alterations in 23 metabolites related to important metabolic pathways, such as those involved in energy production, oxidative stress, and hyperammonemia, among others. Changes in the expression of some unique metabolites, such as β-ketoglutarate, isocitrate, pipecolic acid, pyroglutamate, and adenosine, were also observed, confirming the potential of the metabolomics approach in the search for new biomarkers of AD [42]. Furthermore, metabolomics has enabled significant progress in studying the evolution and progression of the disease from mild cognitive impairment to its advanced stage [31].

### Example Step 4-3:

#### *Metabolomic Study of Serum, Urine, and Bronchoalveolar Lavage Fluid to Study Lung Cancer Pathology*

This study addresses the metabolomic analysis of different fluids: serum, urine, and bronchoalveolar lavage fluid (BALF), this latter of great interest due to the direct contact of this fluid with alveolar tissue. Samples from lung cancer patients, healthy controls, and patients with non-cancerous lung diseases (NCDs) were analyzed, the later as a control group for BALF, due to the invasive nature of sample collection of this fluid. Extracts from the samples were analyzed by GC-MS. The final statistical analysis reveals the existence of metabolites with altered expression in lung cancer when comparing serum and urine samples. Furthermore, alterations were observed in the BALF of patients with lung cancer compared to NCD patients. The analysis of metabolic pathways indicated that several amino acid metabolic pathways were the most affected in lung cancer [32].

### 14.1.5 Step 5: Untargeted Analysis of Proteins Under Metal Contamination and Pathological Health Processes (Proteomics)

**Example Step 5-1:**

*Proteomics in Free-Living Mice* (Mus spretus) *to Monitor Environmental Contamination*

High-throughput proteomic methods based on MALDI-TOF-PMF after proteins separation by sodium dodecyl sulfate polyacrylamide gel electrophoresis and digestion with trypsin. This proteomic method was used to monitor terrestrial ecosystems, comparing free living mice (*M. spretus*) from environments with different degrees of trace metal contamination. The study shows significant changes in protein expression related to the differential presence of contaminants. These proteins play diverse roles related to cytoskeleton dynamics, proteolysis, biotransformation, adaptation to oxidative stress, and metabolism. Animals from different contaminated environments showed differences in their proteomes, with specific increases and decreases in selected protein groups that appear to be regulated in a coordinated manner. The proteomic data were consistent with the responses from metal biomonitoring and conventional biomarkers [33].

### 14.1.6 Step 6: Multi-omics

The systemic approach to diseases has opened a new world of innovative studies that consider them from a broader perspective. Thus, pathologies extend to the entire organism, including organs and fluids, and their interactions [34, 35]. The complexity of the system associated with this conceptual approach also increases the complexity of the tools for its study. Multiomics [36] is the fundamental methodological tool in the study of complex biological systems, such as those defined by alterations associated with exposure to exogenous factors throughout life, which define the exposome [37].

**Example Step 6-1:**

*Integrated Application of Transcriptomics, Proteomics, and Metallomics in Environmental Studies*

This study employs a comprehensive application of omics methodologies—transcriptomics, proteomics, and metallomics—in an environmental monitoring study, using the wild-type mouse (*M. spretus*) as a bioindicator and as a reference for searching gene/protein sequence databases. The results show that absolute expression signatures of transcripts, obtained through reverse transcription and real-time polymerase chain reaction (PCR) of key genes in *M. spretus*, constitute a new and useful assessment criterion for biomonitoring. Furthermore, the suitability of commercially available *M. musculus* oligonucleotide arrays for whole-genome transcriptional profiling in *M. spretus* has been confirmed. Transcriptomic studies indicate considerable similarities in gene sequences between the two mouse species. Therefore, the applicability of proteomic methods in free-living

*M. spretus* and the correlation of peptides with the *M. musculus* database have been verified. On the other hand, the problem has been addressed with a metallomics approach, using SEC-ICP-MS coupling, which allows the study of biomolecule profiles bound to metals. This integration of omics in the study of the *M. musculus/M. spretus* pair, rodent species with high genetic similarity, represents a clear advance in environmental studies related to pollution, since it allows the correlation of results from laboratory exposure experiments (*M. musculus*) with those obtained in field monitoring (*M. spretus*) [38].

### Example Step 6-2:

#### *Integrated Application of Metallomics and Metabolomics in Environmental Studies*

The influence of heavy metal exposure in natural environments was studied using the free-living mouse (*M. spretus*) as a bioindicator. Various geographical areas with varying levels of contamination, from uncontaminated to areas contaminated by mining leachate, were considered. Metallomics techniques were applied to characterize the metallic biomolecules using size-exclusion chromatography coupled with MS (SEC-ICP-MS) and a metabolomics approach based on direct infusion MS (DI-ESI-QTOF-MS). The results were then subjected to statistical analysis using PLS-DA to compare the biological response of the bioindicator mouse in areas with different levels of contamination. Additionally, the activities of key antioxidant enzymes, such as Cu/Zn-SOD, Mn-SOD, CAT, GR, and guaiacol peroxidase, were measured. The results obtained from kidney samples of the mouse (*M. spretus*) revealed significant molecular differences caused by differential exposure to metals, such as the overexpression of metalloproteins (metallothionein, Cu/Zn-superoxide dismutase, and Mn-catalase). Degradation of membrane phospholipids and disturbances in metabolic pathways related to energy metabolism and oxidative stress were also observed [39].

## 14.2 LEVEL 2: *MICROSPATIAL LEVEL*. STUDY OF MOLECULES BEHAVIOR AND INTERACTIONS AT SPATIAL/CELLULAR SCALE

The application of omic methodologies, either separately or in combination through multi-omics approaches, is a powerful way to address the complex problem of the interactions of metals and metalloids in living organisms, and particularly in humans. This interplay is linked to environmental factors and exposures lifelong, being the origin of many pathologies, whose onset and progression have a molecular basis. Therefore, the systematic application of multi-omics is a groundbreaking and valuable tool for addressing this problem. However, despite the enormous progress it represents, it is necessary to approach it from a more holistic perspective. It is essential to consider not only the greater or lesser presence and evolution of the molecules involved, but also the "scenery" in which thousands of molecules interact, with very diverse shapes, sizes, and molecular masses: metabolites, lipids, proteins, transcripts, gens, enzymes, and intra/extracellular formations, as vesicles, exosomes, and others, whose behavior and function are associated not only to their mass and chemical composition but also to the presence of active sites on the molecule with a specific biological function, to which we access through non-invasive samples (blood, urine), relatively invasive samples (cerebrospinal fluid or bronchoalveolar lavage fluid), or invasive ones (biopsies), to obtain information about a person's health status.

On the other hand, we must consider the complexity of their "*in vivo*" worlds, taking into account the impact of their native environments on their functions. This "landscape" is not empty (molecules are not alone!), and there is a multitude of biomolecules interacting with each other [13]. We tend to consider the molecules integrated into the functioning of living things in terms of static, interacting structures, but the process is much more dynamic and complex, like snapshots from a film [40].

### 14.2.1 Techniques to Study Molecular Interactions

Given the importance of characterizing biomolecular interactions, it has been developed different experimental approaches to analyze them, some techniques are only focused at detecting molecular interactions, but others identify interfaces involved or determining affinities and kinetics of these interactions [41]. Some methods rely on the proximity of interacting molecules that produce a signal, such as the in vivo expression of a protein [42], or by change in fluorescent emission using Förster resonance energy transfer (FRET) between fluorophores fused to interacting proteins [43]. Other approaches are based on interactions between biomolecules, considering the changes that occur on a surface, where the biomolecule under study is immobilized to be exposed to molecules that can potentially interact with it. In this sense, the quartz crystal microbalance (QCM) detects interactions through changes in the surface mass of a quartz crystal due to changes in its resonance frequency [44].

Another way to evaluate interactions between biomolecules is to consider the changes they can cause in the optical properties of the surface, such as the refractive index or the interference of reflected light. This forms the basis of surface plasmon resonance (SPR) [45] and biolayer interferography (BLI) [46, 47], respectively.

Other techniques that also immobilize the bonds for the study of interactions between biomolecules are co-immunoprecipitation (Co-IP), which uses antibodies attached to an agarose surface or magnetic beads onto which the antigen is precipitated, along with another protein with which it interacts. In contrast to Co-IP, used to detect protein-protein interactions, chromatin immunoprecipitation (ChIP) allows the detection of protein-DNA interactions [11], and ribonucleoprotein immunoprecipitation (RIP) allows the study of protein-RNA interactions [48].

Comigration techniques do not require immobilization; instead, they utilize the joint migration of interacting molecules, which are then fractionated for identification using MS or Western blot. A critical aspect is fractionation, for which a chromatographic or electrophoretic technique is employed [49]. This technique is commonly used to study protein-DNA interactions by evaluating differences in DNA migration in the presence and absence of a specific protein [50].

Field flow fractionation (FFF) has also been used to study the co-migration of interacting molecular species in solution. However, in this case, instead of a gel or chromatographic column, an electric, magnetic, gravitational, or other force field is used to fractionate molecules by size [51]. This separation method is more suitable for weak molecular interactions compared to the previously mentioned techniques, where the matrix can influence the interactions. Analytical ultracentrifugation (AUC) is another matrix-free technique for separating molecules based on size, although the separation is only temporary while under the influence of the force field. Therefore, AUC requires prior knowledge of the sample composition to identify intermolecular interactions. Therefore, detection systems such as fluorescence have been incorporated into the AUC to expand its capabilities, allowing the analysis of protein-protein interactions in the nanomolar and picomolar range, and in highly concentrated solutions [52, 53]. The AUC with fluorescence detection allows the detection of sedimentation in complex, native-like environments, such as serum [54], or cell lysates [55].

Other tools for studying intermolecular interactions are heat-based, such as mid-scale thermophoresis (MST), which measures the movement of molecules along a local temperature gradient using an infrared laser and fluorescence detection [56], and isothermal titration calorimetry (ITC), which measures small heat changes in a solution to assess bond strength [57]. ITC is used to evaluate protein–ligand and protein–protein interactions. These techniques only allow for the measurement of these interactions under carefully controlled conditions, making them difficult to use under the native conditions of a cell.

Finally, spectroscopic techniques, such as NMR, can also be used to measure protein-protein interactions through chemical shift perturbation analysis, in which the spectra of a labeled protein are collected at a series of varying concentrations of unlabeled ligand [58]. Furthermore, this technique can also provide structural information about the nature of the binding. The use of intracellular NMR allows the study of biomolecules within living cells, enabling the study of proteins in their natural environment [59, 60].

### 14.2.2 Study of the Folding of a Protein in its Natural Environment

To study the effect of molecular crowding at the cellular level on protein conformation, experiments were conducted using various spectroscopic techniques (UV-visible spectroscopy, circular dichroism, dynamic light scattering) to understand the action of the medium and the mechanism of action of two polyethylene glycols (PEGs) of different molecular masses on cytochrome c. The results showed that small PEGs cause a perturbation in protein structure, while larger PEGs have a stabilizing effect on cytochrome c. Furthermore, binding measurements showed that small PEG molecules interact strongly through weak interactions compared to larger PEG molecules, the latter being more governed by the excluded volume effect or preferential protein exclusion. Therefore, these results suggest that protein conformations can be influenced under cellular crowding conditions by interactions that depend on the size of the molecule in the environment. Therefore, both bulk exclusion and soft chemical interactions govern protein conformation and functional activities. The internal architecture of the cellular environment, as evidenced by the size and shape of the clusters, plays an important role [61].

### 14.2.3 Consequences of Molecular Crowding in Very Prevalent Diseases

#### 14.2.3.1 Neurodegenerative Diseases (Alzheimer and Parkinson)

Fibril formation by the Alzheimer's β-amyloid (Aβ) peptide in brain tissue is one of the hallmarks of AD pathology. Understanding the conformational properties and the mechanisms triggering aggregation of the Aβ peptides, at an atomic level of detail, is of crucial importance for the design of effective therapeutic agents. Therefore, it is important to study the conformational transitions and dynamic properties of amyloidogenic peptide fragments (Aβ 10–35) by molecular dynamics simulations, in systems modeling infinite dilution and the presence of macromolecular crowding agents, like the inert polymer dextran T10 [62]. The results show that molecular crowding disrupts fragment dynamics. Fragment diffusivity decreases with increasing concentration of the crowding agent, which favors the formation and appearance of β-bridges and β-chains [63]. Trimethylamine N-oxide (TMAO) is also a molecular crowding agent in beta-amyloid peptide aggregation and tau protein formation. TMAO, is a gut microbiota metabolite derived from dietary choline and carnitine, having recently been identified as a potential risk factor for AD. Therefore, the presence of TMAO could be used in diagnostic and therapeutic procedures for AD, as its association with neuroimaging patterns or plasma levels could be used for the early prediction of AD in individuals without cognitive impairment [64].

Macromolecular crowding even affects the formation of the amyloid plaques themselves, which are highly heterogeneous in content, size and density, composed of masses of fibrils and other cellular material. Given this complex architecture, the microenvironment has been studied with powerful imaging techniques as PET and molecular imaging probes (MIPs). In relation to this, Shoghi-Jadid et al. [65] assert that macromolecular crowding impact the diffusion within plaque regions, which can affect the estimation of amyloid load and consequently the estimation of pathological progression of AD. The molecular characteristics of the key pathogenic Aβ forms are not well understood, being difficult to identify optimal agents that target disease relevant forms of Aβ. In this way, Jin et al. [66] have combined the use of Aβ-rich aqueous extracts from AD patients brain samples as source of human Aβ and live-cell imaging of iPSC-derived human neurons to develop a bioassay suitable for quantify the relative protective effects of anti-Aβ antibodies.

Molecular crowding also has an important role in the molecular transport along the axon of neurons. Many degenerative diseases show increased physical crowding owing to the greater number of stalled organelles and structural changes associated to it. Increased physical crowding may be a significant factor in slowing cargo transport to synapses, contributing to disease progression and culminating in the dying back of the neuronal process. Therefore, it is necessary to examine the sources of physical crowding and actions of molecular motors to avoid crowded areas, in order to understand the subcellular changes that occur in neurodegenerative diseases that can alter physical crowding and influence the progression of the pathology [67].

Regarding the Tau protein, another hallmark of AD, due to its aggregation in the brain forming tangles, Zweckstetter et al. [68] have shown, that in solution Tau exhibits an intrinsic disorder that makes it highly soluble, binding to microtubules. It is unknown, how this conversion from a soluble, functional, and harmless phase into solid neurotoxic deposits begins. Furthermore, the lysine-rich Tau repeats that bind to microtubules undergo liquid-liquid phase separation in solution. This liquid-liquid unmixing leads to the molecular aggregation of Tau elements, promoting amyloid production and driving electrostatic coacervation. Moreover, the three- and four-repeat Tau isoforms differ in their unmixing capacity. Therefore, alternative Tau splicing could regulate the formation of non membrane bound Tau compartments. Furthermore, phosphorylation of Tau repeats promotes liquid-liquid phase separation under cellular protein conditions. The combined data suggest a mechanism in which liquid droplets formed by the positively charged microtubule-binding domain of Tau undergo coacervation with negatively charged molecules to promote amyloid formation.

Finally, in relation to Parkinson's disease neuropathological hallmarks are Lewy bodies and Lewy neurites associates to neuronal inclusions immunopositive for the protein α-synuclein, which shows a crowded environment of membrane therein, including vesicular structures and dysmorphic organelles. These findings were verified by correlative light and electron microscopy and tomography on postmortem of human brain tissue from Parkinson's disease donors. Crowding of organellar components was confirmed by stimulated emission depletion (STED)-based super-resolution microscopy, and high lipid content within α-synuclein immunopositive inclusions was corroborated by confocal imaging, Fourier-transform coherent anti-Stokes Raman scattering infrared imaging and lipidomics [69].

#### 14.2.3.2 Lung Cancer

Macromolecular crowding is also a key variable in the study of other pathological processes, many of which are based on the development of in vitro models. Examples include protein aggregation disorders, vascular diseases, and cancer, particularly lung cancer. In this way, macromolecules promote the deposition of other proteins, such as fibronectin by fibroblasts and fibrillin-1 by pulmonary fibroblasts [70]. Thus, in the analysis of Carcinoembryonic Antigen in serum samples, for the diagnosis of cancerous processes, inaccurate results are obtained when the concentrations of the antigen are high, possibly due to the increase in molecular crowding and protein interaction, since the viscosity of the mixed serum is greater in highly concentrated protein solutions [71].

#### 14.2.3.3 Breast Cancer

Macromolecular crowding also plays an important role in the development of cell-derived array based in vitro tumor models that accurately simulate the tumor microenvironment associated with breast cancer [72]. In this way, in cancer progression, tumor microenvironments become progressively denser and more hypoxic, with cells migrating toward denser, higher oxygen levels as they invade the tumor–stromal boundary. However, although cell invasion is generally considered to depend on collagen density, it was unclear whether prior oxygen conditions alter the phenotype of cell invasion. In this way, [73] have demonstrated that human breast tumor cells exhibit higher rates of invasion following prior exposure to hypoxia compared to normoxic controls, which demonstrates the influence of the microenvironment on the evolution of this cancerous process.

## 14.3 LEVEL 3: *MACROSPATIAL TISSUE LEVEL*. IMAGING STUDY AND DIAGNOSIS OF PATHOLOGICAL PROCESSES ASSOCIATED WITH THE ACTION OF EXPOSURE AND LIFESTYLE THROUGHOUT THE LIFESPAN

The relationship between molecules and tissues/organs has also led to two approaches in the study of pathologies. Let's take AD as an example. Traditionally, it is considered a primary neurodegenerative process affecting the CNS, through clearly identifiable deposits in the brain (β-amyloid peptides and tau protein). However, AD can also be considered a systemic disease, associated with metabolic and biochemical processes that extend beyond the CNS and appear even before the clinically recognizable symptoms. These metabolic abnormalities are not only present in the brain but also in peripheral cells and fluids [34, 74, 75]. Therefore, the question is to what extent the systemic changes represent an effect of a CNS process or reflect a biological process involving both the body and the brain. Various risk and protective factors can influence a person's long-term risk of developing AD. Some components are genetic in nature, but others involve comorbid processes originated outside the CNS, such as diabetes, obesity, inflammation, vascular disorders, physical inactivity, and the action of environmental and exposure factors, whose systemic effects are presumably transferred to the brain [15]. This binomial nature of AD has led to a shift from traditional AD characteristics, based on the amyloid and tau hypothesis [76, 77], toward a more systemic approach that considers metabolic, proteomic, transcriptomic, and metallomic changes in systemic fluids and tissues. Therefore, the range of potential biomarkers [77] for the disease will continue to grow as new studies are developed. Due to the binomial nature of AD, the brain and systemic approaches are complementary, as are their biomarkers. This dual organ centered and systemic approaches can also be translated to other pathologies, particularly the other two we are considering: lung and breast cancer. Therefore, multi-omics studies should be complemented by imaging diagnostic techniques, mainly magnetic resonance imaging (MRI) [78, 79], computed tomography (CT) [79], positron emission tomography (PET) [80], ultrasound [81] and contrast-enhanced mammography [82, 83]. This image-based study of pathologies constitutes Level 3 in the global view of them, which has to be complemented by levels 2 and 3 described above.

## 14.4 LEVEL 4: *LIFELONG POPULATION LEVEL*. IMAGING COHORTS-LIFELONG PATHOLOGICAL PROCESSES UNDER ENVIRONMENTAL ACTION

The action of environmental and exogenous factors on health along the time requires the study of cohorts to assess prodromal symptoms and the onset of diseases. These cohorts should consider age-stratified recruitment, planned to reflect the influence of the environment, habitat (e.g., coastal, mountainous, mining, industrial, urban), occupation, diet, lifestyle, alcohol and/or tobacco consumption, pollution, comorbidities, family history, and other factors if deemed necessary, based on the established objectives. Furthermore, they should cover different age ranges from childhood to old age, and external factors that may affect the onset and progression of the disease being studied throughout life.

Numerous cohorts have been recruited and studied for different purposes, particularly to evaluate the role of all exogenous factors in the development of diseases throughout life. Here, we will only discuss a few cases related to highly prevalent diseases, which we are considering as models.

### 14.4.1 AD Cohorts

Generally, combined psychological tests with amyloid or tau biomarkers analysis in CSF and blood, complementary diagnostic techniques, such as MRI and PET imaging has been used to visualize proteinopathies in brain, centered in the old life. Some typical examples are the following:

- ***Synaptic proteins and neurofilament light (NfL)*** were analyzed in CSF samples at two time points of about 2 years, including subjects with subjective cognitive decline (SCD) or mild cognitive impairment (MCI) and AD patients, from the Amsterdam Dementia Cohort. Trajectories of proteins levels were assessed and associated with cognitive decline evaluated by Mini-Mental test [84].
- ***Blood biomarkers proteins***, together with life behaviors and disease history were used to predict the cognitive function decline using data from Chinese Longitudinal Healthy Longevity Survey [85]. Depressive symptoms may occur along the AD process; it was studied associations between plasma markers of AD (Ab42/40, p-tau181, NfL, and GFAP) and depressive symptoms using questionnaires in individuals without dementia, from the Netherlands Consortium of Dementia Cohorts [86].
- ***The application of morbidity and aging deficit indicators,*** based on the COMPASS-ND study [87], has been linked with adverse outcomes such frailty, functional deficits, accelerated cognitive decline or impairment, and neurodegenerative disease. These indicators were collected through self-report, physical examination, and formal standardized scales, allowing several cohorts comparison from the Canadian Consortium on Neurodegeneration in Aging dataset [88].

### 14.4.2 Lung Cancer Cohorts

#### 14.4.2.1 Study of Urine Metabolites of Polycyclic Aromatic Hydrocarbon in Relation to Lung Cancer in Lifelong Never Smokers (Shanghai Cohort Study)

Exposures to polycyclic aromatic hydrocarbons (PAHs) from various environmental and occupational sources are considered a primary risk factor for lung cancer among lifelong never smokers, based largely on results from epidemiologic studies utilizing self-reported exposure information. It was prospectively investigated levels of urinary metabolites of a PAH and volatile organic compounds in relation to lung cancer risk in a nested case-control study of 82 cases and 83 controls among lifelong never smokers of the Shanghai Cohort Study. The results demonstrated the important role of exposure to PAH in the development of lung cancer among lifelong never smokers [89].

#### 14.4.2.2 Passive Smoking and Lung Cancer in Non-smoking Women

Smoking is a major cause of lung cancer, but the proportion of lung cancer cases among Japanese women who have never smoked is high. The development of lung cancer in nonsmoking Japanese women could be significantly affected by second-hand smoke. A prospective population-based study was conducted to delve deeper into this issue. It was based on two cohorts, the 1990 Cohort I and 1993 Cohort II. The study population was defined as all residents aged 40 to 69 years at baseline. 28,414 women who had never smoked provided baseline information on exposure to tobacco smoke from their husbands, in the workplace, and during childhood. During 13 years of follow-up, 109 women were diagnosed with lung cancer, of whom 82 developed adenocarcinomas. These findings confirm that passive smoking is a risk factor for lung cancer, especially for adenocarcinoma among women [90].

#### 14.4.2.3 Radon-Induced Lung Cancer in Miner Cohorts

The radon-induced carcinogenesis is supported on two-mutation model with clonal expansion of pre-malignant cells, this lung cancer pathogenesis is used in the study to describe of lung cancer mortality data from French and Czech miners with relatively low exposures to radon. Applying a multi-stage model it is possible to transfer radon-induced lung cancer risks results across populations. The inherent age-time dose-rate relationships in the model allow for extrapolation to lifelong exposures to residential radon concentrations. The resulting cumulated (lifetime) risks from continuous exposure to low-level radon concentrations were found to agree with the results [91].

### 14.4.3 Breast Cancer Cohorts

#### 14.4.3.1 Endocrine-Disrupting Chemicals and Breast Cancer

Endocrine-disrupting chemicals (EDCs) are ubiquitous in our daily lives, found in pesticides, plasticizers, pharmaceuticals, personal care products, and also in food products and packaging. Increasing epidemiological evidence suggests that EDCs can affect the development or progression of breast cancer and cause health problems through their effects lifelong, especially when exposure occurs in early life. Numerous cohort studies have been conducted on the effects of EDCs on breast cancer pathology. These studies have identified certain endocrine-disrupting chemicals (EDCs) that increase the risk of breast cancer. Because EDCs are highly persistent in the environment and bioaccumulative, it is important to assess the long-term impacts of exposure to them, especially through multigenerational and transgenerational studies. Since food is a major route of exposure to EDCs, studies on their effects, considering EDCs present in food and food packaging, are necessary in relation to the development of breast cancer. This will allow for the formulation of policies and regulations regarding these substances [18].

#### 14.4.3.2 Alcoholic Beverages and Risk of Breast Cancer

Population studies show a relationship between alcohol consumption and an increased risk of breast cancer. Studies have been conducted based on a cohort of 70,033 women, of whom 2,829 developed breast cancer. The type of alcoholic beverage and interactions with other predictors of breast cancer were considered. The results show a higher risk associated with estrogen receptor-positive tumors, without a significant disparity related to the choice of alcoholic beverage [92].

## 14.5 LEVEL 5: *OVERALL LEVEL*. ARTIFICIAL INTELLIGENCE TO ACHIEVE THE ENTIRE KNOWLEDGE OF PATHOLOGIES LINKED TO THE ENVIRONMENT AND EXPOSURES

Multi-omics [9] has proven more effective in studying environmental problems and predicting diseases compared to individual omics data, demonstrating its translational value for analyzing environmental and health heterogeneity. Despite these advances, challenges remain in the use of massive multi-omics data, particularly in integrating high-dimensional multi-omics data in a statistically rigorous, computationally efficient, and biologically meaningful way [93].

This immense amount of data and information from diverse sources (LEVELS 1–4), its high heterogeneity, its varying levels of precision, and its interpretive options fall outside the mechanisms used to date for analyzing complex problems. The challenge lies in analyzing and drawing conclusions from this "global reality" we have created. Because we are trying to study the interaction of the environment with living beings, and in particular with humans, to understand its influence on health and the onset of diseases, addressing the problem from the smallest to the largest scale in a comprehensive way. Starting from the study of metals and elements (ionomics), chemical species (speciation), metal-biomolecule species (metallomics), metabolites (metabolomics), proteins (proteomics), transcripts (transcriptomics), metagenomics, all at the molecular, cellular, regional, and organ levels (**LEVEL 1**). Continuing with their interactions at the molecular level, considering the limitations imposed by molecular crowding intra- and extracellular and the limitations (confinement) caused by the biological structures (**LEVEL 2**). In addition, it is necessary to consider the information on the temporal evolution of these processes throughout life with that from the macroscopic processes that occur in organs and tissues, making use of imaging techniques (**LEVEL 3**). Supporting the combined use of these tools with the evolution that occurs at the human level over time, under the action of external and internal factors, relying on the information provided by epidemiological studies (**LEVEL 4**). Finally, addressing this "whole" with an appropriate tool that allows us to make the leap from the specific to the global. Therefore, it is necessary to turn to the groundbreaking help of artificial intelligence, which will allow us to make the qualitative leap required to understand the consequences of human interaction with the environment throughout life, and to find the scientific tools to prevent pathological processes or diagnose them early (**LEVEL 5**; Figure 14.1).

### 14.5.1 Exposome the New Frontier to Know the Influence of Exogenous Factors on Population Throughout Life

We have seen that the lifelong action of exogenous factors—pollutants, diet, occupation, lifestyle, and many others—while also considering endogenous factors such as comorbidities (diabetes, cholesterol, hypertension, etc.), family history, additions (tobacco, alcohol) and even the effect of the social environment and personal psychological state, has significant consequences for our health. In this way, it is useful to know whether and to what extent chemical substances penetrated our bodies. In cases of nutrition this biomonitoring is critical to identify excess or deficiency of essential nutrients. In contamination studies the measurement levels of substances in our fluids or tissues, which allows to identify potential health risks. However, these biomonitoring studies are usually focused on a single compound or mixtures measurement at a single time point in lifecycle, but interactions of our body with stressors occurs throughout life, and its intensity varies over time. In addition, these interactions lead to dynamic responses of the entire biological system involving proteome, metabolome, transcriptome, ionome, ant others, which configure the global organism response that constitutes the exposome [94].

#### 14.5.1.1 Exposome Paradigm Case-Study

##### *14.5.1.1.1 Health Impact Assessment of Industrially Contaminated Areas Using the Exposome*

Studying and assessing the health impact on residents of industrially contaminated areas is a significant scientific challenge with multiple human and social implications. Most studies in this area do not provide clear answers regarding the relationship between environmental exposure and disease burden. Therefore, it is difficult to establish appropriate risk management measures. Consequently, it is necessary to improve risk assessment and management through the comprehensive approach that the exposome provides. The analytical tools for the exposome study were based on precise exposure assessment, internal dosimetry, and human biomonitoring based on multi-omics analysis of toxicity pathways, supported by statistics and artificial intelligence. Waste flows and environmental pollution are not considered in isolation, but rather as components of the expotype—the exposure vector to which an individual is exposed over time. Therefore, a multi-pathway and multi-route exposure estimation can be made, establishing a realistic basis for the integrated assessment of health risks and impacts.

Thus, a study was conducted with children from the area near the Fill landfill outside Athens (Greece), who were subjected to clinical tests to detect neurodevelopmental disorders at different stages of childhood. The exposome results, based on multi-omics techniques, were correlated with levels of environmental exposure, also considering the influence of socioeconomic, dietary, and metabolic factors on health risks. The results showed that proximity to a landfill and the resulting soil contamination with metals seriously affect children's neurodevelopment, although other factors, such as parental education level, socioeconomic status, and nutrition, also influence children's neurodevelopment, either positively or negatively [95, 96].

Another important study in relation to children and exposure is the Human Early Life Exposome (HELIX) study, involving several European countries (e.g., France,

Greece, Lithuania, Norway, Spain and the UK), which measure and describe multiple environmental exposures during early life (pregnancy and childhood) using molecular omics signatures and clinical outcomes of child health. In this way, the following tests and analysis were performed during the study: (a) questionnaire information from the mothers about child's diet, physical activity, sleeping patterns, socioeconomic status, social capital of the family, stress of the mother, exposure to environmental tobacco smoke, familiar habits, child's use of mobile phones and other electronic devices, use of green spaces, holidays and sun exposure and puberty development of the child, and other issues; (b) anthropometry and body composition; (c) blood pressure and spirometry test; (d) neurodevelopment test; (e) biomarkers analysis of contamination exposure in blood and urine such as metals and elements, pesticides, and many others contaminants and their degradation products.

To complete the exposome profile, several omics were applied to the samples, such as transcriptomics, methylomics, metabolomics, and proteomics [97].

##### *14.5.1.1.2 Relevance of Exposome to Chronic Kidney Disease Progression and in the Context of Kidney Transplantation and its Outcomes*

The role and importance of the exposome in the development and progression of chronic kidney disease (CDK) is well documented [98, 99], and the importance and benefits of low-calorie, vegetarian-based diets on the outcomes of this disease have also been demonstrated [100, 101]. In this disease, as in others, the influence of the exposome is crucial to its course, as we have seen throughout this chapter; however, we are referring to the external exposome. However, other researchers also emphasize the importance of the internal exposome, which comprises endogenous biological processes triggered or modified by external exposures, including inflammation, oxidative stress, metabolic changes, hormonal alterations, epigenetic modifications, and especially shifts in the microbiome [102–104]. This approach links disease not only to the action of exogenous agents, but also to each organism or patient with their own unique characteristics. One aspect that has been little studied in this field is the impact of the exposome on renal implantation [105].

Patients undergoing kidney transplantation (KRT) are particularly vulnerable to external and internal exposomes. It is worth emphasizing the importance, alongside the action of external factors, of a low-fat, high-vegetable diet, as previously discussed in patients with CDK, due to its influence on transplant success. In this regard, it is important to combine knowledge of the external exposome with the analysis of the internal one and the application of tools such as whole epigenome association studies (EWAS) and multi-omics profiling [105].

### 14.5.2 Understanding the Role of the Environment in Health Throughout Life: Foundations for Methodological Developments and Specific Actions

As we have seen, the influence of our environment, including pollutants (particularly metals), lifestyle, occupation, habitat, diet, physical exercise, and other factors, plays a significant role in our health throughout our lives. Highly prevalent diseases, such as AD, lung cancer, and breast cancer, which we have used as examples and references in this chapter, are powerfully influenced by our environment in their onset and progression. It is important to note that these diseases, like others, although they affect a specific organ, are considered by many experts to be systemic diseases, involving the entire organism, and therefore the pathways of environmental influence are numerous. We also believe that these problems are frequently addressed using a very specific methodological approaches for their study. Thus, when specifically considering the effects of metals on our health, attention is often focused on methodologies that provide information about their presence or concentration in our environment, in our food, in the air we breathe or the water we drink, their level in our blood, liver, or brain—that is, we approach the problem from a reductionist perspective. But the problem is much more complex and intricate, which is why it is necessary to resort to a methodology that more closely approximates reality. Therefore, a multi-level strategy is proposed. This begins with understanding metals, elements, their chemical forms, metal-biomolecule associations, metabolites, proteins, etc., in which we apply omics and multi-omics approaches (Level 1), and progresses to understanding their behavior when they act together in a cell, fluid, or tissue. It also includes the problems associated with molecular crowding and confinement, and how to characterize, measure, and study them (Level 2). On the other hand, when alterations occur at the molecular level, disrupting the pattern associated with a healthy state, and these alterations persist over time, the pathological process manifests macroscopically as tumors or harmful deposits. These are studied using imaging techniques (LEVEL 3). Furthermore, the influence of the environment and its consequences is not an individual but a collective process, requiring study at the population and epidemiological approaches (LEVEL 4). Finally, this multifaceted approach to the problem of environment and health requires a powerful tool to harmonize information from such diverse sources and reach relevant conclusions. Role currently fulfilled by artificial intelligence (LEVEL 5). As we have seen in the various examples and cases considered, this multilevel approach is not yet fully implemented. But this is the path we will be following.

Logically, this multilevel approach cannot be tackled by a single expert, so it requires multidisciplinary study. This requires the participation of analytical chemists to develop and apply multi-omics methodologies, biochemists to study biomolecules and their interactions, physicists to address the problem of interactions between biomolecule clusters in restricted biological spaces, epidemiologists to approach diseases from a population perspective, and, of course, medical specialists in various fields—physiologists, oncologists, neurologists, and others, depending on the problem being addressed—as well as radiologists and imaging specialists. Finally, mathematicians and artificial intelligence specialists are needed to support the interpretation of the results. Therefore, this new frontier in the study of lifelong health–environment interaction can no longer be the work of a single expert and their team; it must be reached together.

Finally, the results of environment–lifelong health studies must reach the society they are intended to serve. This requires the participation and commitment of society representatives, so that healthy environmental regulations and projects, appropriate food standards, and health systems that align with the findings can be established, along with appropriate control policies.

## REFERENCES

1. Stoccoro A, Coppedè F. 2024. Exposure to metals, pesticides, and air pollutants: Focus on resulting DNA methylation changes in neurodegenerative diseases. *Biomolecules* 14:1366.
2. Maret W. 2004. Exploring the zinc proteome. *J. Anal. At. Spectrom* 19:15–19.
3. Gomez-Ariza J L, Elham E Z, González-Fernández M, et al. 2011. Liquid chromatography-inductively coupled plasma-based metallomic approaches to probe health-relevant interactions between xenobiotics and mammalian organisms. *Metallomics* 3:566–577.
4. Outten CE, O'Halloran TV. 2001. Femtomolar sensitivity of metalloregulatory proteins controlling zinc homeostasis. *Science* 292:2488–2492.
5. Mounicou S, Szpunar J, Lobinski R. 2009. Metallomics: The concept and methodology. *Chem. Soc. Rev* 38:1119–1138.
6. Thiele D J, Gitlin J D. 2008. Assembling the pieces. *Nat. Chem. Biol* 4:145–147.
7. O'Halloran T V, Culotta V C. 2000. Metallochaperones, an intracellular shuttle service for metal ions. *J. Biol. Chem* 275:25057–25060.
8. Maret W. 2010. Metalloproteomics, metalloproteomes, and the annotation of metalloproteins. *Metallomics* 2:117–125.
9. Ding E, Wang Y, Liu J, et al. 2022. A review on the application of the exposome paradigm to unveil the environmental determinants of age-related diseases. *Human Genomics* 16:54.
10. Wang L, Qu J, Harari O, et al. 2024. The impact of multi-omics in medicine. *Cell Rep. Med* 5:101742.
11. Collas P. 2010. The current state of chromatin immunoprecipitation. *Mol. Biotechnol* 45:87–100.
12. Yoon J H, Lee D, Lee Ch, et al. 2024. Paradigm shift required for translational research on the brain. *Exp. Mol. Med* 56:1043–1054.
13. Gierasch L M, Gershenson A. 2009. Post-reductionist protein science, or putting Humpty Dumpty back together again. *Nat Chem Biol* 5(11):774–777.
14. Durek P, Walther D. 2008. The integrated analysis of metabolic and protein interaction networks reveals novel molecular organizing principles. *BMC Syst. Biol* 2:100.
15. Alberts B. 1998. The cell as a collection of protein machines: Preparing the next generation of molecular biologists. *Cell.* 92: 291–294.
16. Binette A P, Vachon-Presseau E, Morris J. et al. 2021. β-amyloid and tau pathology associations with personality traits, neuropsychiatric symptoms, and cognitive lifestyle in the preclinical phases of sporadic and autosomal dominant Alzheimer's disease. *Biol. Psychiatry* 89:776–785.
17. Yuan J M, Butler L M, Gao Y T et al. 2014. Urinary metabolites of a polycyclic aromatic hydrocarbon and volatile organic compounds in relation to lung cancer development in lifelong never smokers in the Shanghai Cohort Study. *Carcinogenesis* 35:339–345.
18. Wana M L Y, Coa V A, El-Nezamia H. 2022. Endocrine disrupting chemicals and breast cancer: A systematic review of epidemiological studies. *Crit. Rev. Food Sci. Nutr* 62:6549–6576.
19. Winchester L M, Harshfield E L, Shi L, et al. 2023. Artificial intelligence for biomarker discovery in Alzheimer's disease and dementia. *Alzheimer Dement* 19:5860–5871.
20. Brem A K, Khan Z, M. Ashworth M, et al. 2024. Screening for Alzheimer's disease in primary care using an AI driven screening platform: Design of the PREDICTOM study. *Alzheimer's Dement* 20:e087279.
21. García-Sevillano M A, González-Fernández M, Jara-Biedma R, et al. 2012. Biological response of free-living mouse *Mus spretus* from Doñana National Park under environmental stress based on assessment of metal-binding biomolecules by SEC-ICP-MS. *Anal. Bioanal. Chem* 404:1967–1981.

22. Contreras-Acuña M, García-Barrera T, García-Sevillano M A, Gomez-Ariza J L. 2013. Speciation of arsenic in marine food (*Anemonia sulcata*) by liquid chromatography coupled to inductively coupled plasma mass spectrometry and organic mass spectrometry. *J. Chromatogr A* 1282:133–141.
23. Contreras-Acuña M, García-Barrera T, García-Sevillano M A, Gómez-Ariza J L. 2014. Arsenic metabolites in human serum and urine after seafood (*Anemonia sulcata*) consumption and bioaccessibility assessment using liquid chromatography coupled to inorganic and organic mass spectrometry. *Microchem J* 112:56–64.
24. García-Sevillano M A, García-Barrera T, Navarro F, Gómez-Ariza J L. 2014. Analysis of the biological response of mouse liver (*Mus musculus*) exposed to As2O3 based on integrated -omics approaches. *Metallomics* 5:1644–1655.
25. Suzuki K T, Mandal B K, Ogra Y. 2002. Speciation of arsenic in body fluids. *Talanta* 58:111–119.
26. Watras C J, Bloom N S, Claas S A, et al. 1995. Methylmercury production in the anoxic hypolimnion of a dimictic seepage lake. *Water Air Soil Pollut* 80:735–745.
27. Gomez-Ariza J L, Lorenzo-Garcia F, García-Barrera T. 2005. Guidelines for routine mercury speciation analysis in seafood by gas chromatography coupled to a home-modified AFS detector. Application to the Andalusian coast (south Spain). *Chemosphere* 61:1401–1409.
28. García-Sevillano M A, García-Barrera T, Gómez-Ariza J L. 2013b. Development of a new column switching method for simultaneous speciation of selenometabolites and selenoproteins in human serum. *J. Chromatogr. A* 1318:171–179.
29. Arias-Borrego A, Callejón-Leblic B, Rodríguez-Moro G, et al. 2020. A novel HPLC column switching method coupled to ICP-MS/QTOF for the first determination of selenoprotein P (SELENOP) in human breast milk. *Food Chem* 321:126629.
30. García-Sevillano M A, Contreras-Acuña M, García-Barrera T, et al. 2014a. Metabolomic study in plasma, liver and kidney of mice exposed to inorganic arsenic based on mass spectrometry. *Anal. Bioanal. Chem* 406:1455–1469.
31. González-Domínguez R, Rupérez F J, García-Barrera T, Barbas C, Gómez-Ariza J L. 2016. Metabolomic-driven Elucidation of serum disturbances associated with Alzheimer's disease and mild cognitive impairment. *Curr. Alz. Res* 13:641–653.
32. Callejón-Leblic B, García-Barrera T, Pereira-Vega A, Gómez-Ariza J L. 2019. Metabolomic study of serum, urine and bronchoalveolar lavage fluid based on gas chromatography mass spectrometry to delve into the pathology of lung cancer. *J. Pharm. Biomed. Anal* 163:122–129.
33. Montes-Nieto R, Fuentes-Almagro C A, Bonilla-Valverde D, et al. 2007. Proteomics in free-living *Mus spretus* to monitor terrestrial ecosystems. *Proteomics* 7:4376–4387.
34. Morris J M, Honea R A, Vidoni E D, et al. 2014. Is Alzheimer's disease a systemic disease? *Biochim. Biophys. Acta* 1842:1340–1349.
35. González-Dominguez R, García-Barrera T, Vitorica J, Gómez-Ariza J.L. 2015b. High throughput multiorgan metabolomics in the APP/PS1 mouse model of Alzheimer's disease. *Electrophoresis*. 36: 2237–2249.
36. O'Connor LM, O'Connor BA, Lim SB, Zeng J, Lo ChH. 2023. Integrative multi-omics and systems bioinformatics in translational neuroscience: A data mining perspective. *J. Pharm. Anal* 13:836–850.
37. Wild C P. 2005. Complementing the genome with an "exposome": The outstanding challenge of environmental exposure measurement in molecular epidemiology. *Cancer Epidemiol. Biomark. Prev* 14:1847–1850.
38. González-Fernández M, García-Barrera T, Jurado J, et al. 2008. Integrated application of transcriptomics, proteomics, and metallomics in environmental studies. *Pure Appl. Chem* 80:2609–2626.

39. García-Sevillano M A, García-Barrera T, Navarro F. 2014. Use of metallomics and metabolomics to assess metal pollution in Doñana National Park (SW Spain). *Environ. Sci. Technol* 48:7747–7755.
40. Van Regenmortel M H V. 1999. Molecular recognition in the post-reductionist era. *J. Mol. Recognit* 12:1–2.
41. Watkin S A J, Bennie R Z, Gilke J M, et al. 2023. On the utility of microfluidic systems to study protein interactions: Advantages, challenges, and applications. *Eur. Biophys. J* 52:459–471.
42. Brückner A, Polge C, Lentze N et al. 2009. Yeast two-hybrid, a powerful tool for systems biology. *Int. J. Mol. Sci* 10:2763–2788.
43. Royer C A, Scarlata S F. 2008. Fluorescence approaches to quantifying biomolecular interactions. *Methods Enzymol* 450:79–106.
44. Cheng C I, Chang Y-P, Chu Y-H. 2011. Biomolecular interactions and tools for their recognition: Focus on the quartz crystal microbalance and its diverse surface chemistries and applications. *Chem. Soc. Rev* 41:1947–1971.
45. Wilson W D. 2002. Analyzing biomolecular interactions. *Science* 295:2103–2105.
46. Concepcion J, Witte K, Wartchow C et al. 2009. Label-free detection of biomolecular interactions using biolayer interferometry for kinetic characterization. *Comb. Chem. High T. Scr* 12:791–800.
47. Shah N B, Duncan T M. 2014. Bio-layer interferometry for measuring kinetics of protein-protein interactions and allosteric ligand effects. *J. Vis. Exp* 84:e51383.
48. Keene J D, Komisarow J M, Friedersdorf M B. 2006, RIP-Chip: the isolation and identification of mRNAs, microRNAs and protein components of ribonucleoprotein complexes from cell extracts. *Nat. Protoc* 1:302–307.
49. Rattray D G, Foster L J. 2019. Dynamics of protein complex components. *Curr. Opin. Chem. Biol* 48:81–85.
50. Hellman LM, Fried M G. 2007. Electrophoretic mobility shift assay (EMSA) for detecting protein–nucleic acid interactions. *Nat. Protoc* 2:1849–1861.
51. Chou RY-T, Pollastrini J, Dillon T M et al. 2012. Field-flow fractionation for assessing biomolecular interactions in solution. In: Williams SKR, Caldwell K D (eds) *Field-Flow Fractionation In Biopolymer Analysis*, pp 113–126. Springer, Vienna.
52. Zhao H, Mayer ML, Schuck P. 2014. Analysis of protein interactions with picomolar binding affinity by fluorescence-detected sedimentation velocity. *Anal. Chem* 86:3181–3187.
53. Chaturvedi S K, Zhao H, Schuck P. 2017. Sedimentation of reversibly interacting macromolecules with changes in fluorescence quantum yield. *Biophys. J* 112:1374–1382.
54. Krayukhina E, Noda M, Ishii K et al. 2017. Analytical ultracentrifugation with fluorescence detection system reveals differences in complex formation between recombinant human TNF and different biological TNF antagonists in various environments. *Mabs* 9:664–679.
55. Kroe R R, Laue T M. 2009. NUTS and BOLTS: Applications of fluorescence-detected sedimentation. *Anal. Biochem* 390:1–13.
56. Seidel SAI, Dijkman PM, Lea WA et al. 2013. Microscale thermophoresis quantifies biomolecular interactions under previously challenging conditions. *Methods* 59:301–315.
57. Ladbury JE, Chowdhry BZ. 1996. Sensing the heat: The application of isothermal titration calorimetry to thermodynamic studies of biomolecular interactions. *Chem. Biol* 3:791–801.
58. Purslow J A, Khatiwada B, Bayro M J, Venditti V. 2020. NMR methods for structural characterization of protein-protein complexes. *Frontiers Mol. Biosc* 7:9.
59. Barbieri L, Luchinat E, Banci L. 2015. Protein interaction patterns in different cellular environments are revealed by in-cell NMR. *Sci Rep-Uk* 5:14456.

60. Li D, Zhang Y, He Y et al. 2017. Protein-protein interaction analysis in crude bacterial lysates using combinational method of 19F site specific incorporation and 19F NMR. *Protein. Cell* 8:149–154.
61. Parray ZA, Ahmad F, Chaudhary A A, et al. 2022. Size-dependent interplay of volume exclusion versus soft interactions: Cytochrome c in macromolecular crowded environment. *Front. Mol. Biosci* 9:849683.
62. Hatters D M, Minton A P, Howlett G J. 2002. Macromolecular crowding accelerates amyloid formation by human apolipoprotein C-II. *JBC* 277:7824–7830.
63. Li X, Mehler E L. 2006. Simulation of molecular crowding effects on an Alzheimer's β-amyloid peptide. *Cell. Biochem. Biophys* 46:123–141.
64. Buawangpong N, Pinyopornpanish K, Siri-Angkul N, et al. 2022. The role of trimethylamine-N-oxide in the development of Alzheimer's disease. *J. Cell Physiol* 237:1661–1685.
65. Shoghi-Jadid K, Barrio J R, Kepe V, Huang S-Ch. 2006. Exploring a mathematical model for the kinetics of b-amyloid molecular imaging probes through a critical analysis of plaque pathology. *Mol Imaging Biol* 8:151–162.
66. Jin M, O'Nuallain B, Hong W, et al. 2018. An in vitro paradigm to assess potential anti-Aβ antibodies for Alzheimer's disease. *Nat. Commun* 9:2676.
67. Sabharwal V, Koushika S P. 2019. Crowd control: Effects of physical crowding on cargo movement in healthy and diseased neurons. *Front. Cell. Neurosci* 13:470.
68. Ambadipudi S, Reddy J D, Biernat J, Mandelkow E, Zweckstetter M. 2019. Residue-specific identification of phase separation hot spots of Alzheimer's-related protein tau. *Chem. Sci* 10:6503–6507.
69. Shahmoradian S H, Lewis A J, Genoud Ch, et al. 2019. Lewy pathology in Parkinson's disease consists of crowded organelles and lipid membranes. *Nat. Neurosci* 22:1099–1109.
70. Lim J M, Supianto M, Kim T Y, et al. 2023. Fluorescent lateral flow assay with carbon nanodot conjugates for carcinoembryonic antigen. *BioChip J* 17:93–103.
71. Galush W J, Le L N, Moore J M. 2012. Viscosity behavior of high concentration protein mixtures. *J. Pharm. Sci* 101:1012–1020.
72. Shologu N, Gurdal M, Szegezdi E, et al. 2022. Macromolecular crowding in the development of a three-dimensional organotypic human breast cancer model. *Biomaterials* 287:121642.
73. Almeida J A, Avila D B, Longmore G D, Pathak A. 2025. History of hypoxia exposure aids future cell invasion according to cell type and collagen density. *MBC* 36:1–13.
74. Cortes C J, Thyfault J P, Wilkins H M. 2023. Editorial: Systemic implications of Alzheimer's disease. *Front. Aging Neurosci* 15:1219987.
75. Zeng W, Yang F, Shen W L, et al. 2022. Interactions between the central nervous system and peripheral metabolic organs. *Sci. China Life Sci* 65:1929–1958.
76. Armstrong R A. 2014. A critical analysis of the 'amyloid cascade hypothesis'. *Folia Neuropathol* 52:211–225.
77. Afrin M R, Upadhyaya P G, Hashim A et al. 2025. Advanced biomarkers: Beyond amyloid and tau: Emerging non-traditional biomarkers for Alzheimer's diagnosis and progression. *Ageing Res. Rev* 108:102736.
78. Liu Y, Nie Y. 2023. Imaging diagnosis analysis and literature review of intrapulmonary solitary fibrous tumor. *Quant. Imaging Med. Surg* 13:8303–8312.
79. Christenson S A, Smith B M, Bafadhel M, Putcha N. 2022. Chronic obstructive pulmonary disease. *Lancet* 399:2227–2242.
80. Mu Y, Gui J, Lang Zh, et al. 2020. Information feedback of 18F-FDG PET/CT computer imaging combined with tumor markers on recurrence and metastasis of non-small cell lung cancer. *J. Infect. Public Health* 13:1336–1341.

81. Sakakibara J T, Nagashima H, Fujimoto M, et al. 2023. A review of MRI (CT)/US fusion imaging in treatment of breast cancer. *J. Med. Ultrason* 50:367–373.
82. Prieto-Falcón MC, Jurado-Gómez MC, Gomez de las Heras C, Rubio F. 2022. Mamografía con Contraste. Madrid. Panamericana Ed.
83. García-Carriazo M, Gómez de las Heras C, Mármol-Vázquez P, Ramos- Solís MF. 2016 Estudio de la insuficiencia venosa crónica mediante ecografía Doppler y realización de cartografía venosa. *Radiología* 58:7–15.
84. Duits F H, Nilsson J, Zetterberg H, et al. 2024. Serial cerebrospinal fluid sampling reveals trajectories of potential synaptic biomarkers in early stages of Alzheimer's disease. *J. Alz. Dis* 100:S103–S114.
85. Duits F H, Nilsson J, Zetterberg H, et al. 2025. Changes in dementia risk along with onset age of depression: A longitudinal cohort study of elderly depressed patients. *BMC Psychiatry* 25:247.
86. Twait E L, Kamarioti M, Verbeck I M W, et al., NCDC Consortium. 2024. Depressive symptoms and plasma markers of Alzheimer's disease and neurodegeneration: A coordinated meta-analysis of 8 cohort studies. *Am. J. Geriatric Psychiatry* 32:1153.
87. Bohn L, Drouin Sh M, McFall G P, et al. 2023. Machine learning analyses identify multimodal frailty factors that selectively discriminate four cohorts in the Alzheimer's disease spectrum: A COMPASS-ND study. *BMC Geriatrics* 23:837.
88. Fah H, Bohn L, Greiner R, Dixon R A. 2025. Comparing machine learning classifier models in discriminating cognitively unimpaired older adults from three clinical cohorts in the Alzheimer's disease spectrum: Demonstration analyses in the COMPASS-ND study. *Front. Aging Neurosci* 17:1542514.
89. Yuan J M, Butler L M, Gao Y T et al. 2014. Urinary metabolites of a polycyclic aromatic hydrocarbon and volatile organic compounds in relation to lung cancer development in lifelong never smokers in the Shanghai Cohort Study. *Carcinogenesis* 35:339–345.
90. Kurahashi N, Inoue M, Liu Y. et al. 2008. Passive smoking and lung cancer in Japanese non-smoking women: A prospective study. *Int. J. Cancer* 122:653–657.
91. Brugmans M J P, Rispens S M, Bijwaard H, et al. 2004. Radon-induced lung cancer in French and Czech miner cohorts described with a two-mutation cancer model. *Radiat. Environ. Biophys* 43:153–163.
92. Li Y, Baer D, Friedman G D, et al. 2009. Wine, liquor, beer and risk of breast cancer in a large population. *Eur. J. Cancer* 45:843–850.
93. Wang L, Qu J, Harari O, et al. 2024. The impact of multi-omics in medicine. *Cell Rep. Med* 5:101742.
94. Zuccherato-Bocato M, Bianchi-Ximenez J P, Hoffmann C, Barbosa F. 2019. An overview of the current progress, challenges, and prospects of human biomonitoring and exposome studies. *J. Toxicol. Environ. Health. B Crit. Rev* 22:131–156.
95. Sarigiannis D A. 2017. Assessing the impact of hazardous waste on children's health: The exposome paradigm. *Environ. Res* 158:531–541.
96. Sarigiannis D A, Karakitsios S P. 2018. Addressing complexity of health impact assessment in industrially contaminated sites via the exposome paradigm. *Epidemiol. Prev* 42 Suppl 1:37–48.
97. Maitre L, de Bont J, Casas M, et al. 2018. Human Early Life Exposome (HELIX) study: a European population-based exposome cohort. *BMJ Open* 8:e021311.
98. Alvarenga L, Ribeiro M, Cardozo L, et al. 2025. The exposome and the kidney: A silent dialogue shaping chronic kidney disease. *J. Xenobiotics* 15:73–92.
99. Liang K H, Colombijn J M T, Verhaar M C, et al. 2024. The general external exposome and the development or progression of chronic kidney disease: A systematic review and meta-analyses. *Environ. Pollut* 358:124509.
100. Kaimori J Y, Sakaguchi Y, Oka T, Isaka Y. 2025. Plant-dominant low-protein diets: A promising dietary strategy for mitigating disease progression in people with chronic kidney disease—A comprehensive review. *Nutrients* 17:643.

101. Hu L, Borelli G, Gessaroli E, et al. 2025. Individualized diets in patients with kidney disease and kidney transplants: A narrative review. *Life (Basel)* 15:896.
102. Vermeulen R, Schymanski E L, Barabási A L, Miller G W. 2020.The exposome and health: Where chemistry meets biology. *Science* 367:392–396.
103. Niedzwiecki M M, Walker D I, Vermeulen R, et al. 2019. The exposome: Molecules to populations. *Annu. Rev. Pharmacol. Toxicol* 59:107–112.
104. Walker D I, Valvi D, Rothman N, et al. 2019. The metabolome: A key measure for exposome research in epidemiology. *Curr. Epidemiol. Rep* 6:93–103.
105. Guldan M, Al-Shiab R, Rustamov A et al. 2025. The exposome era in kidney transplantation: A new frontier in graft outcomes and precision medicine. *Clin. Transplant* 39:e70384.

# Index